LEARNING, LANGUAGE AND INVENTION

Essays presented to Francis Maddison

LEARNING, LANGUAGE AND INVENTION: Essays presented to Francis Maddison

Edited by W.D. Hackmann & A.J. Turner

Variorum and the Société Internationale de l'Astrolabe
Aldershot & Paris, 1994

ASTROLABICA N° 6
Collection Published by the Société Internationale de l'Astrolabe

Published by VARIORUM
Ashgate Publishing Limited
Gower House, Croft Road
Aldershot, Hampshire GU11 3HR
Great Britain

Ashgate Publishing Company
Old Post Road
Brookfield, Vermont 05036
USA

and Société Internationale de l'Astrolabe
24 rue du Buisson Richard
78600 Le Mesnil-le Roi
France

Btitish Library Cataloguing in Publication Data
Learning, Language and Invention: Essays
presented to Francis Maddison
I. Hackmann, W.D. II. Turner, A.J.
ISBN 0-86078-467-3

U.S. Library of Congress Cataloging in Publication Data
Learning, Language and Invention: Essays presented
to Francis Maddison / edited by W.D. Hackmann & A.J. Turner
p. cm. Includes bibliographical references.
ISBN 0-86078-467-3
I. Maddison, Francis Romeril. II. Hackmann, Willem Dirk.
III. Turner, Anthony John.
AC5.L38 1994
081—dc20 94-27295
CIP

This book is printed on acid free paper.

Typeset by Stanford Desktop Publishing Services, Milton Keynes

Printed in Great Britain at the University Press, Cambridge

Contents

Editors' Introduction

Solicited for two posts in Oxford at virtually the same time, Francis Maddison in 1953 chose that of assistant curator in the Museum of the History of Science despite the fact that it offered a lower salary. The Museum's collections are extensive and cross-cultural, offering ample materials to appeal to an historian's sense of the past, an archæologist's delight in objects, and a philologist's fascination for the linguistic problems that inscriptions on objects can offer. All three elements have combined to characterize Francis Maddison's scholarly work, but his published contributions are further controlled by the discipline of bibliographical and archival description. Indeed it may be argued that it is his introduction of the rigour of bibliographical description to the cataloguing of scientific artefacts, notably in the *Supplement* to the Billmeir collection catalogue, but also in the hundreds of detailed labels that he produced for the Museum's display, that Francis Maddison has had a fruitful influence on the study of those objects of material culture that may be designated as mathematical or 'scientific' instruments.

Born in Hounslow in 1927, Francis Maddison was educated at Hounslow College (an independent Grammar school where he became librarian and Head of School) and Exeter College, Oxford. Matriculating to read Modern Languages (French), he soon changed to Modern History, although archæology continued to exercise a strong fascination. President of the Oxford University Archæological Society in 1948, he had the embarrassment of organizing a meeting which only he and the speaker attended, but nonetheless learnt to cut flint in the Palæolithic manner from R.J.C. Atkinson under whose supervision he directed excavations at Cricklade and Dorchester in 1948. A member of the British School at Rome Expedition to Leptis Magna, Tripolitania, in 1949, he was assistant archivist in the Glamorgan County Record Office (February 1950–March 1952), and in that of Warwickshire (April 1952–March 1953), where with Dorothy Styles and Anthony Wood, he was responsible for mounting an exhibition concerning the life and work of Sir William Dugdale in 1953. The catalogue published on this occasion (Bibliography no. 5) remains a fundamental source for anyone interested in Dugdale, and the Maddison hallmark of systematic description combined with some discursive but relevant commentary is already fully displayed. It is also characteristic that Maddison designed the catalogue himself.

While preparing the Dugdale exhibition, Maddison had met C.H. Josten, curator of the Museum of the History of Science, Oxford, who was already collecting materials for his monumental study of Elias Ashmole, Dugdale's father-in-law. Soon afterwards Josten encouraged Maddison to apply for the assistant curatorship then vacant in the Museum of the History of Science, thus initiating his association with the Museum. Return to Oxford and work in the Museum meant that Maddison now had a very wide range of material to stimulate his diverse interests. Immediately study and redisplay of the remarkable collection of over one hundred astrolabes led to the development of his interest in Arabic. Already in 1955 a forty-page typescript study of 'Some Early Islamic Astrolabes' had been prepared although it was not submitted for publication. Seminal however was the *Supplement* that he prepared in 1957 to C.H. Josten's *A Catalogue of Scientific Instruments from the 13th to the 19th Centuries, the Collection of J.A. Billmeir* (Oxford, 1954). Billmeir's collection was on loan display in the Museum of the History of Science, and Maddison's *Supplement* describes additions made to it between 1954 and 1957. Twice as large as the parent catalogue although it describes fewer instruments (121 as opposed to 154) the *Supplement* (Bibliography no. 8) is a model of detailed descriptive cataloguing, which set new standards in the field and also displayed the wealth of historical information that could be derived from instruments when they were studied in combination with contemporary sources and with other surviving contemporary examples.

In the following years archæological and archival researches interwove with a continuing interest in teaching methods and popularization. On 17 January 1963 a patent was taken out in association with Raymond E. Maddison for the teaching machine TALPA 2 (Bibliography no. 15); a general survey of instruments was made for popular consumption in *The Connoisseur Concise Encyclopedia of Antiques* (Bibliography no. 12) and a series of accompanying leaflets were prepared for the Advisory Council on Education to accompany an ITA television programme 'Towards 2000' (1964–5). Neither this however, nor increasing demands for consultative services from Sotheby's and Christie's auction houses, hindered further basic research. Careful archival delving led to the establishment of a virtually complete biographical list of the curators and their assistants in Ashmole's Museum which, although never published, has yet proved invaluable to subsequent students of the history of the Ashmolean. Investigation of some remarkable instruments among the Museum's holdings led to the publication of important studies (Bibliography nos 13, 16, 21), while in 1963 an essential bibliographical tool for studying the development of scientific instruments was provided which dealt with both the major sources and modern studies in all languages (Bibliography no. 17).

In 1964 Maddison succeeded Josten as curator of the Museum of the History of Science, developing its collections, increasing the number of staff and ably navigating it through difficult times of recession and the onslaughts of an increasingly demanding and bureaucratic university administration. At the same time Maddison's interests were widening further to embrace navigation, particularly that practised in the Iberian peninsula at the time of the discoveries, and new languages such as Armenian and

Georgian. From the early 1970s onwards he collaborated with the late Alain Brieux in the major research project of a complete repertoire of Islamic and Hindu instrument-makers. This, now nearing completion, is already partly in press.

The mere existence of a *festschrift* is evidence of esteem for its recipient as it is for his stature in his profession. An examination of its contents will give some idea of his interests. What such a volume cannot do is to express the affection in which its recipient is held or the debts that its contributors may have to him. Thanks to the generosity with which Maddison has disposed of his time in answering questions posed, in teaching, and in passing on pearls of information relevant to his colleagues' researches, indebtedness to him is as widespread as it is varied. The contributors to this volume have been able to make their own acknowledgements. It falls to the editors to express those of all others as well as their own, and also perhaps to avow, like the Canon's Yeoman, who realized

> With this Chanoun I dwelt have seven yeer,
> And of his science am I never the neer. (720–1)

a certain *désespoir* when confronted by the apparently endless well of knowledge upon which their colleague can draw.

We are glad to acknowledge here the pertinent comments and information imparted by A.V. Simcock to this introduction, the enthusiasm of the contributors and their patience with our exigencies, and the subventions from the Société Internationale de l'Astrolabe and from Rod and Madge Webster, The Adler Planetarium, Chicago, which have made the volume possible.

W.D. Hackmann — *Oxford & Le Mesnil-le-Roi*
A.J. Turner — *March 1994*

Francis Romeril Maddison

Bibliography of the Published Works of Francis Romeril Maddison

1. 'An Ox-bow Lake', *Hounslow College Magazine*, no. 4, Trinity Term 1938, p. 13. (Signed; 'Madison [*sic*] i, Form II').
2. 'Geographical Basis', pp. 7–8, and 'Farming', pp. 8–12, in (G.G.Barker (ed.)), *Hounslow College. Sixth Form Rural Survey (in Dorset) 1942. Record and Report*, Hounslow, 1942.
3. (With John Tyrer Egg), 'The Effigy of John Crocker at Exeter College', *Oxoniensa*, xiii, 1948, pp. 82–84 & plate VII, figs B & C.
4. (Associated editor), *Notes on Archaeological Technique*, Ashmolean Museum, 1950.
5. (With Dorothy Styles and Anthony Wood), *Sir William Dugdale 1605–1686. A List of his Printed Works and of his Portraits with Notes on his Life and the Manuscript Sources.* (Published on the occasion of the 'Exhibition of the Life and Works of Sir William Dugdale', Warwick: July & August, 1953), Warwick, 1953, 92 pp., illus.
6. (With Dorothy Styles and Anthony Wood), *Exhibition of the Life and Works of Sir William Dugdale, 1605–1686, Historian of Warwickshire (County Museum, Warwick, 15 July–31 August 1953). Description of the Exhibits Produced as a Supplement to* Sir William Dugdale, 1605–1686. A List of his Printed Works and of his Portraits with Notes on his Life and the Manuscript Sources (Warwick, 1953), *and Containing Additions and Corrections to that Publication*, Warwick, 1953, 49 pp. (duplicated).
7. (With R.E.W. Maddison), 'Joseph Priestley and the Birmingham Riots', *Notes and Records of the Royal Society of London*, xii, no. 1, 1956, pp. 98–113.
8. *A Supplement to a Catalogue of Scientific Instruments in the Collection of J.A. Billmeir, Esq., C.B.E. Exhibited by the Museum of the History of Science, Oxford*, Oxford & London (Published by Frank Partridge & Sons Ltd, 144–146 New Bond Street, London, W.1), 1957, 95 + (4) pp., plates XVII–XXXVII.
9. Review of L.A. Mayer, *Islamic Astrolabists and their Works*, in *Oriental Art*, n.s., iii, no. 3, 1957, p. 115.
10. 'An Eighteenth-Century Orrery by Thomas Heath and some Earlier Orreries', *The Connoisseur*, cxli, no. 569, 1958, pp. 163–167.

11. '(Obituary of) Professor L.A. Mayer', *Oriental Art*, n.s., v, no. 2, 1959, p. 76.
12. 'Scientific Instruments', in L.G.G. Ramsay (ed.), *The Connoisseur Concise Encyclopædia of Antiques*, vol. 5, London, 1961, pp. 193–203 & plates 149–156; and in various reprints of this work, sometimes anonymously, as in the one volume, *The Connoisseur Complete Encyclopædia of Antiques*, London, 1975, pp. 580–591.
13. 'A 15th Century Islamic Spherical Astrolabe', *Physis. Rivista di storia della scienza*, iv, fasc. 2, 1962, pp. 101–109, illus.
14. (With Bryan [*sic, recte* Brian] Scott and Alan Kent), 'An Early Medieval Water-Clock', *Antiquarian Horology*, iii, no. 12, 1962, pp. 348–352.
15. (With Raymond E. Maddison), 'TALPA 2' *(Triple Association Pedagogical Apparatus)*, teaching machine, subject of Patent Application no. 2073 of 1963, 17 January 1963: 'Pedagogical Apparatus', by Francis Romeril Maddison and Raymond Emil Maddison, and Assignment, 17 January 1963, of the invention to the Reader's Digest Association Limited, London.
16. (With Emmanuel Poulle), 'Un Equatoire de Franciscus Sarosius', *Physis. Rivista di storia della scienza*, v, fasc. 1, 1963, pp. 43–64.
17. 'Early Astronomical and Mathematical Instruments. A Brief Survey of Sources and Modern Studies', *History of Science*, ii, 1963, pp. 17–50; reprinted (paper-bound xerox) by the Librairie Alain Brieux, Paris.
18. Review of W.E. Knowles Middleton, *The History of the Barometer*, Baltimore, 1964, in *Journal of the History of Medicine*, xx, no. 1, 1965, pp. 70–72.
19. 'The Development of Clock Technology and the Problem of Finding Longitude at Sea, *c.* 1500–*c.*1800', in an experimental course devised by the Advisory Council for Education, in connection with a series of ITA television programmes entitled 'Towards 2000', separate fascicules, published between 1 August 1964 and 31 July 1965.
20. (With Silvio A. Bedini), *Mechanical Universe. The Astrarium of Giovanni de' Dondi*, Washington DC (Smithsonian Institution), 1966; bound (hardback & paperback) versions of *Transactions of the American Philosophical Society held at Philadelphia for Promoting Useful Knowledge*, n.s., lvi, part 5, 1966, 69 pp., illus.
21. *Hugo Helt and the Rojas Astrolabe Projection* (Agrupamento de Estudos de Cartografia Antiga XII), Coimbra, 1966; reprinted from *Revista da Faculdade de Ciências* (Coimbra), 61 pp., illus., folding table, and *addendum* and *corrigenda* sheet issued by F.R.M.; translated into Spanish by Claudio de Elejabeitia and printed, with a few additions, as the 'Estudio preliminar' to Hugo Helt, *Declaración y uso del reloj español. Edición facsimil con transcripción actualizada* (TEMPVS FVGIT, Monografías españoles de Relojeria VIII), Madrid–Valencia, 1984, pp. 13–56.
22. *Intervento* on the history of the mechanical clock, pp. 35–38, in the discussion of J.D. North, 'Relations between Mathematics and Astronomy before the Scientific Revolution', in A.C. Crombie, J.D. North, and M. Schramm, 'Physics and Astronomy', *Atti del Primo Convegno internazionale di Ricognizione delle Fonti per la Storia della Scienza italiana: i secoli XIV–XVI (Pisa, 14–16 Settembre 1966)*

(Pubblicazioni di Storia della Scienza della Domus Galilaena, sezione V, vol. 1), Florence, 1967, 48 pp.

23. (With R.E.W. Maddison), English translation of Henri Michel, *Les Instruments Scientifiques dans l'art et l'histoire*, Rhode-St-Genèse, 1966 as *Scientific Instruments in Art and History*, London, 1967.
24. 'Rapport de la Commission des Instruments Scientifiques', pp. 23–25 of the reprint of 'Union Internationale d'Histoire et de Philosophie des Sciences. Assemblée Générale, Paris, 24–31 août 1968', *Archives internationales d'histoire des sciences*, xxii, nos 86–87, 1969, 36 pp.
25. *Medieval Scientific Instruments and the Development of Navigational Instruments in the XVth and XVIth Centuries* (Agrupamamento de Estudos de Cartografia Antiga XXX), Coimbra, 1969, 60 pp., illus.; 'reprint' from *Revista da Universidade de Coimbra*, vol. 24, but this was delayed, so the late Professor Armando Cortesao had printed 10 copies of *I Reuniao Internacional de Historia de Nautica (23–26 outubro, 1968). Discursos e communicadoes*, Coimbra, November 1970, where it appears on pp. 115–72; reprinted (paper-bound xerox) by the Librairie Alain Brieux, Paris.
26. (With Francis Bédarida), *Maison française d'Oxford. Exposition. Cinq siècles de l'art française des instruments scientifiques: XVe–XIXe siècle, 9 novembre–4 dècembre 1970*, Oxford, 1970, 30 pp., & *corrigenda* sheet.
27. 'Dondi, Giovanni', in Charles Coulton Gillespie (ed.), *Dictionary of Scientific Biography*, New York, 16 vols, 1970–1980, vol. 4, 1971, pp. 164–165.
28. (With A.J. Turner), *University of Oxford, Museum of the History of Science. Catalogue 2. Watches*, Oxford, 1973, 91 pp., illus.
29. 'A Sceptical View of the *Tartar Relation*', in Helen Wallis, F.R. Maddison, G.D. Painter, D.B. Quinn, R.M. Perkins, G.R. Crone, A.D. Baynes-Cope, Walter C. and Lucy B. McCrone, 'The Strange Case of the Vinland Map. A Symposium', *The Geographical Journal*, cxl, part 2, 1974, pp. 187–191.
30. (With Alain Brieux), 'Bastûlus or Nastûlus? A Note on the Name of an Early Islamic Astrolabist', *Archives internationales d'histoire des sciences*, xxiv, 1974, pp. 157–160.
31. 'Preface' to facsimile reprint, Paris (Librairie Alain Brieux), 1976, of Henri Michel, *Traité de l'astrolabe*, Brussels, 1948.
32. (With Anthony J. Turner), *Catalogue of an Exhibition, 'Science and Technology in Islam', held at the Science Museum, London, April–August 1976, in association with the Festival of Islam*, Paris, 1976; limited circulation of paper-bound xerographic copies of most of the text.
33. Edited (with Margaret Pelling and Charles Webster), *Linacre Studies. Essays on the Life and Work of Thomas Linacre c. 1460–1524*, Oxford, 1977, 416 pp., 20 plates, and wrote the 'Editors' Introduction', and 'The Life of Thomas Linacre: a Brief Chronology', pp. xiii–liii.
34. Descriptions of exhibits: nos 2.33 'Compasso di proporzione inglese del tipo Hood', and 6.17 'Egnazio Danti–astrolabio', in (§2) 'La rinascita della Scienza'; and nos

2.22 'Il giusto eliotropio', 2.23 'Zaffiro per gli occhi', 2.24 'Sfera di cristallo', 2.25 'Matrice di un sigillo magico', 2.26 'Amuletti italiani', 2.27 'Due dischi in ottone italiani', 2.31 'Astrolabio di origine ignota', 2.32 'Astrolabio astrologico italiano', 2.33 'Orologio e quadrante orario italiano secondo Regiomontano', 2.34 'Quadrante italiano di Giovanni Antonio Magini (1555–1617)', 2.36 'Annulo solare italiano', and 3.5.17 'John Dee – Tavola sacra', in 'Astrologia, magia e alchemia nel Rinascimento fiorentino ed europeo', of *Consiglio d'Europa. Sedicesima Esposizione Europea di Arte, Scienza e Cultura. Firenze e la Toscana dei Medici nell'Europa del Cinquecento*, Florence, 1980, illus.

35. Descriptions of instruments, nos 90–234, pp. 90–189, in *Instruments scientifiques/Livres anciens... Collection Leonard Linton, Point Lookout, N.Y., et Central Resources Corporation, New York, N.Y.... et Objets appartenant à divers amateurs... dont la vente aux enchères publiques aura lieu au Nouveau Drouot, 9 rue Drouot, 75009 Paris, 9 et 10 octobre 1980. Mes Etienne Liber et Alain Castor, Commissaires-Priseurs... 3, rue Rossini... M. Alain Brieux, Expert près des Douanes, 48, rue Jacob..., Paris*, 1980; illus.
36. (Anon.) Text, published in English and Arabic, on history and use of the astrolabe, for a booklet to accompany each of 50 small astrolabes of gold, made by Asprey and Company, London, based on but not copied from one in the Museum of the History of Science, Oxford, for Benson & Hedges Special Filter (cigarettes) – 'Discover Gold', 1981.
37. 'Foreword', pp. xv–xvi, to C.F.C. Beeson, *Perpignan 1356. The Making of a Clock and Bell for the King's Castle* (edited by Charles W. Aked; Antiquarian Horological Society, Monograph no. 23), London, 1982.
38. Descriptions of astrolabes (no. 296 on pp. 284–285, & no. 302 on pp. 292–293) in Toby Falk (ed.), *Treasures of Islam. Published in Association with the Musée d'Art et d'Histoire, Geneva*, London, 1985, illus.; and Arabic and ? other editions.
39. 'Early Mathematical Wheelwork. Byzantine Calendrical Gearing', *Nature*, cccxiv, no. 6009, 1985, pp. 316–317, illus.
40. 'Al-Jazarī's Combination Lock: Two Contemporary Examples', *Studies in Oriental Art*, no. 1, 1985, pp. 141–157, illus.
41. Description of plate 174, p. 194, 'Astrolabe', in Robert Anderson and Ibrahim Fawzy (eds), *Egypt in 1800. Scenes from Napoleon's* 'Description de l'Egypte', London, 1987.
42. 'The Planispheric Astrolabe. A Brief Modern Bibliographical Guide', *The Horological Journal*, cxxx, no. 4, 1987, pp. 14–15.
43. Descriptions of the Islamic astronomical instruments in *Louisiana Revy* (published by the Louisiana Museum, Humlebæk, Copenhagen), xxvii, no. 3, 1987: 'Art from the World of Islam 8th–18th Century', which contains the catalogue of the exhibition with this title, held at the Museum, 7 March–14 June 1987, illus.; also Danish edition.
44. Review of Sharon Gibbs with George Saliba, *Planispheric Astrolabes from the National Museum of American History* (Smithsonian Studies in History and

Technology 45), Washington 1984, in *Bulletin of the Scientific Instrument Society*, xii, 1987, pp. 18–19.

45. (With Moshe J. Brown), 'A Unique Judaeo-Arabic Astrolabe, probably from a Jewish Community in Muslim Spain or possibly in North Africa...probably circa 1300...', *Important Judaica...Sale: Thursday 15 December 1988...Christie's Amsterdam B.V., Cornelis Schuytstraat 57, 1071 JG Amsterdam...*, Amsterdam, 1988, pp. 88–95, lot 247, illus.
46. Descriptions of items nos 474 'Lamina Universelle', 481, 483, 485, 486 'Astrolabe(s) plani-sphérique(s), 487 'Disque horaire et quadrant', and 490 'Astrolabe-quadrant', in *De l'Empire romain aux villes impériales... 6000 ans d'art au Maroc ... Musée du Petit Palais*, Paris, 1990, illus.
47. 'Symbols of Identity: Armorial Bearings, Badges and Colours (of Linacre College)', *Linacre News*, no. 2, 1991, pp. 9–10.
48. Articles on 'Armillary Sphere', 'Astrolabe', 'Compass', 'Marine Compass', 'Cross-staff', 'Kamal', 'Lead and Line', 'Lodestone', 'Navigation', 'Instruments of Navigation', 'Quadrant', and 'Timeglass' in Silvio A. Bedini (ed.), *The Christopher Columbus Encyclopaedia*, 2 vols, New York, 1992, illus.
49. 'Tradition and Innovation: Columbus' First Voyage and Portuguese Navigation in the Fifteenth Century', pp. 89–94, and descriptions of exhibits nos 121–123, pp. 221–225 & 5 (Martin Bylica's instruments (15th century) and another astrolabe), in Jay A. Levenson (ed.), *Circa 1492. Art in the Age of Exploration* (Catalogue for an exhibition at the National Gallery of Art, Washington DC, 1992–1993), New Haven, Conn. & London, 1992, illus.
50. 'The Barber's Astrolabe', *Interdisciplinary Science Reviews*, xvii, 1992, pp. 349–355.
51. 'A Consequence of Discovery: Astronomical Navigation in Fifteenth Century Portugal', in T.F. Earle and Stephen Parkinson (eds), *Studies in the Portuguese Discoveries I, Proceedings of the First Colloquium of the Centre for the Study of the Portuguese Discoveries* (Linacre College, Oxford), Warminster, 1992, pp. 71–110, illus.
52. (With Willem Hackmann), 'L'Appareil de Foucault pour la détermination de la vitesse de la lumière. Note sur l'exemplaire de 1855 conservé à Oxford' (title in French, but published in English), *Bulletin de l'Académie du Second Empire*, ix, 1993, pp. 9–11.

The Transmission of Decorated Wooden Ceilings in the Early Islamic World

J.W. Allan

In the Islamic Near East, as in Europe, surviving wooden ceilings of medieval date are few and far between. The ravages of fire, woodworm and beetle, the almost total destruction of the material culture of the Abbasid heartlands, the medieval Islamic propensity for destroying one's predecessor's official residence: these are some of the factors that have militated against the survival of one of the richest segments of early Muslim material culture. Spain, of course, is well-known for its *artesonado* ceilings, and a brief visit to the Alhambra is enough to impress on the visitor the richness of the western Islamic tradition. But even here the evidence is limited to the very latest period of Muslim occupation of the peninsula, and no wooden ceilings survive from before the 13th century. All the more important, therefore, has been the survey work carried out over the last two decades by Lewcock, Finster and others in the north Yemen. For the survival in the Yemen of at least five, largely intact, early Islamic ceilings of superb quality, and the existence of a number of other fragmentary examples, means that scholars for the first time can begin to study the history of early Islamic ceilings, and start to trace their evolution and development. For this reason, the surviving Yemeni ceilings will be used in this paper as a focus for a more general study of the early Islamic tradition.

It is, however, impossible to talk of Islamic ceilings without first drawing attention to their forebears in the Near East, especially those of the Greco–Roman and early Christian worlds. For sumptuous ceilings were scarcely an Islamic invention, even if medieval Islamic craftsmen were highly accomplished in the art. Greco–Roman coffered ceilings have only survived in their stone forms, but their outstanding craftsmanship gives some idea of the splendour which their wooden forerunners and contemporaries must also have displayed. As early as the last quarter of the 5th century BC, the Temple of Apollo Epicurius at Bassae had coffered stone ceilings,[1] and Greek

In this article an alif maqṣūrah will be transliterated with a macron rather than an acute accent.

[1] B. Fletcher, *A History of Architecture on the Comparative Method*, London, 1963, pp. 123–125.

From *Learning, Language and Invention: Essays presented to Francis Maddison*, ed. W.D. Hackmann and A.J. Turner. Published by Variorum, Ashgate Publishing Ltd, Gower House, Croft Road, Aldershot, Hampshire, GU11 3HR, Great Britain.

peristyle ceilings were regularly of this type: square or rectangular panels of carved stone or marble. Naos ceilings on the other hand were probably of coffered wood. In the Roman period the tradition of coffering was extended to the underside of domes, half-domes and vaults, and the stone or marble coffering on temple peristyles was at times elaborated into geometric patterns. Timber ceilings were coffered, or alternatively flat and covered with painted stucco designs in low relief.

The full glory of Roman ceilings is to be seen at two sites in greater Syria, both of them well-known to the Islamic world, and indeed endowed with medieval Islamic fortifications: Baalbeck and Palmyra. At Palmyra in the 1st century AD, in the great Temple of Bel, are two extraordinary *tours de force* in the monolithic ceilings of the north and south cellas. The northern one has a central domical vault decorated with planetary figures, with a rectangle of ceiling at either side divided into diamond-shaped, square and domical coffers. The southern one has a central domical vault with a large acanthus fleuron projecting down from the centre within a broad key-pattern band surrounded by an elaborate octagonal system of coffering.[2] This was the ceiling that so excited Wood and Dawkins during their visit of 1756 and made such an impact on English architecture of the second half of the 18th century. Another coffered ceiling is to be found in the annexe room, where a square domical vault is surrounded by hexagons. From the entrance comes the remains of a ceiling with a central square coffer, surrounded by four triangles within a diamond, within another square subdivided into squares and pairs of diamonds, while eight other geometric ceiling patterns have been restored from surviving fragments.[3] Other small buildings at Palmyra also contain coffered ceilings. Their various designs include all-over hexagons or rectangles; rosettes combined with diamonds, hexagons or ovals; diamonds and crosses; diamonds and triangles; interlacing circles.[4] A reconstructed ceiling in the Temple of Bel consists of two stars each made up of eight, diamond-shaped projections.[5]

The 2nd century AD ceilings of the Temple of Bacchus at Baalbeck are no less impressive and show a further range of designs utilized by the Roman masons. In the peristyle the ceiling design centres on a hexagon within an interlaced six-pointed star;[6] on the vault of the south exedra of the basement is an all-over hexagon-and-triangle design;[7] in the pronaos are two square ceilings, one with a central circular coffer surrounded by four irregular hexagons, with cruciform coffers in the corners, the other with a central octagon, within an interlaced eight-pointed star, projected beyond a

[2] M. Seyrig, R. Amy and E. Will, *Le temple de bel à Palmyre*, Paris, 1975, pls 27, 34, 58, 78; K.A.C Creswell, *Early Muslim Architecture*, Oxford, 1969, vol.1, part 1, fig. 125 (hereafter cited as *EMA 1*).

[3] Seyrig *et al. Op.cit.* (n. 2), pls 29, 62, 87, 100, 110, 121.

[4] T. Wiegand (ed.), *Palmyra*, Berlin, 1932, Taf. 30, 34, 44; B. Filarska, 'Etudes sur le décor architectural à Palmyre', in K. Michalowski, *Etudes palmyrennes*, vol. 2, Warsaw, 1967, figs 122–124.

[5] Seyrig *et al. Op.cit.* (n. 2), fig. 82.

[6] T. Wiegand, *Baalback*, 4 vols, Berlin/Leipzig, 1921, Taf. 40–48, Abb. 31; Creswell *EMA 1* (n. 2), vol. 1, part 1, figs 116–117.

[7] Wiegand *Op.cit.* (n. 6), Taf. 133–4; Creswell *EMA 1* (n. 2), vol.1, part 1, figs 114-115.

circle circumscribed around it into another eight-pointed star.[8] From elsewhere on the site come fragments with all-over hexagon and rhomb patterns.[9]

Although such ceilings are very rare outside these two great Roman sites, there is ample evidence that they were once more common. The ruins of Husn Niha temple in Syria include an elaborate coffered ceiling with busts within diamonds,[10] while that of Mejel Anjar includes a design of hexagons and triangles virtually identical with that in Baalbeck.[11] A further example of hexagonal coffering comes from the Western Tomb in 'Amman.[12] Such ceilings were clearly widespread in classical times, and tens of buildings so decorated were probably standing at the time of the Islamic conquests.

Of the ceilings of pre-Islamic Iraq and Iran we know virtually nothing. That there must have been a tradition, and an elaborate one, is shown by the great colonnaded halls (*apadana*'s) of Persepolis, dating from Achaemenid times. It is inconceivable that these were not covered with lavishly decorated wooden ceilings, though, as a result of the burning of Persepolis by Alexander the Great's troops, they have been lost for well over two thousand years. The capitals found at Taq-i Bustān in Fars[13] show that the Sasanians too sometimes built colonnaded structures; presumably these would also have had decorated wooden ceilings. On the other hand, the emphasis on undecorated stone or brick vaulting in surviving Sasanian monuments of Iraq and Iran suggests that the tradition of coffered ceilings was far more limited than in the Greco–Roman world.

Turning now to Islamic Yemen, a word on certain forms of structure inherited by the early Islamic architects is called for. The first of these structures has columns equally spaced in both directions supporting the ceiling directly, without the use of arcades. This would be characterized in the Achaemenid and later Persian world as an *apadana*. However, that would hardly be appropriate in south-west Arabia, despite a period of Yemeni subservience to a Sasanian satrap. The form also seems to have existed in Aksumite Ethiopia.[14] Unfortunately there seems to be no precise term available for it in the geographical and cultural context of Arabia.

In her most recent work on Yemeni Islamic architecture, Finster has drawn attention to the fact that this type of structure was well-known in Arabia in pre-Islamic times,[15]

---

[8] Wiegand *Op.cit.* (n. 6), Taf. 67–68.

[9] B. Finster, 'Die Freitagsmoschee von Šibām-Kaukabān', in *Baghdader Mitteilungen*, x, 1979, Taf. 86.

[10] D. Krencker and W. Zschietzschmann, *Romische Tempel in Syrien*, Berlin/Leipzig, 1938, vol. 2, Taf. 60.

[11] Krencker and Zschietzschmann *Op.cit.* (n. 10), vol. 1, Abb. 286.

[12] Creswell *EMA 1* (n. 2), part 2, fig. 498.

[13] A.U. Pope (ed.), *A Survey of Persian Art*, Oxford, 1938–39, pls 153A & B.

[14] The relevance of Aksumite ceilings to those of early Islamic Arabia is uncertain, though some of their forms have been reconstructed by D. Buxton and D. Matthews, 'The reconstruction of vanished Aksumite buildings', in *Rassegna di Studi Etiopici*, xxv, 1971–1972, pp. 53–77. Among the forms suggested is the type under discussion here – Buxton and Matthews, fig. 18.

[15] J. Schmidt. 'Zur altsudarabischen Tempelarchitektur', in *Archäologische Berichte aus dem Yemen* (henceforward *ABADY*), i, pp. 161–169.

and has described the early Islamic mosques associated with it as 'cubical'.[16] Ceilings of such buildings will inevitably divide themselves into squares, so that their weight is transmitted directly down the columns. When such a structure is conceived on a grand scale, for example the Great Apadana at Persepolis, the squares so formed will be so large that many smaller beams will have to be used to subdivide the space. But in smaller buildings of this type such squares can be filled with a single square coffer.

The second type of pre-Islamic structure which deserves notice also utilized regularly spaced columns. Here, however, the architraves were laid in one direction only, and the spaces between were covered by beams set at right angles to the architraves. Sometimes these beams might be contiguous, at other times they might be spaced out. In the latter case the areas between were covered by a further set of beams, set at right angles to the larger beams and in alignment therefore with the architraves.[17] Stone ceilings will not feature directly in the rest of this article, but it is worth noting the existence of a small mosque at Tamūr, in the north Yemen, in which the stone ceiling follows the latter form.[18] The historical prevalence of these types of ceilings in stone or wood is suggested by their popularity in the smaller urban mosques of Saudi Arabia, even to this day.[19]

When the Arabs conquered Iraq and Iran on the one hand, and Egypt and Syria on the other, they inevitably came into contact with two highly developed civilizations, those of the Sasanians and the Byzantines. Each had its own cultural traditions, and in this particular context, its own tradition or traditions of ceiling construction. In the east, the *apadana* style, which had been widely used since Achaemenid times, was destined to continue. Its precise development is lost for ever due to the destruction of the evidence, but groups of wooden mosques of Safavid date, such as those of Bunāb and Marāgheh,[20]and the much better known Safavid palaces, like those of the Chehel Sutūn and 'Ālī Qapū in Isfahan,[21] show its continuing vitality throughout the Islamic centuries as an essential element of wooden architecture.

In the Byzantine world, at the time of the Arab conquests, the most striking form of wooden roof was that associated with the basilica. The early Christian basilica had a wide central nave and two or more side aisles. The nave was covered by a pitched

---

[16] B. Finster, 'Cubical Yemeni Mosques', in *Proceedings of the Seminar for Arabian Studies*, xxi, 1991, pp. 49–68.

[17] J. Schmidt, 'Bericht über die Yemen-Expedition 1977 des Deutschen Archäologischen Instituts', in *ABADY*, i, 1982, Taf. 46, and 'Der Stadttempel von Ma'in', in *ABADY*, i, 1982, Taf. 65.

[18] B. Finster, 'Die Moschee von Tamūr', in *ABADY*, i, 1982, Taf. 82d.

[19] e.g. G.R.D. King, *The Historical Mosques of Saudi Arabia*, London, 1986, illustrations on pp. 53, 96, 127.

[20] e.g. the Masjid-e Mehrābād, dated 951/1544–5, and the contemporary Masjid-e Meidān (M.A. Mukhlis, *Fihrist-e banāhā-ye tārīkhī-ye azerbaijān-e sharqī*, Tehrān, 1371 pp. 81–84). A Qajar example in the same town is the Masjid-e Ismāīl Beg (Mukhlis, pp. 84–87), and others include the Masjid-e Sheikh Mu'azz al-Dīn in Marāgheh, and the Masjid-e Jame in 'Ajabshir (Mukhlis, pp. 166–168 and 171).

[21] Pope *Op.cit.* (n. 13), pp. 473–474.

roof of timber, of very simple, open construction, and the beams, rafters and planks were often painted. The roofs of the aisles were constructed on the same principles.

Associated with basilicas were *atria*, or courtyards, surrounded by an ambulatory. Continuing a tradition going back to the stoas of the ancient Greeks, such ambulatories consisted of porticos of flat ceilings on closely-set beams, the beams being supported on the one side by the architrave above the colonnade, and on the other by the back wall of the structure. These beams thus divided the ceiling into narrow rectangular areas at right angles to the line of the colonnade. The relevance of Byzantine ceilings of these types will soon become apparent.

Turning now to the Islamic Yemen, the earliest surviving ceiling is that of the west *riwāq* (portico) of the Great Mosque in San'ā'.[22] Here, the flat roof of the building is supported on arcades, and the aisles thus created are straddled by a series of great, cased beams, creating a series of narrow rectangular ceiling spaces at right angles to the arcades. These rectangular spaces are divided into square coffers by smaller beams set at right angles. Each coffer has a simple moulded frame and contains a diamond-shaped inset. The smaller beams are decorated with a single carved rosette, while the centre of each coffer is decorated with a carved and painted rosette pattern surrounded by a white pearl border. The other parts of the coffers and beams are painted with a variety of vegetal motifs. The visual result of the carved and painted decoration from the ground is a regular series of squares containing a circle within a diamond.

Structurally, this ceiling is of Byzantine origin. Not only do the motifs used in the coffer paintings suggest that the artists were brought by al-Walīd from Umayyad Syria, but the use of arcades shows the introduction of arches to a tradition which had hitherto been without them. That the earliest ceiling in the Great Mosque in San'ā' should derive from a Byzantine form need not come as a surprise. al-Walīd I (AD 705-715), from whose reign this *riwāq* almost certainly dates, was resident in Syria, and is likely to have commissioned Syrian or Egyptian architects for his San'ā' project.

Perhaps more surprising is the south *riwāq* of the mosque at Shibām, at the foot of the Kawkabān escarpment.[23] For here the very tall columns are surmounted directly by great beams, and there is no arcading. Nevertheless, the ceiling structure is similar to that of San'ā', and the simple coffering and painted decoration are so similar to those of San'ā' that there can be little doubt that the south *riwāq* of Shibām-Kawkabān too dates from al-Walīd's reign. Indeed, it is probably the work of the same craftsmen. The influence of the Byzantine tradition on Yemeni ceilings under al-Walīd is therefore very strong.

The survival of the Shibām-Kawkabān *riwāq* is of great importance for reconstructing the ceiling of one of the most important buildings in the Arabian peninsula, the Prophet's Mosque in Medina, as rebuilt by al-Walīd I. Destroyed in the fire of AD 1256, the

[22] B. Finster, 'Die Freitagsmoschee von San'ā'. Vorlaufiger Bericht, I. Teil', in *Baghdader Mitteilungen*, ix, 1978, pp. 92–133.

[23] Finster *Op.cit.* (n. 9), pp. 193–228.

general style of the ceiling is suggested by a passage in Ibn 'Abd ar-Rabbih's book, *Al-'Iqd al-Farīd*, written in the early 10th century:

> The capitals are gilded, and on them are carved and gilded architraves; then on the architraves are the ceilings, also carved and gilded.[24]

Ibn 'Abd ar-Rabbih's description testifies to a columned structure with architraves, but does not, unfortunately, speak of the shape of whatever coffering decorated the ceilings. Two pieces of evidence are valuable here, however. First is the fact that the ceiling of the bay in front of the mihrab was said to be like a buckler. That must mean it was domical, and domical vaults or coffers have yet to be found in cubical Yemeni mosques. Secondly, the sources tell us that al-Walīd sent Greek and Coptic craftsmen from Syria and Egypt to undertake the work.[25] We may therefore surmise with confidence that, despite the lack of arcades, the ceiling was of the Byzantine portico type, corresponding to those of the west *riwāq* of the contemporary mosque at Shibām-Kawkabān.

In the ceilings of San'ā' and Shibām-Kawkabān we may also have the solution to a hitherto unexplained comment of at-Ṭabarī about the Great Mosque at Kufa, rebuilt in AD 670.[26] He writes that the ceiling resembled the ceilings in Byzantine churches. Now Ibn Jubayr noted that the columns in height were like ships' masts, and that they had no arches above them. Given that there were thus no arcades, any comparison with Byzantine work has always seemed inappropriate, the more so since the Arabic text of at-Ṭabarī appears to suggest that the ceiling came from a palace formerly belonging to one of the Persian kings.[27] If his comment is remotely accurate, however, at-Ṭabarī must surely be indicating that the ceiling of the Great Mosque, like that at Medina, followed the Byzantine tradition in that it was arranged in aisles, each aisle covered by a series of beams at right angles to the architraves, with coffering of some sort between them, the whole elaborately painted and gilded.[28]

In contrast, the ceiling of the Ka'aba was almost certainly of the type found in Yemeni cubical mosques. For, as Finster has pointed out, its original form, with two rows of three columns each, is identical to those of the two outstanding early examples of the cubical mosque in the Yemen, the Masjid al-'Abbās, and the mosque at Ṣarḥah. The Ka'aba was later reduced in size, and one row of columns was removed, which explains its present shape. We also know that the ceiling was richly decorated, and that five alabaster slabs were let into it, to allow light to filter through.[29]

---

[24] J. Sauvaget, *La mosquee omeyyade de Medine*, Paris, 1947, p. 74.

[25] Creswell, *A Short Account of Early Muslim Architecture*, revised and supplemented by J.W. Allan, Aldershot 1989, p. 43. It should be noted that there is a mistake on p. 45 of this book, last paragraph: the roof was 25 cubits (not 25 metres) above the floor, i.e. 2 cubits above the ceiling.

[26] Abu Djafar Mohammed ibn Djarir at-Tabari, *Annales*, ed. M.J. de Goeje, Leiden, 1893, vol. 1, p. 2489; *The History of al-Tabari*, vol. 13, translated and annotated by G.H.A. Juynboll, New York, 1989, p. 69.

[27] Juynboll's translation (n. 26), p. 69, n. 246.

[28] This also seems to have been the original form of the ceilings in the outer ambulatory of the Dome of the Rock, see Creswell/Allan *Op.cit.* (n. 25), p. 30.

[29] Finster *Op.cit.* (n. 16), pp. 54–55.

The next earliest ceiling in the Great Mosque in San'ā' is that of the east *riwāq* [Figure 1]. This was built in 262/875–6, just four years before the Mosque of Ibn Ṭūlūn, by the Ṣulayḥid Ibrāhīm ibn Muḥammad ibn Yu'fir.[30] Structurally it corresponds to the ceiling of al-Walīd I, but decoratively it is much more elaborate. The casing of the large beams is not only painted but richly carved [Figure 2]. These carved designs are sometimes rectilinear, sometimes curvilinear and almost always show sequences of large and small cartouches filled with vegetal motifs. The smaller beams too have more elaborate carved centres, and the frame of the coffering has three recesses carved along each side, each recess filled with a pine-cone in the round; in the corners are carved rosettes. Finally, carving is not confined to the central rosette of each coffer, but is also used for the triangular spaces between its edge and that of the surrounding square.

The form of the coffer frame reflects the continuing Syrian influence in San'ā' architecture, for it is a straightforward variant of the ovolo moulding of classical antiquity. Such mouldings originally supported, and later simply decorated, square or rectangular ceiling panels of the Roman period.[31] The carved and painted designs of the ceiling, on the other hand, are to be linked with the art of Abbasid Iraq, most closely resembling those of the wooden minbar in the Great Mosque at Qairawan. Whether the latter was made by an Iraqi craftsman in Qairawan or imported in carved panels from Iraq itself, the Iraqi connection is attested by the literary sources, as is its date of AD 862–3.[32]

Iraqi influence in the decoration of the San'ā' ceiling does not seem to have been accompanied by any structural ideas. Not only are the mouldings of the coffers derived from the Mediterranean world, but the most characteristic feature of Abbasid ceilings is lacking: the use of rectangular coffers, which were almost certainly alternated with square ones. The evidence for this from Iraq itself is confined to one such rectangular coffer elaborately carved, found in the Jawsaq al-Khāqānī in Samarra[33] [Figure 3]. However, the complete, if simplified, style appears in the ceiling of the Mosque of Ibn Ṭūlūn in Cairo[34] [Figure 4], a building which almost entirely depended on the culture of Abbasid Iraq for its form, construction and decoration. This must surely have been the source of the Fatimid style, illustrated by the remains of part of

[30] B. Finster and J. Schmidt, 'Die Freitagmoschee von San'ā'. Vorläufiger Bericht, 2. Teil. Der Östriwāq', in *Baghdader Mitteilungen*, x, 1979, pp. 179–181.

[31] Seyrig *et al. Op.cit.* (n. 2), pl. 62.

[32] K.A.C. Creswell, *Early Muslim Architecture*, Oxford, 1940, vol. 2, pp. 317–319 (henceforward *EMA 2*).

[33] E. Herzfeld, *Der Wandschmuck der Bauten von Samarra und seine Ornamentik*, Berlin, 1923, pp. 76–77 orn. 110. The only other evidence for ceilings in imperial Abbasid buildings comes from the Great Mosque of al-Mutawakkil. There, early photographs make clear that the colonnades on which the architraves must have rested ran at right angles to the qibla wall in the sanctuary, and at right angles to the adjacent part of the surrounding wall in the *riwāqs* (E. Herzfeld, *Erster vorläufiger Bericht über die Ausgrabungen von Samarra*, Berlin, 1929, Taf. II). The assumption must therefore be that the ceilings were constructed in the same general way as those of al-Walid in San'ā' and Shibām, though no trace of the ceilings themselves survived.

[34] Creswell *EMA 2* (n. 32), p. 349, fig. 256.

the ceiling of their western palace in Cairo, now included in a rather confused state in the *māristān* (hospital) of the late-13th-century Mamluk Sultan Qalā'ūn,[35] and other fragments in the Museum of Islamic Art[36] [Figure 5].

The next intact Yemeni ceiling, chronologically, is the most splendid of all the early ones, and outstanding by any standards. This is the sanctuary ceiling of the mosque at Shibām[37] [Figure 6]. The mosque is unfortunately undated, but its structure and the ceiling of the southern *riwāq*, mentioned above, probably date from al-Walīd's reign. The sanctuary ceiling, however, is later. Popular opinion accords Ya'fur ibn 'Abd ar-Raḥmān, the founder of the Yu'firid dynasty (d. AD 872), the honour of having founded the mosque in Shibām, his capital, but this could well refer to such a major embellishment as the addition of the sanctuary ceiling coffers. Certainly, the second half of the 9th century seems an appropriate dating for them. Excluding the coffering itself, the carved and painted designs find their closest parallels in the decoration of the mihrab and minbar of the Great Mosque in Qairawan. These, as mentioned above, reflect the Abbasid Iraqi style, and were placed in position in the mosque in AD 862–3.

Without a doubt, such a sumptuous set of coffers would have required a patron of considerable resources in terms of finance and craftsmen. The ceiling is divided into twenty-six separate areas of coffering, which are divided from each other by a further eighteen narrow rectangles of carved geometric work. For the coffers, only two different grids are used, one of squares, the other of equilateral triangles, but visually the ceiling gives an immensely rich impression. The designs include the following combinations: squares and elongated hexagons; large octagons with small diamonds; large and small cusped circles; stars and crosses; octagons, irregular hexagons and squares; hexagons, rhombs and diamonds; hexagons and triangles; hexagons, squares, triangles and irregular duo-decagons; hexagons and interlacing circles; and a number of visually complex patterns of rosettes in interlacing bands.

Structurally, there appear to be three different methods used for building these ceilings. The commonest, used for two-thirds of them, utilizes three or four planks running longitudinally over the rectangular space. Circular, hexagonal, octagonal or rosette-shaped holes are cut in these planks for the small coffers of the design, and then the other geometric forms are applied as pieces of shaped wood to the surface of the planks, leaving narrow gaps betweeen neighbouring shapes so that they are visually distinct. The geometric forms usually cross the plank joints, showing that a whole unit was constructed as one before being placed in position. Four of the ceilings (nos 2, 37, 40, 41) appear to be constructed rather differently. Six (or in one case five) narrow beams have been laid latitudinally, and have been joined together by short sections of longitudinal beam. The spaces resulting from this construction have then been framed with applied pieces of wood, and further geometric shapes of wood have finally been

[35] V. Meinecke-Berg, 'Materialien zu fatimidischen Holzdekorationen in Kairo I): Holzdecken aus dem fatimidischen Westpalast in Kairo', in *Mitteilungen des deutschen archäologischen Instituts. Abteilung Kairo*, xxxxvii, 1991, pp. 227–233.

[36] E. Pauty, 'Un dispositif de plafond fatimite', in *Bulletin de l'Institut d'Egypte*, xv, 1932–33, pp. 99–107.

[37] Finster *Op.cit.* (n. 9), pp. 193–228.

Figure 1. San'ā' Great Mosque, east riwāq *(Photo: J.W. Allan and R. Hillenbrand)*

Figure 2. San'ā' Great Mosque, east riwāq, *detail of coffering (Photo: J.W. Allan and R. Hillenbrand)*

Figure 3. Coffer from the Jawsaq al-Khāqānī, Samarra (after Herzfeld)

Figure 4. Cairo, ceiling of the Mosque of Ibn Ṭūlūn (after Creswell/Comité de Conservation)

Figure 5. Cairo, design of the ceiling of the western Fatimid palace (after Meinecke-Berg)

Figure 6. *Shibām-Kawkabān, mosque ceiling, general view (Photo: J.W. Allan and R. Hillenbrand)*

Figure 7. Shibām-Kawkabān, mosque, ceilings nos 39 and 40 (Photo: J.W. Allan and R. Hillenbrand)

Figure 8. Marrakesh, ceiling of the Kutubīya Mosque (after Terrasse)

Figure 9. *Granada, Generalife, ceiling of the north portico (Photo: J.W. Allan)*

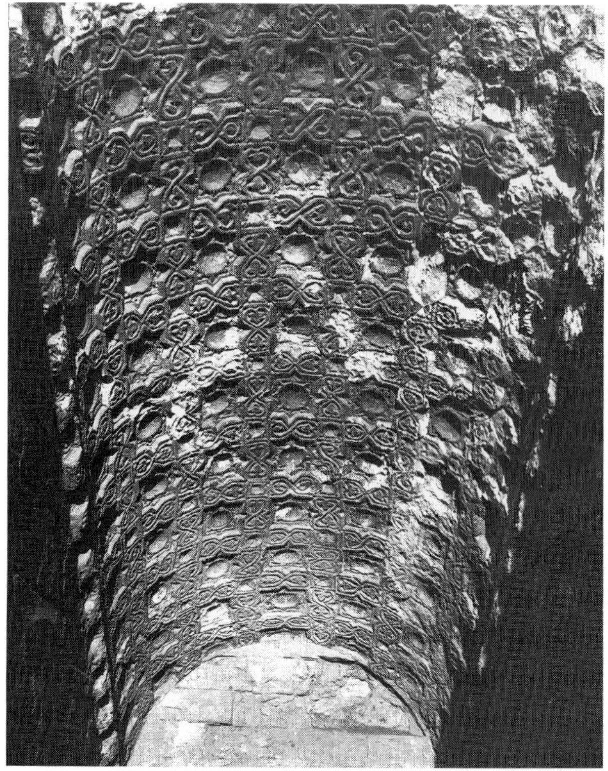

Figure 10. *Cairo, Burg aẓ-Ẓafar, stone ceiling (Photo: Creswell Photographic Archive, Ashmolean Museum)*

applied to the surface. The most complex method occurs in ceiling no. 39 [Figure 7] (and possibly too in no. 18). Here the star patterns are made of fairly thick battens and actually form the structure of the ceiling. Upon them sits a set of painted planks, forming the background.

That such a magnificent set of coffered ceilings should only survive in Shibām in the Yemen must be but an accident of history. There can be no doubt that in its day it would have been just one of many such ceilings in palaces and mosques, not just in the Yemen but in other more important areas of the Islamic world. Neither can there be any doubt that it continued an equally rich classical tradition, for many of these designs are already present in the remarkable ceilings of 2nd-century Palmyra. And while there are no other earlier Islamic ceilings of this type surviving, there is plentiful evidence of the use of the patterns in other media, and of the way such patterns moved freely between media. For example, some at least of the more complex patterns at Shibām are already to be found in earlier Islamic buildings in stucco, as for example in the stuccos of the palace of Khirbat al-Mafjar, dating from the AD 740s.[38] It is highly likely that the porticos around the courtyards of some of the Umayyad princely residences in greater Syria would have been covered with just such coffers. The ceilings in Shibām, though evidence of the marvellous quality of craftsmanship available in the Yemen in the 9th century, should not therefore be taken as evidence of the Yemen's predominance in this field. Rather, they give us a taste of an otherwise unknown form of ceiling which must have been among the most impressive parts of the great architectural complexes of early Islam.

Simple structural analysis of the Shibām ceilings is important. For it shows that in the 9th century in Yemen and the central Islamic lands the so-called *artesonado* ceiling was already developing. The *artesonado* ceiling is a type of ceiling formed of joists and lathes, which have been shaped and then interlaced. Its earliest appearance in the west is in the Kutubīya mosque at Marrakesh [Figure 8], dating from the middle of the 12th century,[39] where it gives the impression of intersecting squares and diamonds. Because it has a long and spectacular history in western Islam and Christian Spain from the 12th century onwards, it might be thought that the *artesonado* ceiling was a western Islamic phenomenon. However, one particular feature of the Marrakesh ceiling suggests that it may ultimately be derived from the east: the decoration between the rafters. This consists of a very simple arrangement of alternating octagons and rectangular octagons which follows the ceiling tradition associated with the Mosque of Ibn Ṭūlūn, mentioned above, ultimately deriving from Abbasid Iraq. They certainly have nothing in common with the indigenous Maghrebi tradition exemplified by the 9th century ceilings of Qairawan[40] and Cordoba.[41] The carpenters who built the ceiling

[38] R.W. Hamilton, *Khirbat al-Mafjar*, Oxford, 1959, figs 238–250.

[39] H. Terrasse, *L'art hispano-mauresque des origines au XIIIe siècle*, Paris, 1932, pl. LXVI.

[40] G. Marcais, 'Plafonds peints du IXme siècle à la Grande Mosquée de Kairouan', in *Revue des arts asiatiques*, 1935, pp. 1–8.

[41] F. Hernandez, 'La techumbre de la Gran Mesquita de Cordoba', in *Archivo Español de Arte y Arqueología*, 1928, pp. 191–225; M. Gomez-Moreno, 'El arte árabe español hasta los almohades. Arte mozarábe', *Ars Hispaniae*, iii, Madrid, 1951, fig. 134; G. Marcais, *L'architecture musulmane d'Occident*, Paris, 1954, pp. 148–149.

at Qairawan followed their predecessors at the mosque of Ibn Ṭūlūn in using planks to cover the unsquared tree-trunk beams, but the decoration was purely painted. This tradition was continued when alterations were made in the 11th century. At Cordoba, on the other hand, flat relief decoration was applied before painting.

Hence it may well be that the *artesonado* style and technique was introduced to the Maghreb from Egypt at this period. It is of course true that there are no extant ceilings of the *artesonado* type among the surviving wooden ceilings of Cairo, but an interesting indication that they may once have existed is an 81 x 53 cm fragment of a grill from the mausoleum of Sayyida Nafisa, dating probably from one of the restorations carried out by the Caliph al-Ḥāfiẓ in AD 1138 or 1147.[42] The technique used here is the same as that used for the Kutubīya ceiling, although it is in every way cruder. There is also a further feature of the Kutubīya which points to an eastern source: a decorative frieze of cusped arches which runs around the base of the ceiling, a design which occurs in other Maghrebi buildings.[43] The remains of a very similar ornament is attached to the fragments of 10th–11th century ceiling in the south-west corner of the sanctuary of the Great Mosque in San'ā'.[44] There is little chance of San'ā' influencing Marrakesh directly. More likely is a common source in Abbasid Iraq, where a frieze of blind arcading might well have been a regular feature at the base of a ceiling.

More evidence for a link between the ceiling traditions of Spain and the Maghreb on the one hand, and the Near East on the other, is provided by two other Spanish ceilings. In the north portico of the 'Court of the Lions' in the Alhambra, in Granada, is an area of coffering which seems to be an elaboration of one of the coffering designs used both in the Great Mosque in San'ā',[45] and at Shibām.[46] In the north portico of the Generalife in Granada is a ceiling with small domical vaults within a geometric surround[47] [Figure 9] which looks very old-fashioned within the context of 14th–15th century *artesonado* ceilings in western Islam. So too does a spectacular ceiling of similar design in the Alcázar in Seville. These are surely part of an older Near Eastern tradition. An important intermediary is provided by the stone vault of one of the sloping staircases of the Burg aẓ-Ẓafar in Cairo, dating from c. AD 1176[48] [Figure 10]. This latter vault must be based on a contemporary, wooden structure, showing that the form must have been known in Fatimid Egypt as well.

Returning now to the Yemen, the climax of the Yemeni ceiling tradition is the ceiling of the royal mosque of Ẓafār Dhībīn, part of a complex of buildings begun by the

[42] E. Pauty, *Les bois sculptés jusqu'à l'époque ayyoubide*, Cairo, 1931, pl. XCVI, no. 1615.

[43] Another western Islamic example of this form of arcade, along with the inscription which formed part of the same frieze, comes from the Madrasa in Ceuta, see E.P. Lopez, *El arte en el sur de Al-Andalus*, Seville, 1988, pl. 173.

[44] B. Finster, 'Die grosse Moschee von San'ā'. Dritter Vorläufiger Bericht. Das vier Westjoche in Ḥaram', *ABADY* (n. 15), i, 1982, Taf. 79a & b.

[45] Finster *Op.cit.* (n. 44), Taf. 76.

[46] Finster *Op.cit.* (n. 9), Taf. 69.

[47] L. Torres-Balbas, *Ars Hispaniae*, vol. 4, *Arte Almohade. Arte Nazari. Arte Mudéjar*, Madrid, 1949, fig. 127.

[48] K.A.C. Creswell, *The Muslim Architecture of Egypt*, Oxford, 1952–59, vol. 2, pl. 19c (henceforward *MAE*).

Imām Manṣūr Billāh in AD 1204, work continuing until his death in 1217.[49] Sadly, only four of the naves of the mosque sanctuary retain their ceilings, and even here they are badly damaged or incomplete, but what remains is nothing if not impressive. The naves concerned are the two to the west of the mihrab and the two to its east, the ceilings of the mihrab nave and the rest of the sanctuary having almost totally disappeared.

The sanctuary of the mosque has eleven naves, the central one being wider than the rest. The ceiling is supported on ten rows of very tall columns supporting large, cased beams running at right angles to the qibla wall. The nave second from the mihrab on the west is in the Byzantine portico style: smaller beams cross it at regular intervals, creating narrow rectangular areas parallel to the qibla wall. Each is divided by two short beams into three square spaces. The central square has a diamond-shaped recess in it, while the side squares bear flat painted designs only. The nave on the immediate west of the mihrab has the two bays nearest the qibla wall still intact. Each consists of two large square units with ovolo frames. One is in the form of a recessed interlaced eight-pointed star with a central circular recess, the other in the form of a recessed diamond with semicircular protrusions in the middle of each side, and a central octagonal recess. Both have broad strapwork designs in relief and background designs of arabesques, all done in painted plaster.

Such designs, though superb, have a long history in Yemen and in Egypt. Square coffers bearing an interlaced eight-pointed star occur, for example, at the mosque at Jibla,[50] which is probably a converted palace, built by the Queen Arwā bint Aḥmad, wife of the Ṣulayḥid ruler al-Mukarram. The palace was built in 1065–6, and the mosque dates from 1088–9.

The diamond with protruding semicircles used at Ẓafār Dhībīn occurs in an earlier context at Hawlān near San'ā', in the ceiling of the cubical mosque of al-'Abbās[51] [Figure 11]. The mosque was built by a local ruler, one Mūsā ibn Muḥammad al-Qutbī, in AD 1126. Outwardly insignificant, it contains the finest coffered ceiling of the cubical mosques of Yemen. Diamond-shaped coffers, square ones, star-shaped ones, and diamonds with protruding semicircles are all to be found here, and the gilded designs on the beams are equally sumptuous.

Only one bay of the nave second to the east of the mihrab at Ẓafār Dhībīn survives [Figure 12]. It is divided into two square coffers, both filled with geometric designs. One consists of stars and crosses,[52] the other of octagons with square and irregular hexagons between. Both appear to be constructed like the second group of ceilings at Shibām. In other words there is a structure of beams running in both directions to which are attached both the frames of the more deeply recessed octagons or stars, and the other geometric forms which decorate the surface.

---

[49] B. Finster, 'Survey islamischer Bau- und Kunstdenkmäler im Yemen. Erster vorläufiger Bericht', in *ABADY* (n. 15), i, 1982, pp. 253–269.

[50] Finster *Op. cit.* (n. 49), pp. 232–240, and especially Taf. 96.

[51] B. Finster, 'Die Masgid al-'Abbās im Haulān', *ABADY* (n. 15), iii, 1986, Taf. 63a.

[52] Finster *Op.cit.* (n. 49), Taf. 114b.

The most elaborate of the ceilings to survive, however, is that of the nave to the east of the mihrab, covering all three bays[53] [Figure 13]. In the centre is a wooden painted dome within an octagonal surround. It is set in a square frame with ovolo moulding. From this a muqarnas ceiling runs both north and south. Longitudinally it is divided into four lines of small domical vaults, each within an octagonal surround, a long wooden stalactite hanging from each corner. At either end of the nave is a small area of ceiling in a rather different design, elongated hexagonal battens forming a cruciform structure overlaid with painted planks.[54] These two areas are constructed like the third group of Shibām ceilings, and are clearly related to the latter.

The most surprising feature of this ceiling, and indeed of the ceilings as a whole, is the muqarnas work. Excluding the actual stalactites, the design follows one of the ceiling patterns from Shibām no. 41), but the only other example of stalactites is to be found in the cubical mosque of Ṣarḥa, which, however, could be later than Ẓafār and is on an incomparably smaller scale. Nor is anything as grand to be found in the surviving ceilings of other Islamic lands. The nearest equivalents geographically and stylistically are two ceilings in Cairo. The first is a very small wooden ceiling in what was either a doorway or window of the mausoleum of the Imām Shāfiʿī, dating from AD 1211.[55] Much later, but on a scale comparable to Ẓafār Dhībīn, is the ceiling of the early 14th century mosque of an-Nāṣir Muḥammad on the Cairo citadel.[56] Both these ceilings have the same layout as the Ẓafār ceiling, but neither has the stalactites. Is such stalactite work therefore to be seen as an invention of royal craftsmen working in Yemen in the early 13th century?

While such a solution is tempting, it does not stand up to detailed scrutiny. In the first place, it should be noted that the Ẓafār ceiling does not occupy the central nave of the mosque, but a nave of no more or less significance than those already described. What originally occupied the central nave is open to question. As a result, it could scarcely have been considered the most original of designs available. Moreover, although ceilings of this form are lacking elsewhere, there is indirect evidence that they must have existed. While none of the three great surviving stucco muqarnas domes of the 11th–12th centuries – those of the Imām Dūr in Iraq (1085–1090), and of the Māristān and Mausoleum of Nūr al-Dīn in Damascus (1154 and 1172)[57] – contain actual stalactites, a pair do occur at the top of the elaborate muqarnas vaulting of the facade of the Māristān of Nūr al-Dīn.[58] In addition, all four of these vaults have at their apex a small domical vault, recalling the way the ceiling at Ẓafār has a similar focus. Moreover, the polylobed keel arches of two early-12th-century mihrabs from Fatimid Cairo, Sayyida Ruqayya (1133) and Muḥammad al-Ḥaṣawātī

[53] Finster *Op.cit.* (n. 49), Taf. 115a.

[54] Finster *Op.cit.* (n. 49), Taf. 114a.

[55] Creswell *MAE* (n. 48), vol.2, pl. 26a.

[56] There is a rather poor photograph showing part of this ceiling in *The Mosques of Egypt*, Giza, 1949, vol. 1, pl. 58.

[57] E. Herzfeld, 'Damascus: studies in architecture – I', in *Ars Islamica*, ix, 1942, figs 44–45, 57.

[58] Herzfeld *Op.cit.* (n. 57), fig. 43.

(c. 1125–1150), suggest that elaborate muqarnas work was also used in Egypt at this period,[59] while the pendant keystones in the portals of the Kilijīya and ʿĀdilīya madrasas in Damascus could also indicate such a tradition.[60] If such work had been developed in stucco and imitated in stone, it is hard to see why it would not also have been developed in woodworking, for wood is the more appropriate medium for large areas of flat ceiling. We must therefore conclude that the stalactites of the Ẓafār Dhībīn ceiling were not Yemeni inventions but part of a much wider tradition of wooden ceiling design, of which examples may well have existed up to a hundred years earlier in Egypt, Syria and Iraq.

This study of the ceilings of Ẓafār Dhībīn brings in its wake the need for a brief discussion of the role of craftsmen in ceiling manufacture. As shown above, there are a wide variety of carpentry techniques involved in these Yemeni ceilings, and workers in stucco and painters are also prerequisites. That the stucco workers and painters would have been local Yemeni craftsmen seems likely, at least in the later period under consideration. Stucco work of one variety or another is found in virtually every Yemeni mosque, while the painted designs of the later ceilings, although they can be related to other Islamic designs of their period, are clearly different from the latter, suggesting the existence of a local school or schools. But the fact that the most elaborate ceilings do not occur in every mosque, even when as at Jibla they are royal foundations, suggests that the carpenters who could manufacture ceilings like those of the Shibām sanctuary or Ẓafār Dhībīn were not always available. It is, moreover, difficult to believe that there were enough building projects and wealth available in the Yemen to keep carpenters of this quality in full-time business year after year, let alone century after century. It therefore seems all the more likely that such craftsmen were itinerant, and that ceilings of the various types that survive in the Yemen would have been found in Baghdad, Damascus and Cairo as well.

The Yemeni focus of this article has inevitably left certain types of early Islamic ceiling out of the discussion. In particular, it is important to return to the Fatimid domain, and to the stone coffers which decorate the exterior of the Bāb al-Futūḥ[61] [Figure 14]. Two features of these coffers are noteworthy. First is their geometric form. The central coffer has a star within an octagon in a square frame; each of the two adjacent coffers has a square with projecting semicircles in the centre of each side, enclosing a rectangle, also within a square frame. Neither is unique in Egypt. The star design occurs in stone in a vault in the Burg aẓ-Ẓafar (AD 1176),[62] while a stone coffer with a design of a square and projecting semicircles is to be found in a small oratory in the western minaret of the Mosque of al-Ḥākim (AD 1003).[63] More interesting, however, is the

[59] Creswell *MAE* (n. 48), vol. 1, pls 88a-b.

[60] E. Herzfeld, 'Damascus: studies in architecture – III', in *Ars Islamica*, xi-xii, 1946, figs 90–91.

[61] Creswell *MAE* (n. 48), vol. 1, fig. 85. These were the focus for an article by the present author, which has been awaiting publication since 1987 in the Proceedings of the Conference on Fatimid Art, held at the School of Oriental and African Studies.

[62] Creswell *MAE* (n. 48), vol. 2, pl. 19c.

[63] Creswell *MAE* (n. 48), vol. 1, pl. 31c.

fact that they both occur in the painted wooden ceilings of the Yemen, as mentioned above. An interlaced eight-pointed star pattern is used in the ceiling of the 11th-century mosque at Jibla, for example,[64] while the square with projecting semi-circles is found at the mosque of al-'Abbās [Figure 11].

Even more important, both designs occur in the ceiling of the mosque at Ẓafār Dhībīn, and here they have hollow centres surmounted by another piece of planking.[65] Now the three central coffers on the Bāb al-Futūḥ also have an open centre. They seem today to be covered with loose pieces of stone. Originally they could have been either covered with single flat stones, or left open. If they were left open, they may reflect a wooden coffer with a very specific function, the admittance of light. In the Yemen, for example, a square or rectangular coffer with an alabaster sheet over its open centre is sometimes found above a mihrab, providing a source of light for the focal point of the prayer ritual in the mosque. Two such coffers, probably of 9th-century date, are to be found in the ceiling of the mosque at Shibām-Kawkabān.[66] The same function is provided by the alabaster sheets which surmount the more elaborate domical vaults in the Great Mosque at San'ā'.[67] The Ka'aba also had alabaster 'windows' set into its ceiling, as mentioned above. Yemen had no monopoly of alabaster, and it is hard to imagine the great Abbasid buildings without such light sources in their inner areas. However, the coffers on the Bāb al-Futūḥ are on a comparatively small scale, and are situated on the exterior of the gateway in broad daylight, so that an origin of this type may be inappropriate.

A more likely alternative takes us back to Shibām-Kawkabān, and its extraordinarily elaborate late-9th-century ceiling. The numerous examples there of designs with overall patterns of recessed elements suggests that elaborate ceilings of this type were well-known, and this three-dimensional quality was much used and admired. It is then all the more likely that when the Fatimids started to build their new capital in Cairo, they would have been employing the best carpenters available in the Near East, and that their palaces, like that of Sitt al-Mulk, daughter of the Caliph al-'Azīz, would have been decorated with ceilings of this quality. The coffers on the Bāb al-Futūḥ would then be the stonemasons' reference to the sumptuous palatial ceilings nearby.

At this point it is appropriate to note the legacy of the Fatimid ceilings, in particular that of the ceiling of the western palace [Figure 5]. For it is undoubtedly from this that a characteristic Mamluk ceiling style is derived, best exemplified by the ceiling of the Great Corridor in the complex of Qalā'ūn in Cairo (AD 1284–5)[68] [Figure 15]. The Great Corridor is straddled by a series of wooden beams, and the narrow areas between the beams are covered by wooden planks running in alignment with the corridor [Figure 16]. Cut through each set of planks are seven stars, usually half a star in the side of two contiguous planks. On certain sections of planking, however, the place

---

[64] Finster *Op.cit.* (n. 49), pl. 96a & c.

[65] Finster *Op.cit.* (n. 49), Taf. 115b, 116, 117b.

[66] Finster *Op.cit.* (n. 9), pls 79–80.

[67] Finster *Op.cit.* (n. 22), pl. 29.

[68] Creswell *MAE* (n. 48), vol. 2, pl. 67b.

Figure 11. Hawlān, mosque of al-'Abbās, coffer (Photo: J.W. Allan)

Figure 12. Ẓafār Dhībīn, mosque, ceiling of the second nave to the east of the mihrab (Photo: J. W. Allan)

Figure 13. Ẓafār Dhībīn, mosque, ceiling of the nave to the east of the mihrab (Photo: J.W. Allan)

Figure 14. Cairo, design of the square coffers of Bāb al-Futūḥ (after Creswell)

Figure 16. Cairo, complex of Qalā'ūn, corridor ceiling detail (Photo: J.W. Allan)

Figure 15. Cairo, complex of Qalāʾūn, corridor (Photo: J.W. Allan)

Figure 17. Radā', Amiriyya Madrasa, stucco ceiling (Photo: J.W. Allan)

Figure 18. Radā', Amariyya Madrasa, stucco ceiling, detail (Photo: J.W. Allan)

of the three central stars is taken by a narrow rectangle with stellar ends. A variant occurs in the small ceilings of the ambulatory of Qalā'ūn's mausoleum,[69] while a late Fatimid example occurs in the Dayr al-Banāt, a Coptic building in old Cairo. In its general appearance, this style of ceiling was destined to remain standard in Mamluk architecture. However, it is rare to find three-dimensional examples, and even where they do occur, as in the mosque of Umm Sulṭān Sha'bān (1368–9), the holes are round not stellar in shape, and the star design is applied in paint. In the vast majority of the surviving ceilings of this type the star designs are painted only, though sometimes on a low relief plaster base. Examples are ceilings in the house of Beshtak (1334–39), the mosque of Aṣlam al-Silāḥdār (1344–45), the khanqa of Shaykhū (1354), and the mosque of al-Mu'ayyad Shaykh (1415–20).

The impact of Fatimid ceilings is seen not only in those of the dynasties which succeeded them in Egypt, but much further afield, in Sicily. For in Christian buildings in Palermo, also, the pierced stellar designs are to be found: for example, in the 13th-century La Steri palace,[70] and the late-14th-century Palazzo Chiaramonte.[71] The only difference here is that the stars are set in individual coffers between the main beams, rather than arranged in longer rows. What is perhaps a further variant of the Fatimid/early Mamluk style is to be seen in the ceiling of the choir of Monreale cathedral.[72] At precisely what date the idea travelled across the Mediterranean is unclear, though it is presumably sometime in the later Fatimid period.

It would be absurd to write an article such as this without mention of the most famous of all 'Islamic' ceilings, that of the Cappella Palatina, and without asking the question how it fits into the picture now established. The Cappella Palatina ceiling is quite distinct in its form from the *artesonado* tradition as developed in the Maghreb and Spain, though it does seem to have something in common with the ceiling of the north portico of the Generalife on the one hand, and the Ẓafār Dhībīn stalactite ceiling on the other. In practice, however, it is probably not closely related to either. Palerman architecture of the early Norman period is most closely allied to that of Ifriqiyya, for example the buildings of Suse or the Qala' of the Bannī Ḥammād.[73] The closest parallels to the Cappella Palatina's ceiling are likewise North African, the stucco domes of Almohad mosques of the Maghreb: the Kutubīya Mosque at Marrakesh, the mosque dedicated to Ibn Tūmart at Tinmal, and the Qarawiyyīn Mosque in Fez. Whether there was also a lively tradition of wooden ceilings in Ifriqiyya and the Maghreb at this period, following the stucco forms, or whether the stucco forms in fact followed such a wooden tradition, is impossible at present to say. However, the lack of

---

[69] Creswell *MAE* (n. 48), vol. 2, pl. 73b.

[70] G. Spatrisano, *La Steri di Palermo e l'architettura Siciliana de trecente*, Palermo, 1972, pls 60, 62.

[71] A. Colasanti, *Volte e soffitti Italiani*, New York, 1943, pl. 4–7.

[72] D. Gravina, *Il duomo di Monreale*, Palermo, 1859–69, vol. 2, Taf. 18. The ceilings of the cathedral, with the exception of those of the choir, were replaced after the fire of 1811.

[73] J. Johns, 'The Norman Kings of Sicily and the Fatimid Caliphate', in *Anglo-Norman Studies*, xv, 1993, p. 140, draws attention to the connection between the plans and architectural decoration of the Norman and early Ifriqiyyan palaces.

connection between the Cappella Palatina ceiling and the wooden ones of Fatimid and Mamluk Egypt on the one hand, and the *artesonado* ceilings of Spain and the Maghreb on the other, suggest that it may well relate to an Ifriqiyyan–Maghrebi palace tradition, now lost.

It is perhaps worth mentioning here that stucco points to a further tradition of coffered ceiling manufacture in early Islamic times about which virtually nothing is known. The only surviving example appears to be the ceiling of the north portico of the early-16th-century Amiriyya Madrasa in Radā', again in the Yemen [Figures 17 and 18]. This is suspended from narrowly spaced wooden beams which form the basis of the latitudinal stucco work. Small stucco beams set longitudinally between each of these pairs make up the series of small square coffers. The coffers, the recessed rectangles on each section of beam, and the recessed circles at each joint were all originally filled with pierced stucco designs of lace-like fineness, though the decoration now only survives in a few of the rectangles. The circles supported small inverted domes of pierced stucco work. In the centre of the ceiling above the door into the sanctuary are the remains of a large square stucco coffer. Sadly its central area is missing so its original form will never be known. It cannot be a coincidence that precisely this layout occurs in the stucco decoration from the walls of the Samarra houses excavated by Herzfeld.[74] One is forced to conclude that the Radā' ceiling is part of a tradition by then at least six hundred and fifty years old.

In one of the most important books on Islamic art published in recent years, Terry Allen[75] devotes a chapter to 'Transformations and correspondences in Abbasid architecture'. Here he argues that Abbasid Iraq provided the models of which we now have only the surviving regional variations as evidence, and he cites such features as the nine-dome plan, framing inscriptions, the muqarnas dome, and domes of intersecting arches. To these I believe we should now add painted wooden ceilings. For the evidence from the Yemen, Egypt and the Maghreb seems to me to be overwhelmingly in favour of Abbasid Iraqi models for the majority, and certainly the most important, of the ceiling designs known from contemporary or later provincial examples. Of that great Abbasid civilization we have so little direct material evidence from Iraq itself. How exciting that we can rebuild in our minds some at least of its ceilings, from the scattered evidence of Marrakesh, Cairo, and above all Shibām and Ẓafār Dhībīn!

[74] Creswell *EMA 2* (n. 32), pl. 78c, House XIa.

[75] T. Allen, *Five Essays on Islamic Art*, n.p., 1988.

Eruditionis bona pars: Oxford and the European Languages around 1600

Giles Barber

'Scis bonam eruditionis partem esse scire rerum vocabula', Erasmus wrote in his *De recte pronuntione*, first published in 1528.[1] Sir Thomas Bodley, the founder of Oxford's major university library as we know it today, was a much travelled diplomat and would have agreed strongly with Erasmus. Born in 1544 at Exeter, he was a member of a strongly Protestant Devonshire family which, fearing religious persecution under Queen Mary, fled to the continent in 1556. Thomas Bodley therefore found himself, at the age of twelve, travelling in Germany and finally settling at Geneva where he studied theology under Calvin and Beza, Greek under Beroaldus, and Hebrew with Antoine Chevalier. The latter was an accomplishment he was particularly proud of and one which allowed him to give his library a motto drawing on this cultural tradition.[2] With the accession of Queen Elizabeth, Bodley returned to England and entered Magdalen College, Oxford, graduating in 1563. He became a Fellow of Merton College, lectured in Greek and undertook further studies in Natural Philosophy, but soon accepted university office, becoming Proctor and Public Orator. In 1576 he 'waxed desirous to travell beyond the seas, for attayning to the knowledge of some speciall modern Tongues' and he visited Germany, Italy and France. In 1585 he entered the diplomatic service, being employed on missions to assist the Protestant interest in Denmark, Germany and France, finally becoming, in that momentous year 1588, England's permanent representative at The Hague. Politics led to his retirement in 1597 however and in his own words he 'concluded at last to set up my staffe at the Library doore in Oxford'.[3]

---

[1] Erasmus, *Opera omnia*, Amsterdam, 1973, vol. 1, section 4, line 983.

[2] See G. Barber, 'Quarta perennis', *Bodleian Library Record*, vii, 1962, pp. 46–50; and C. Roth, 'Sir Thomas Bodley – Hebraist', *ibid.*, pp. 242–251.

[3] For the biography of Sir Thomas Bodley see Bodley's own account, often reprinted; the *Dictionary of National Biography*, vol. 2, pp. 756–759, and I.G. Philip, *The Bodleian Library in the Seventeenth and Eighteenth Centuries*, Oxford, 1983, ch. 1.

From *Learning, Language and Invention: Essays presented to Francis Maddison*, ed. W.D. Hackmann and A.J. Turner. Published by Variorum, Ashgate Publishing Ltd, Gower House, Croft Road, Aldershot, Hampshire, GU11 3HR, Great Britain.

In this famous passage Bodley goes on to say that his enterprise would not have been successful without 'some purse-ability', a 'very great store of honourable friends' and leisure: but the first qualification he mentions is, in a modest manner, 'some kinde of knowledge, as well in the learned and modern tongues'. He is thought to have spoken French, German, Dutch and possibly Spanish, while a knowledge of Italian is suggested by his personal ownership of two works by Boccaccio, the 1544 *Vita di Dante* and a 1569 *Corbaccio*.[4] Bodley's interest in languages, probably originally motivated by scriptural reasons to some degree, was clearly both academic and practical; nor was it limited to what he himself knew, for in commissioning the consul of the English merchants in Aleppo to collect books 'in the Syriacke, Arabicke, Turkish and Persian tongues, or in any other language of those Esterne nations' he believed that 'in processe of time, by the extraordinarie diligence of some one or other student they may bee readily understoode'. He also insisted that the librarian of 'his' library should be a linguist. The recipient of the present volume, Francis Maddison, clearly shares both Sir Thomas' linguistic ability and his interest in languages and it is therefore the aim of this paper to explore, through the Bodleian's contemporary resources, the knowledge of the vernacular European languages in Oxford around 1600.

Oxford has always received travellers, including a fair number of persons of heterodox opinions, from all parts of Europe, and in the book world alone it can be noted that its earliest printer, Theodoric Rood, came from Cologne. In the early sixteenth century Cambridge had been more successful than Oxford in challenging London's monopoly of the book trade: their 1534 charter allowing them to appoint three stationers and printers, alien or native. Oxford on the other hand had to rely on aliens who were more restricted in their actions, but who often had close trade relations with foreign countries. An early Oxford bookseller of German origin, John Thorne or Dorne, some of whose accounts survive, clearly had regular links with colleagues overseas. Many books came from Rouen or Antwerp by boat, a method which allowed them to come up the Thames as far as Burcot. The Elizabethan settlement, while it left the University library non-existent, allowed colleges and individuals to expand their libraries and although there were restrictions on direct importation in the 1560s, there was an active trade in second-hand books. Later another bookseller, Nicolas Clifton, held stock including 'a scattering of vernacular works in French and Italian: Boccaccio in French, "sonetti" and "canzoni", the *Heptameron*, a French 'life of the Trojans', and an Italian grammar'.[5] However, signs of local composition and printing in foreign languages in Oxford itself do not come before the proper establishment of printing in the city by Joseph Barnes, under the protection of the Chancellor of the University, Robert Dudley, first Earl of Leicester, in 1584. This was equally the time of the presence

[4] Sir Edmund Craster, *History of the Bodleian Library 1845–1945*, Oxford, 1952, pp. 76, 282.

[5] See C. Blagden, *The Stationers' Company*, London, 1960, ch. 6; R.J. Roberts, 'Importing Books for Oxford in the Sixteenth and Seventeenth Centuries', address to the Oxford Bibliographical Society, April 1994; and *The History of the University of Oxford*, vol. 3, ed. J. McConica, Oxford, 1986, p. 706.

in Oxford of Giordano Bruno, who however did not publish his Italian works there, but in London (1594).

The English interest in Spanish culture has been attributed to Queen Catherine of Aragon and certainly both Oxford and Cambridge produced influential sons who knew Spanish. Some eight years before Giordano Bruno's visit the controversial divine Antonio del Corro had, also through Leicester's influence, reached Oxford after leaving Spain in 1557 and moving around Europe quarrelling with the Calvinist Churches of Switzerland and France. After some difficulty he eventually became established for a few years at Christ Church and during this time got Barnes to publish his small anonymous book, *Reglas gramaticales para aprender la lengua española y francesa* in 1586. This had originally been written for the use of Prince Henry of Navarre in 1559. The work, printed in Oxford, is known with both an Oxford imprint and a Paris one and continued to be advertised at the Frankfurt book sales until the 1620s. It was accompanied by another work in Spanish, equally published by Barnes the same year but, for very different reasons, with a totally false Paris imprint. This was the work of a dissident Catholic, Alfonso de Valdés, whose *Dialogo en que particularmente se tratan las cosas acaecidas en Roma*, which blamed the Sack of Rome on the Pope, was the first of a long series of Spanish evangelical literature to be printed in England for distribution on the Continent. These books were not only the first books in a continental European language printed at Oxford, but also the first books in Spanish to be printed in England.[6]

A few years later Barnes went on to produce other grammars of this nature, namely Petrus Morletus' *Ianitrix sive institutio ad perfectam linguae Gallicae cognitionem acquirendam* (STC 18114), in 1596 and then two works by John Sanford, Chaplain of Magdalen College 1593–1616 and possibly the corrector of Corro's *Reglas*, namely *Le guichet françois. Sive janicula et brevis introductio ad linguam gallicam* (STC 21736) in 1604 and *A Grammar to the Italian Tongue* (STC 21735) in 1605. Evidence of linguistic ability is not easy to come by since it was neither required nor examined by the university, but there were clearly both native speakers and others about with a knowledge of the European languages. Fashionable young men wished to learn at least the rudiments although, as John Florio puts it, once they had a few words of Italian, French, or Spanish their enthusiasm quickly waned. It has been recorded that Sir Henry Wotton, who matriculated in 1584, studied Italian; that Merton, eleven years later, was paying an 'Italo cuidam' for teaching Italian and Spanish; and it has been suggested that George Sandys (St Mary Hall 1589) was taught, while at Oxford, by John Florio.[7] This seems unlikely since Florio, the son of an Italian Protestant refugee, appears to have arrived at Oxford sometime after 1578, matriculated as a poor commoner in 1581, married Samuel Daniel's sister, had a daughter by her in 1585, but had probably left for London in 1583. He could however have tutored Emanuel Barnes (Magdalen

[6] F. Madan, *Oxford books, vol.1: The Early Oxford Press*, Oxford, 1895.

[7] See McConica *Op. cit.* (n. 5), pp. 61, 696 and F. Yates, *John Florio*, London, 1934, p. 54.

1577), the son of the Bishop of Durham. Also at the same period, 1583–85, Giordano Bruno was paying his controversial visit to Oxford.

Other printed evidences of a knowledge of, and an interest in, foreign languages in the early seventeenth century include the collections of poems published in 1613 to mark the wedding of Frederick, the Elector Palatine, with Elizabeth, the eldest daughter of James I, the death of Sir Thomas Bodley and that of the son of the re-founder of Exeter College, Sir William Petre. Each of the first two books, *Epithalamia; sive lusus Palatini in nuptias Friderici comitis Palatini ad Rhenum et Elisabethae, Jacobi regis filiae* (STC 19022) and *Iusta funebria Ptolomaei Oxoniensis Thomae Bodleii* (STC 3194) contains two items in Italian, while *Threni Exoniensium in obitum Ioannis Petrei* (STC 19044) has one in French. There is also a strange work which somehow failed to be ready in time for the royal wedding and, dedicated to James I, had to appear later in the year. This was *Raccolta, d'alcune rime, del Cavaliere Lodovico Petrucci, Nobile Toscano, in piu luoghi, e tempi composte, & a diversi Principi dedicate; con Selva delle sue Persecutioni* (STC 19814). The volume contains Italian poems with Latin translations, a poem on the death of Sir Thomas Bodley, various testimonials and a brief account of the author's stay at New College (despite the fact that he seems to have matriculated at St Edmund Hall in 1612). There is also a long list of errata, starting notably with errors on the title-page!

During the years in which Sir Thomas Bodley was busy re-founding the university library he wrote regularly to his librarian, Thomas James, and these fascinating letters show his detailed attention to all aspects of library life and to the building up of its stock. In 1601 Bodley deliberately seeks to have Lomazzo's *Trattato dell'arte della pittura* in Italian as well as in English and in thanking the Regius Professor of Hebrew for a psalter he notes, 'but if it be that of Nebiensis, with five languages, besides two translations in latin, I have it already printed at Genua 1516'. Later he borrowed several dictionaries (nomenclators) for James's perusal and there is of course much evidence of his pressure on his booksellers to acquire foreign material of all sorts and even to travel abroad to do so. John Norton imported from the Frankfurt book fair, while John Bill went in 1602 to Paris and toured extensively in Italy, bringing back catalogues and books of all sorts.[8] He also visited Spain briefly in 1604, bringing back that rare bibliographical treasure, the first edition of *Don Quixote.* However Bodley was only concerned with serious works and in a famous passage condemns 'idle bookes & riffe raffes', although later this contemporary of Shakespeare can write:

> Happely some plaies may be worthy the keeping: but hardly one in fortie. For it is not alike in Englishe plaies, & others of other nations: because they are most esteemed, for learning the languages & many compiled, by men of great fame, for wisdome & learning, which is seldom or never seene among us'.[9]

The main evidence of an interest in vernacular European texts at this period must nevertheless be the earliest printed catalogue of the Bodleian, published by Barnes in

[8] G.W. Wheeler, *The Earliest Catalogues of the Bodleian Library*, Oxford, 1928, p. 55.

[9] *Letters of Sir Thomas Bodley to Thomas James*, ed. G.W. Wheeler, Oxford, 1926, especially p. 222.

1605.[10] Compiled by James – like Bodley, another Marian exile – the catalogue is an expanded shelf-list, taking the library faculty by faculty: theology, jurisprudence, medicine and arts. Within each section the books were catalogued as they were shelved, in roughly alphabetical order by author; the folios, chained on the main stall bookshelves, coming first in each letter group, while the smaller formats, kept in the 'closets', follow all together afterwards. Dictionaries and grammars as well as literary texts therefore appear in two places in the catalogue among the 'Libri Artium'.

The 1605 catalogue naturally contains the Bible in a number of different European languages: a Danish edition of 1550, French ones of 1580 and 1597, Luther's German text of 1580, Italian editions of 1517 and 1545, Spanish ones of 1553 and 1602, and even a version in Slavonic. While works in German are few there were a number of historical items in both French and Italian. From the days of the early English humanists like Linacre and Grocyn, Oxford scholars had made the journey to Italy and had studied there. It is not surprising therefore that although both Bodley and his first librarian saw the library as primarily a bastion of Protestant controversy nevertheless continental European works in the vernacular were present there in some numbers. It has been suggested that Spanish and French works outnumber Italian in all sections except the Arts, but that in total there were over five hundred books in the latter language.[11] Here literature was evidently particularly well represented. There were four editions of Dante (1484, 1512, 1544 and 1568), various sixteenth-century editions of vernacular works by Petrarch, a manuscript of Boccaccio's 'des cas des nobles hommes & femmes' as well as the *Decamerone* of 1522 and 1552. Sannazaro was represented by the *Arcadia* of 1549, his sonnets and *Opera* of 1593, Ariosto by an *Orlando furioso* of 1548 and other works, Aretino by his comedies of 1535, not to mention works by Annibale Caro, Doni, Dolce, Parabosco, Tasso, Trissino and others, from the 'Mondo al rovescio' and some 'Lettere amoroso' to Buoni's *Tesoro di proverbi*. Among the French books there were Commines' memoirs of 1559, Marot's works of 1551, Desportes' of 1581, Ronsard's of 1584, and Montaigne's *Essais* of 1598, not to mention a work by Pontus de Tyard and a work entitled *Les esguillons d'amour*. On the Spanish side the Bodleian could provide two copies of the 1595 *Celestina*, Alemán's *Guzmán de Alfarache* of 1603 and two or three other works. A number of Italian works in particular occur in the appendix pages at the end of the catalogue and since several date from the first years of the seventeenth century it seems likely that they represent the fruit of John Bill's 1602 tour of Italy.

This chapter is however more particularly concerned with foreign language dictionaries and grammar books. These occur throughout the Arts sections of the catalogue and while there are many relating to Greek, Hebrew, Latin and other

[10] T. James, *Catalogus librorum bibliothecae publicae quam vir ornatissimus Thomas Bodley… instituit*, Oxford, 1605, and *The First Printed Catalogue of the Bodleian Library, 1605: a facsimile*, Oxford, 1986. Mr M. Heaney has kindly assisted with various identifications.

[11] J.L. Lievsay, *The Englishman's Italian Books, 1550–1700*, Philadelphia, 1969, pp. 34–35. For much information on early Spanish books in Oxford, see the exhibition catalogue, *Hispanic manuscripts and books before 1700, published in connection with the Primer Congreso internacional de Hispanistas*, Oxford, 1962.

languages more specifically associated with biblical studies, the following list has concentrated rather on those which relate to the learning or study of the vernacular European languages.

The 1605 catalogue contained some fifteen thousand works and, as the size of the Appendix suggests, the library was growing fast. The year in which it appeared also saw the first publication of the first part of Cervantes' *Don Quixote* and the acquisition by the library of that work, an act of remarkable prescience and good luck. However, at first members of the Arts faculty were not admitted to the library which was conceived of as a place for the three superior faculties (theology, medicine and law) only.[12] By 1603 nevertheless, they could gain some limited access provided that they had a college recommendation, that they came in academical costume, that they only consulted Arts books – and that they did not incommode theological students.

During Sir Thomas Bodley's lifetime the entrance to the library was, as it had been from the beginning, by means of two stairs on either side of the west end of what is now termed Duke Humfrey's Library. It seems likely that the classified shelves therefore started their sequence from that end, although theology, as the queen of the sciences, may always have had the preferable southern side. The growth of the library caused rapid expansion and the smaller formats, neither foreseen nor liked by the founder, were at first kept in the locked studies at the far (that is east) end of the library, although a gallery over the west, entrance, end had soon to be put up. Several years after the publication of the catalogue and just before the death of the founder the area called Arts End was built on at the eastern extremity of Duke Humfrey's in 1610–12. This new area allowed for a considerable expansion of the library stock and while the senior faculties' (theology, medicine and law) stock was redisposed around the central portion of the library, represented by Duke Humfrey's, the Arts faculty books were relegated, so to speak, to the new shelves then still at the far end from the entrance. Shortly after Sir Thomas' death however the university, engaged in planning the further development of the 'schools' and library around the new quadrangle which Bodley's bequest funded in part, provided staircases at the east end, which soon became the main entrance. The Arts books became therefore the first to be seen on entering the library.[13]

The growth of the library also forced the librarian to produce an updated edition of the catalogue. This appeared in 1620, the year in which James retired. Comparison of the two editions naturally allows the acquisitions of the intervening years to be identified, but for the purposes of the present article another document is perhaps of equal interest. The first librarian, Thomas James, was more liberal in his approach than the founder; he started the compilation of a manuscript subject catalogue as soon as the 1605 printed volume was finished, and on his retirement in 1620 turned to completing the Arts section of this (finished around 1623): 'Ad Spectatae probitatis

[12] Wheeler *Op. cit.* (n. 8), p. 102.

[13] S. Gillam, *The Divinity School and Duke Humfry's Library*, Oxford, 1988, and W.D. Macray, *Annals of the Bodleian Library Oxford*, 2nd edn, Oxford, 1895.

& optimae spei Tyrunculos vel candidatos in Facultate Artium', as he dedicated the manuscript.[14] This text concentrates on dictionaries, grammars and language learning books rather than literature, and the items which it gives and which do not figure in the 1605 catalogue have been included in the list below. The identification of foreign material in the college libraries at this date has not been attempted, but one can note that New College had received a number of French books in 1584, that Sir Arthur Throckmorton's continental books reached Magdalen in 1626 and Lord Herbert of Cherbury's were received by Jesus in 1648.[15]

There are therefore in the early seventeenth century four successive lists: the first and main part of the 1605 catalogue, the appendix containing the latest accessions, the later, manuscript, subject index, and the 1620 catalogue. The following list places all relevant works in one alphabetical sequence by their authors' names, gives a short title and imprint, records the style of binding and any information on it, states in which source the item first appears, and finally gives its present shelf-mark. Unidentified anonymous items are under the first word of their title.

1. Accademia della Crusca: *Vocabolario*. Venice: 1612, folio.
   Brown calf, two ties, two staple holes. ?Oxford binding.
   9 on foredge.
   1620 *sub* Crusca. *Olim* C 1. 16 Art. C 2. 9. Art.
2. Accarigi, Alberto: *Vocabolario, grammatica, et orthographia de la lingua volgare con ispositioni di molti luoghi di Dante, del Petrarca, et del Boccaccio*. Cento: 1543, quarto.
   Contemporary limp vellum, two ties, shelf-mark written on spine [Figure 1].
   1605 pp. 283. 4° A 9 Art.
3. Alberti, Lorenzo: *Teutsch Grammatick oder Sprach-Kunst certissima ratio discendae linguae Alemanorum sive Germanorum*. Strasburg: 1573, octavo.
   Contemporary limp vellum, shelf-mark written on spine.
   1605 pp. 284. 8° A 13 Art.
4. Alexander, *Grammaticus*: (Opus interpretatum a L. de Guaschis). Venice: per Andream Catherensis & I. de Leodio, 1483, folio.
   GW 989. Early-19th-century calf, gilt Bodleian stamp.
   No trace of earlier shelf-marks.
   1605 pp. 277. Auct O 4.3.
5. Alunno, Francesco: *Le richezze della lingua volgare*. Venice: 1551, folio.
   Contemporary paneled calf, blind stamped decoration, two ties, two sets of staple holes, nos. 12 and 3 on foredge. Signature of Sir Michaell [*sic*] Dormer on title-page. See also no. 28.
   Olim A 7. 12 Art. A 1. 3 Art.
   1605 pp. 281. Auct 2 R 2. 12.

[14] Wheeler *Op. cit.* (n. 8), pp. 94–116. The original does not survive but a copy is in the library as MS Rawl. D. 984, which has been quoted as evidence that the work was in the library by 1623. I am indebted to Mr W.H. Clennell for my knowledge of this source.

[15] P. Morgan, *Oxford Libraries outside the Bodleian*, Oxford, 1973, pp. 50, 66, 90.

6. Antonio, of Lebrija: *Vocabularius Nebrissensis; lexicon, id est dictionarium nuperrime ex Hispanice in Gallicum traductum eloquium*. Paris: 1523, octavo.
Brown calf, panels with blind fillets, 4 ties, title written on foredge. Rebacked.
1620 *sub* Nebrissa. *Olim* O 1. 11 Med. 4° D 18 Jur.
7. Antonio, of Lebrija: *Vocabularium Nebrissense; ex Latino sermone in Siciliensem et Hispaniensem traductum. Adjunctis L. C. Scobaris additionibus*. Venice: 1570, folio.
Bound in vellum, staple holes. Heavily damaged and nearly turned out in 1865 – see F. Madan's manuscript note on repair.
1605 pp. 359. G 4. 9 Art.
8. Antonio, of Lebrija: *Dictionarium quadruplex*. Granada: 1572, folio [Latin/Spanish, Spanish/Latin and two geographical indices].
Brown morocco, 3 blind fillet panels with ornaments at corners, two circle ornaments in centre around small blind coat of arms – probably a Spanish binding.
1605 pp. 359. *Olim* N 2. 3 Art. G 4. 5 Art.
9. Antonio, of Lebrija: *Dictionarium quadruplex opera J. Lopez Serrani auctum*. Antequera: 1600, folio.
Blind ruled calf, two ties, two staples, 15 on foredge.
1605 pp. 619. *Olim* N 2. 9 Art. D 7. 15 Art.
10. *Arte de la lengua Mosca de los Indios del nuevo reyno de Granada, en las Indias Occidentales.* [?Seville: ?1603], duodecimo.
Bound with no. 77.
1605 Not listed; 1620 *sub* Mosca. 8° V 23 (2) Art.
11. Benesovinus, Matthaeus: *Grammatica Bohemica*. Prague: 1577, octavo.
Contemporary limp vellum, two ties, shelf-mark written on spine and lower cover.
1605 pp. 295. 8° B 22 Art.
12. Berlaimont, Noel de: *Der new Barlamont oder Gemeine Gesprache zu Teutsch und Frantsosisch beschrieben... les devis familiers de Berlamont*. Cologne: 1587, octavo.
Bound with no. 63.
1605 pp. 296. 8° O 3 (2) Art.
13. [Bukvar]: *Slovenskaya rekomaya grammatika*. [Ostrog]: 1598, octavo.
Rebound but retaining the original wooden boards and a strip of the blind-stamped leather. Arch shelf-mark at top of title-page. Presented by Sir Richard Lee 1602.
1605 (pp. 349 as *Moscoviticum Abecedarium*), 1620 as Arch A 157. The 1605 catalogue suggests that there were two items, described there as *Grammatika Moscovitica*, and given the shelf-marks Arch A 157 and 158, which is confirmed by an early handlist. The second 'copy', possibly in fact not the same work, has not been located. It should not be confused with Zizanii's *Grammatika Slovenska* (no. 82 below), which was also a gift from Sir Richard Lee, but which was Arch A 153. 8° R 19 Art. BS.

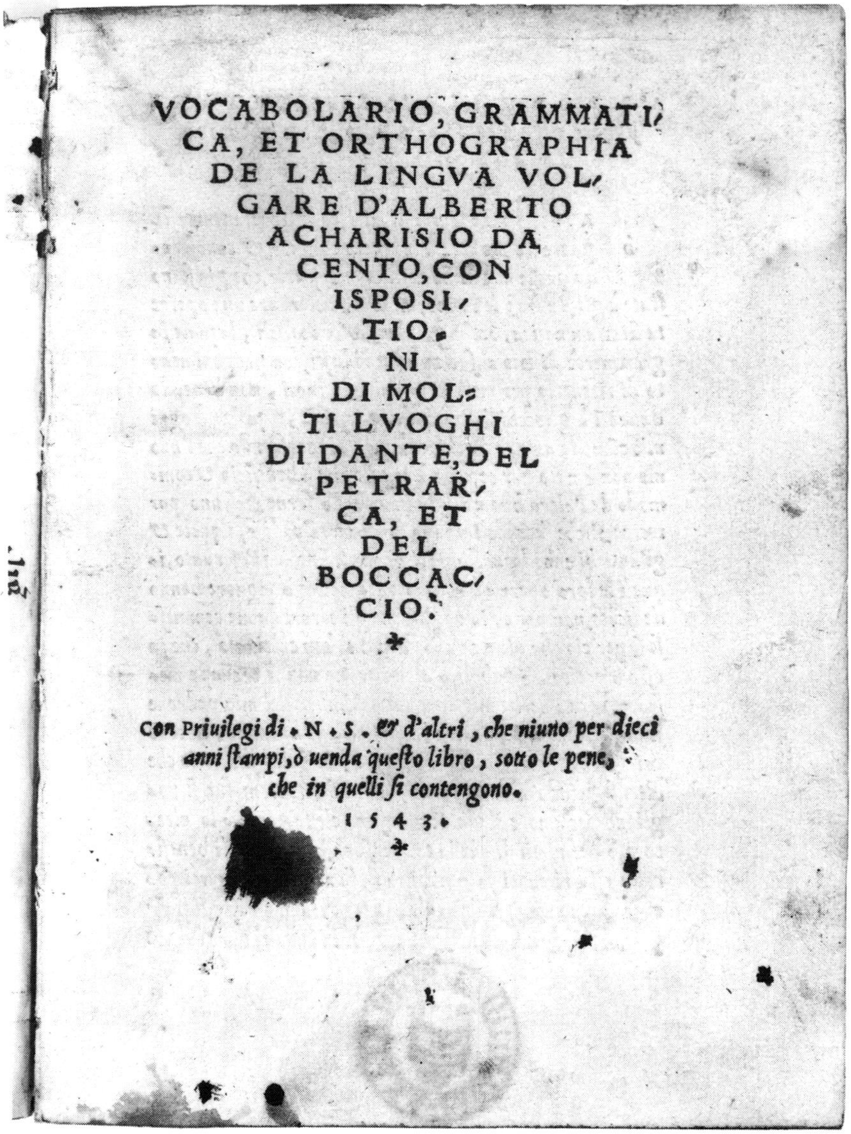
VOCABOLARIO, GRAMMATI-
CA, ET ORTHOGRAPHIA
DE LA LINGVA VOL-
GARE D'ALBERTO
ACHARISIO DA
CENTO, CON
ISPOSI-
TIO-
NI
DI MOL-
TI LVOGHI
DI DANTE, DEL
PETRAR-
CA, ET
DEL
BOCCAC-
CIO.

Con Priuilegi di. N. S. & d'altri, che niuno per dieci
anni ſtampi, ò uenda queſto libro, sotto le pene,
che in quelli ſi contengono.
1543.

Figure 1. A. Accarigi, Vocabolario, grammatica, et orthographia de la lingua volgare. *Cento, 1543. Bodleian Library (4° A 9 Art)*

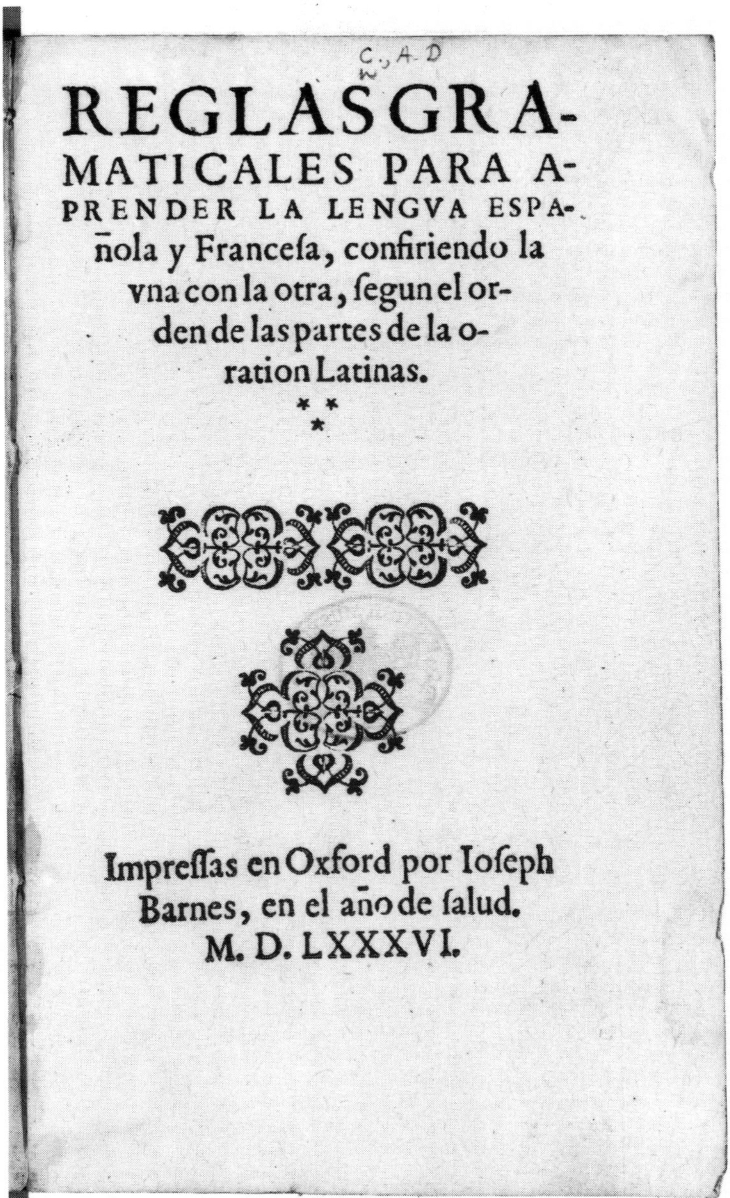

REGLAS GRA-
MATICALES PARA A-
PRENDER LA LENGVA ESPA-
ñola y Franceſa, confiriendo la
vna con la otra, ſegun el or-
den de las partes de la o-
ration Latinas.

Impreſſas en Oxford por Ioſeph
Barnes, en el año de ſalud.
M. D. LXXXVI.

Figure 2. A. del Corro, The Spanish Grammar. *Oxford, 1586. British Library (C66 f.33)*

POLONICÆ
GRAMMATICES
INSTITVTIO.

In eorum gratiam, qui eius linguæ elegantiam cito & facile addiscere cupiunt.

CRACOVIÆ.
Apud Mathiam Wirzbiętam, Typographum Regium.
1568.

Figure 3. P. Stojenski, Polonicae grammatices institutio. *Cracow, 1568. Bodleian Library (8° P23 (1) Art)*

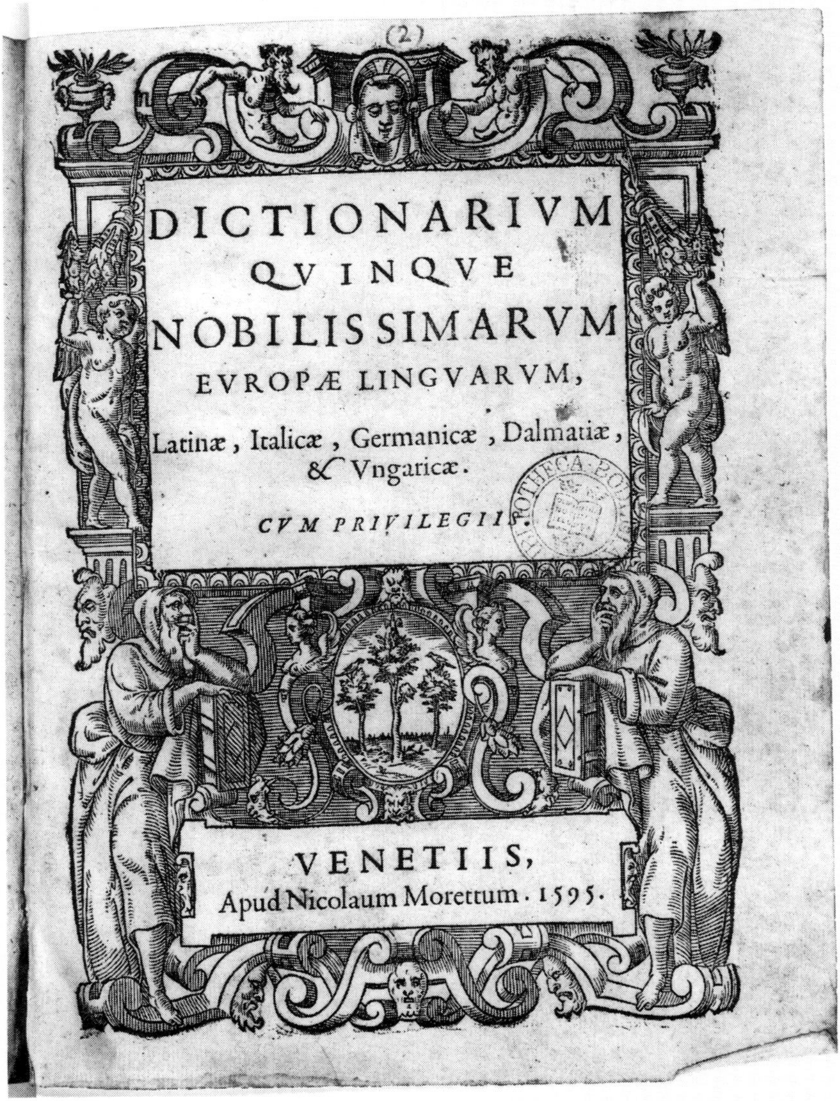
DICTIONARIVM
QVINQVE
NOBILISSIMARVM
EVROPÆ LINGVARVM,
Latinæ, Italicæ, Germanicæ, Dalmatiæ,
& Vngaricæ.
CVM PRIVILEGIIS.
VENETIIS,
Apud Nicolaum Morettum. 1595.

Figure 4. *F. Verantius,* Dictionarium quinque nobilissimarum Europae Linguarum. *Venice, 1595. Bodleian Library (4° G 12(2) Art)*

14. Calepinus, Ambrosius: *Dictionarium decem linguarum*. Lyons: 1586, folio.
    1605 pp. 298. C 2. 2 Art.; 1620 but not in 1738. Possibly replaced by the 1616 edition.
15. Casas, Christoval de las: *Vocabulario de las dos lenguas Toscana y Castellana*. Venice: 1597, octavo.
    This edition not in Palau.
    MS Rawl D 984. Not in 1605, 1620. The present 1570 copy (4° F 33 Med.) bought 1825.
16. Claius, Joannes: *Grammatica Germanicae linguae.* Leipzig: 1578, octavo.
    Limp vellum, overlapped edges, two ties, shelf-mark on spine.
    1605 pp. 306. 8° C 30 Art. 8° C 94 Art.
    Reprinted 1894 in *Ältere deutsche Grammatiken*, no.2.
17. Corro, Antonio del: *The Spanish grammar...with a dictionarie adjoyned to it*. London: 1590, quarto.
    Brown reversed calf, triple blind fillet, two ties, printed waste. For the Oxford edition of 1586, see introduction above. A facsimile edition, *Reglas gramaticales para aprender la lengua española y francesa*, ed. by L. Nieto, was published at Madrid in 1988. A specially bound copy of this facsimile was presented to Her Majesty Queen Sophia of Spain on the occasion of her visit to Oxford in 1989 [Figure 2].
    MS Rawl D 984. Not traced in 1605 or 1620. 4° C 71 Art.
18. [Corso, Rainaldo]: *Fondamenti del parlar Thoscano.* Venice: 1550, octavo.
    Vellum, probably Italian; lettered down the spine. MS notes.
    1620 *sub* Corso. 8° C 106 Art.
19. Cotgrave, Randal: *A dictionarie of the French and English tongues.* London: 1611, folio.
    1620 (as Fo. 28). Presumably replaced by the Selden copy, now C 2.11 Art Seld. For the Bodleian's substitution of the Selden copy of books for any earlier held one, see page 52.
20. Dasypodius, Petrus: *Dictionarium Latinogermanicum et Germanolatinum*. Iam quarto recognitum. Strasburg: (c. 1570), octavo.
    Contemporary vellum, overlapped edges, shelf-mark on overlap. Signature of Modestin Himbracht 1579.
    1605 pp. 313. 8° D 8 Art.
21. *Dictionarium Latino-Gallicum acc. Graecae dictiones suis quaeque Latinis respondentes.* Paris: 1576, folio.
    1620 D 3. 3 Art. 1674 & 1738 as 1570.
    Not identified 1993.
22. *Dictionarius seu nomenclatura quatuor linguarum Latine Italicae, Polonice, & Theutonice.* Cracow: [c. 1575], duodecimo.
    Bound with no. 44. 8° V 12 (2) Art.

23. *Dictionarium tetraglotton, seu Voces Latinae omnes, et Graecae his respondentes, cum Gallica & Teutonica (quam passim Flandricam vocant) earum interpretatione.* Franeker: 1610, octavo.
Brown leather, red edges, shelf-mark on foredge and spine.
1620. 8° D 40 Art.
24. *Dictionnaire François-Latin*, augmenté outre les précédentes impressions d'infinies dictons françoises, spécialement des mots de marine, vénerie, & faulconnerie. Recueilli des observations de plusieurs hommes doctes: entre autres de M. Nicot [Based on the dictionary of Robert Estienne and edited by J. Du Puy]. Paris: 1573, folio.
Brown calf, gilt fillet and oval centrepiece, two ties, staple hole. Probably a French binding. Rebacked.
1605 pp. 359. *Olim* N.2. 2 Art. H 2. 7 Art.
25. Doergangk, Henricus: *Institutiones in linguam Gallicam*. Cologne: 1604, octavo.
Limp vellum, shelf-mark on foredge. Bound with the following item.
1605 pp. 599. 8° D 15 Art. 8° D 16 (1) Art.
26. Doergangk, Henricus: *Institutiones in linguam Italicam*. Cologne: 1604, octavo.
Limp vellum, shelf-mark on foredge. Bound after the previous item.
1605 pp. 599. 8° D 16 (2) Art.
27. Dolce, Ludovico: *Espositione di tutti i vocaboli, et luoghi difficili ... [nel] Ariosto.* Venice: 1550, quarto.
Vellum with overlapped edges, shelf-mark on spine and overlap.
1620. 4° A 12 (3) Art.
28. Dolce, Ludovico: *Osservationi nella volgar lingua*. Venice: 1550, octavo.
Nineteenth-century binding. Signature of Sir Michaell [*sic*] Dormer. See also no. 5.
1620. 8° D 6 (3) Art.
29. Donatus, Aelius: *Elementa una cum traditione Polonica et Germanica*. Cracow: 1560, octavo.
White leather on bevelled wooden boards. Probably a Polish binding.
No shelf-mark on binding.
1605 pp. 314. 8° D 17 Art.
30. *Ein schoene Lere mit kurczen sinnen begriffen Deutsch und Polnisch beiden sprach zu reden*. Cracow: 1584, duodecimo.
Bound with no. 44.
Not found in early catalogues nor in *VD 16*.
8° V 12 (3) Art.
31. Emmelius, Helfricus: *Nomenclator quadrilinguis, Latino Germanico-graeco gallicus.* (Basle: 1600), octavo.
Bound in vellum. Two copies of the work.
1620. 8° E 18 (3,4) Art.
32. Emmelius, Helfricus: *Sylva quinque linguis vocabulorum et phrasium Germanicae, Latinae, Graecae, Hebraicae, Gallicae linguae*. Basle: (1592), octavo. Bound with the previous item.
8° E 18 (5) Art.

33. [Estienne, Robert]: *Dictionarium Latinogallicum*. Ex hoc, Latini sermonis cum Gallico idiomate consensum ita. Editio postrema. [Latin, Greek, French]. Paris: 1570, folio.
Calf, two ties, staple hole, 5 on foredge, rebacked.
1620 *sub* Dictionarium. E 3.5. Art.
34. Fabrini, Giovanni: *Teorica de la lingua dove s'insegna a trasmutare tutte le lingue ne la lingua Latina*. Di nuova ristampata. Venice: 1588, octavo.
Vellum, two ties, shelf-mark on spine.
1620. 8° F 31 Art.
35. Fenice, Giovanni Antonio (Pierre Canal): *Dictionnaire françois et italien*. (Geneva) J. Chouet: 1598, octavo.
Brown calf, centrepiece ornament, two ties, shelf-mark on foredge.
1605 pp. 307. 8° C 8 Art.
36. Fenice, Giovanni Antonio (Pierre Canal): *Dittionario italiano e francese* [reverse of above]. (Geneva) J. Chouet: 1598, octavo. Bound with the above.
1605 pp. 307. 8° C 8 Art.
37. Florio, John: *A worlde of wordes, or most copious dictionarie in Italian and English.* London: 1598, folio.
1605 pp. 603. Present in 1620 catalogue but absent from the 1674 one. Presumably replaced by the 1611 edition.
38. Florio, John: *Queen Anna's new World of words, or Dictionarie of the Italian and English tongues.* London: 1611, folio.
1620 (as Fo. 6), probably replaced by Selden copy (F 2. 26 Art Seld).
39. Frischlinus, Nicodemus: *Nomenclator trilinguis, graeco-latinogermanicus*. Frankfurt: 1600, octavo.
Vellum with overlapped edges, shelf-mark on foredge.
1605 pp. 602. 8° F 23 (2) Art.
40. Frisius, Joannes: *Dictionarium Latino-Germanicum*. Ed. nuova. Zurich: 1556, folio.
Fine binding, bevelled wooden boards, two roll frames of saints with an ornamental frame between, staple hole, 1 on foredge.
1605 pp. 601. T 7. 1 Jur.
41. Gabriele, Giacomo: *Regole grammaticali.* Venice: 1545, quarto.
Brown calf, medallion roll panel, two ties, low staple mark, shelf-mark on foredge. Bound with no. 72.
1605 pp. 309 (*sub* Cabriele, 4° P 2. 11 Art). 4° P 63 (1) Art.
42. Gelli, Giovanni Battista: *Dialogo sopra la difficulta delle ordinare la lingua che si parla & scrive in Firenze.* Florence: 1551, octavo.
1605 & 1620. Entered separately but in fact part of no. 43 below.
8° G 17 Art.
43. Giambullari, Pierfrancesco: *De la lingua che si parla & scrive in Firenze e un dialogo di Gio. Bat. Gelli sopra la difficulta delle ordinare detta lingua*. Florence: (1551), octavo.

Bound in vellum with overlapped edges, two ties, green edges, shelf-mark on spine. Italian manuscript notes at the beginning.
1605 pp. 331. 8° G 17 Art.

44. Heyden, Sebaldus: *Formulae puerilium colloquiorum pro primis tyronibus, ex comicorum campo hinc inde collectae ac Germanico, Polonico, idiomate illustratae.* Cracow: 1584, duodecimo.
Bound with three other relevant items, nos 22, 30, 71.
Vellum with overlapped edges, two ties, shelf-mark on spine.
1605 pp. 336. 8° V 12 (5) Art.
45. Hulsius, Levinus: *Dictionarium Teutsch und Frantzosisch.* Frankfurt: 1607, octavo. Brown calf, blind fillet and centrepiece, two ties, red edges. Probably an Oxford binding.
1620. 8° H 49 Art.
46. Hulsius, Levinus: *Dictionaire François-Allemand & Allemand-François.* Der 3. Truck. Frankfurt: 1607, octavo.
Bound as preceding item. MS Rawl D 984. Not listed separately in 1620 but evidently part of no. 45 above. 8° H 50 Art.
47. Junius, Hadrianus: *Nomenclator octolinguis omnium rerum propria nomina continens.* Nunc renovatus. Accessit huic postremae ed. alter Nomenclator H. Gerembergij opera ac studio. Geneva: 1602, octavo.
1605 pp. 339. 8° 1. 4 Art.
First published in 1567 by Adriaen de Jonge (1511–75), it is based on the Latin but contains words and expressions translated into Greek, German, Dutch, French, Italian, Spanish and sometimes English as well. There were numerous sixteenth-century editions. See C.S.M. Rademaker, 'De Nomenclator van Hadrianus Junius', *Hermeneus,* xxxix, 1968, pp. 217–27.
48. Kiel, Cornelius van: *Etymologicon linguae Teutonicae, sive Dictionarium Teutonico-Latinum.* [Alcmar?]: 1599, quarto.
1620 as 4° K 2 Art, 1674. Probably replaced by the 1613 or 1620 editions.
49. Kiel, Cornelius van: *Etymologicum Teutonicae linguae, sive Dictionarium Teutonico-Latinum*. Quartae editio. Alcmar: 1613, octavo. 4° K 1 Art.
50. Kiel, Cornelius van: *Etymologicon linguae Teutonicae, sive Dictionarum Teutonico–Latinum, cum auctario Potterii*. Middleburg: 1620, octavo.
Probably replaced the 1599 edition. 8° Z 199 Th.
51. La Faye, Abraham de: *Linguae Gallicae et Italicae hortulus amoenissimus. Plaisant jardinet, ensemble vii. dialogues communs françois italiens.* Halle: 1608, quarto.
1620. 8° F 40 Art.
52. Maaler, Josue: *Die teutsch spraach. Alle worter, namen, un arten zu reden in hochteutscher spraach, mit gutem Latein vertolmetscht. Dictionarium Germanico-latinum novum, hoc est linguae Teutonicae, superioris praesertim, thesaurus*. Zurich: 1561, quarto.
Reversed calf, diamond stamp in centre, two ties, staple hole, shelf-mark on foredge. Signature of Thomas Skeffington.
1605 pp. 368. 4° W 10 Art.

53. [Maupas, Charles]: *Grammaire françoise contenant règles très certaines et addresse très aseurée à la naive connoissance & pur usage de nostre langue: En faveur des estrangers qui en seront désireux.* Par C[harles].M[aupas].B.L. Blois: 1607, octavo. The *privilège* is to Charles Maupas, *chirugien* of Blois. Vellum, two ties, shelf-mark on foredge.
MS Rawl D 984, 1738 *sub* Maupas. 8° G 82 Art.
54. Megiserus, Jerome: *Dictionarium quatuor linguarum videlicet Germanicae, Latinae, Illyricae (quae vulgo Sclavonica appellatur), & Italicae sive Hetruscae* [columns headed Deutsch, Latine, Windisch, Italice]. Gratz: 1592, octavo.
1605 pp. 617. 8° G 42 (2) Art.
55. Minsheu, John: [Ductor in linguas] *The guide into tongues. Vocabularum hispanolatinum*. London: 1617, folio.
1620. S 5. 14 Art. Not present 1993. Probably replaced by the 1623 edition (*olim* A 1.36 Art. M 4. 12 Art.).
56. Molina, Alonso de: *Vocabulario en lengua castellana y mexicana*. [2 parts] Mexico: 1571, folio.
1605 pp. 615 (Q.M.5.11), 1620. The copy now present (S. Seld. d 46) and previously in other Seld. shelf-marks was undoubtedly Selden's and has his inscription. The original copy was presumably disposed of as a duplicate when Selden's came.
57. Molnar, Albert: *Dictionarium Latinoungaricum*. Nuremberg: 1604, octavo. Vellum with overlapped edges, two ties, shelf-mark on foredge and spine. Bound with second part.
1605 pp. 617. 8° M 28 (1) Art.
58. Molnar, Albert: *Dictionarium Ungarico-Latinum*. Nuremberg: 1604, octavo. Bound with part 1 above.
1605 pp. 617. 8° N 28 (2) Art.
59. Monosinus, Angelus: *Flores Italicae linguae libri novem*. Venice: 1604, quarto. Brown leather, blind fillet, two ties, staple hole.
1605 pp. 615. BB 10. 1 Med.
60. Montemerlo, Giovanni Stefano: *Delle phrasi toscane lib. xii.* Venice: 1566, folio. Brown calf, two ties, staple hole, numbered 4 on foredge, blue edges.
1605 pp. 352, *olim* M 6. 7 Art. G 4. 4 Art.
61. Murmelius, Joannes: *Dictionarium variarum rerum, cum Germ. at[que] Polon. interpretatione tam pueris tam adultis utilissimus.* [Cracow: 1584], octavo.
Bound in vellum with overlapped edges, two ties, shelf-mark on upper cover and spine. Bound with no. 76.
1605 pp. 352. 8° P 23 (2) Art.
62. Nicot, Jean: *Thresor de la langue françoise, tant ancienne que moderne. Auquel sont ajoutés les mots propres de marine, vénerie & faulconnerie cydevant ramassez par Aimar de Ranconnet. Avec la grammaire francoyse.* Paris: 1606, folio.
1620 *sub* Ranconnet as F° 53. M 9.10. Jur.

63. Oelingerius, Albertus: *Vnderricht der hoch teutschen Spraach: Grammatica seu institutio verae Germanicae linguae*. Strasburg: 1573, octavo.
Bound in vellum with overlapped edges, shelf-mark on foredge. Bound with no.12.
1605 pp. 363. 8° O 3 (1) Art.
64. Oudin, Cesar: *Refranes o proverbios espanoles traduzidos en lengua francesa*. Paris: 1605, duodecimo.
Vellum spine, marbled boards of early nineteenth century.
1620. 8° R 27 Art.
65. Peletier, Jacques: *Dialogue de l'ortografe françoise*. Paris: 1550, octavo.
Manuscript wrapper.
MS Rawl D 984 (as 8° F 30). 8° S 239 Art.
66. Percyval, Richard: *A dictionarie in Spanish and English*. London: 1599, folio.
1674 (*sub* Minsheu) as M 3. 1 Art. Probably replaced by later edition under name of Percyval or Minsheu. The work is now present only in the Douce and Lister copies.
Not identified 1993.
67. Percyval, Richard: *A dictionarie in Spanish and English*. London: 1623, folio.
Reversed calf, two ties, staple hole, 12 on foredge.
Olim A 1.36 Art.
No evidence before 1674. M 4. 12 Art.
68. [Pergamino, Giacomo]: *Il memoriale della lingua*. Venice: 1602, folio.
Plain brown calf, staple hole, two ties, 2 on foredge.
1605 pp. 368. H 8. 2 Art.
69. Pomi, David de' (Semah saddik): *Dittionario nouo Hebraico dechiarato in tre lingue* [Hebrew, Latin, Italian]. Venice: 1587, folio.
Brown calf, blind fillet, staple hole, 8 & 6 on foredge.
1605 pp. 366. *Olim* G 2. 8 Art Seld. HH 6 Art Seld.
70. Pontanus, Johannes Isacius: *Glossarium Prisco-Gallicum* [in his *Itinerarium Galliae Narbonensis*]. Leiden: 1606, duodecimo.
1620 (8° R 27 Art.) but probably replaced by Selden copy (8° I 17Art Seld.).
71. [Pressius, P.]: *Wokabulàr wno-/we sprawený a rozssýrený* [By P. Pressius according to the dedicatory poem] *Vocabularium trilingue pro usu scholarum diligenter & accurate editum. Nunc vero tertia editione ... auctum*. Prague: 1581, duodecimo. [Latin, Czech, German]. Bound with no. 44.
1605 pp. 408. 8° V 12 (1) Art.
72. Priscianese, Francesco: *Della lingua romana*. Venice: 1540, quarto. Bound after no. 41 above. Brown calf, medallion roll panel, two ties, low staple holes, shelf-mark on foredge.
1605 pp. 365. 4° P 64 (2) Art.
73. *Promptuarium linguae Lat. Gr. & Gall*. Antwerp: 1591.
Probably an edition of the *Promptuarium Latinae linguae...una cum Gallica earum interpretatione* [followed by *Les mots françois*] published by Plantin in 1568.
MS Rawl. D 984 but not traced in 1605, 1620, etc.

74. Ruscelli, Girolamo: *Commentarii della lingua italiana libri vii di nuovo posti in luce* [by Vincenzo Ruscelli]. Venice: 1581, octavo. Brown leather, blind centrepiece ornament, two ties, staple hole, shelf-mark on foredge.
 1605 pp. 378 as R 3. 13 Art. 4° R 27 Art.
75. Salazar, Ambrosio de: *Espejo general de la grammatica en dialogos* [In French and Spanish]. Rouen: 1622, octavo.
 Vellum with overlapped edges, shelf-mark on spine and foredge.
 MS Rawl D 984, 1674. 8° S 130 Art.
76. [Stojenski (Statorius), Piotr]: *Polonicae grammatices institutio.* [Cracow: 1568], duodecimo.
 Bound with no. 61 [Figure 3].
 1605 pp. 364. 8° P 23 (1) Art.
77. Torres Rubío, Diego de: *Grammatica y vocabolario en la lengua general del Peru, llamada Quicha, y en la lengua española*. Seville: 1603, octavo.
 Vellum, two ties, shelf-mark on foredge and spine. Bound with *Arte de la lengua Mosca* and *Doctrina Christiana en lengua Mosca de los Indios del nuevo reyno de Granada, en las Indias Occidentales.*
 1605 pp. 637. 8° V 23 Art.
78. *Trium linguarum dictionarium Teutonicae, Latinae, Gallicae.* Franeker: 1587, octavo.
 Vellum, overlapped edges, two ties, modern lettering up spine.
 1605 pp. 313. 8° D. 3 Art.
79. [Verantius, Faustus]: *Dictionarium quinque nobilissimarum Europae linguarum* [Latin, Italian, German, Dalmatian, Hungarian]. Venice: 1595, quarto.
 1605 pp. 311. D 1. ° Art., 1620 [Figure 4]. 4° G 12 (2) Art.
80. *Vocabulario nuovo, con il quale da se stessi, si puo benissimo imperare diversi linguaggi, cioe, Italiano, Greco, Turcho, Todesco*. Nuovamente ristampato. Venice: 1599, octavo.
 Gilt floral paper on boards, shelf-mark on foredge.
 1605 pp. 637. 8° M 33 Th.
81. Vulcanius, Bonaventura: *Thesaurus vtriusq; Linguae.* Leyden: 1600, folio.
 1605 pp. 398 & 1620 as T 3. 2 Art.
 Not present 1993.
82. Zizanii, Lavrentii: *Grammatika slovenska...novo s"st. L.Z.*. Vilna: 1596, octavo. Bound with the same author's *Nauka ku chitanyu, rozumenyu pisma slovenskogo.* [Vilna: 1596], octavo [imperfect].
 MS Rawl D 984 as Arch A 153.
 Nineteenth-century half-leather binding. Presented in 1602 by Sir Richard Lee. Arch shelf-mark on title-page. 8° R 17 Art. BS.

The four sources used therefore include between them eighty-two works, ranging in date from 1483 to, effectively, 1622, but with the majority dating between 1550 and 1605. The foreign books were published in many parts of Europe and one even in

Mexico. Venice alone accounts for over a quarter of the total and is particularly strong in the earlier period as one might expect. East European and German-speaking towns are well represented, including Basle, Cologne, Cracow, Frankfurt, Halle, Leipzig, Nuremberg, Prague, Strasburg, Tübingen, and Zurich. The Low Countries and the French-speaking world feature Alcmar, Antwerp, Blois, Geneva, Lyons, Paris, and Rouen. Spanish publications came from Antequera, Granada, and Seville. At this period the folio volume is still found in all countries (although less in Germany), but the format of over half the items is octavo or less, this trend being particularly strong in the Germanic countries. Overall therefore these volumes are very representative of European publishing in the late sixteenth century and also demonstrate how this particular market, that for practical, vernacular language books, was, as one might expect at this period, a very healthy and active one.

The range of languages covered is also wide, there being several multilingual dictionaries, the largest of which, claiming ten languages, covered Low Dutch, English, French, German, Greek, Hebrew, Italian, Latin, Portuguese, and Spanish. The best represented language in these Bodleian holdings is clearly Italian, which is particularly strong in books of the earlier part of the period. German and French come almost equally second, although it is evident that the Bodleian's French holdings are stronger in the years after 1600 rather than before. Spanish has less than half of these two last languages but still comes well before any other. Several books cover Polish and interest in Eastern Europe was evidently strong, other works relating to Hungarian, Slavonic, Modern Greek and Turkish, not to mention other Middle, or even Far Eastern languages which are not covered by the present survey. The sixteenth-century dictionaries, whether bi-, tri- or multi-lingual, mostly contain Latin as a key language, sometimes accompanied by Greek and a vernacular, but after 1600 the pattern changes: Latin becomes much rarer, dictionaries are largely bilingual and that of the Accademia della Crusca heralds the age of the monolingual dictionary, officially promoted as an act of national policy and unification. Grammars feature as frequently as dictionaries and, in the Italian field, the vocabulary and style of particular writers is the subject of study. Sir Thomas might of course not have approved of the attention paid to Italian literature, since he belonged in many ways to the generation of Ascham's Scholemaster, for whom the 'merrie books of Italie' were more wont to make Roman converts than all the duller tomes of Louvain. On the learning side however and in all languages set-situation dialogues continued to be popular.

The size of the Appendix to the 1605 catalogue suggests a sudden rush of material, possibly the result of John Bill's European trip, but another factor may have been Bodley's success in persuading people to present books to the library. Gifts naturally included manuscripts as well as printed books and a number were equally in foreign languages, the prize item in these early years being perhaps the 'Romance of Alexander' (MS Bod. 264). It is however striking that foreign works feature fairly frequently among the first donations. As early as 1601 Bodley's friend Savile, the Warden of Merton, presented a manuscript 'Evangelia lingua Moscovitica' and in the following year Sir Richard Lee, a former ambassador in those parts, gave items in Finnish, Slavonic and

Persian.[16] Numerous Spanish and Portuguese works came from the Earl of Essex as a result of the raid on the town of Faro in 1596. Sir Michael Dormer, a Cambridge graduate but a resident of Great Milton near Oxford, gave Italian books and two with his signature feature in the list above. Thomas James' speech to the king in 1605 could indeed boast that the library then contained items in at least thirty languages and the Preface to the 1620 catalogue claimed that something approaching a quarter of the 16000 books then thought to be in the library were in French, Italian or Spanish.[17] James also pointed out that the library was attracting numerous foreign readers, principally Italians, French, Germans, Poles, and Swedes.

Interestingly the bindings of the books also contribute to our knowledge of Bodley's acquisitions in this field. Many books have of course had to be rebound by the library over the years, but a sizeable number remain in their early bindings, often with their original shelf-marks still visible inside them, and may thus, through a localization of the binding, indicate where they were bought. Outstandingly the Donatus would appear to be in a Polish binding, the 1572 Antonio of Lebrija in a Spanish one, and the Frisius in a German one, but the very uniformity of vellum with yapp edges on the majority of the smaller formats suggests that they were bound in this style, then popular in England, on arrival in Oxford. The larger formats were evidently chained and retain their staple holes and their number on their foredge. Even smaller items were however also originally shelved spine in/foredge out and equally retain their foredge number, demonstrating that they are the original early-sixteenth-century copies.

This latter point is important in the present context for assessing the strength of the library's stock at a particular period, for evidently the library went on acquiring such books at later dates. The existence of the 1605 catalogue and indeed of its successors of 1620, 1674, 1738 and 1843 allow snapshots of the library's collections at those dates. They also allow one to trace the appearance or disappearance of a particular book, this evidence having sometimes to be corroborated by examination of the actual volumes to check that their shelf-marks tally with those shown in the catalogues. Sir Thomas Bodley himself had been conscious of the rate of growth of the library and, as Ian Philip puts it, 'Bodley laid it down in his first draft statutes that the Curators at their annual Visitation should consider changing editions and disposing of duplicates, the money so obtained to be used for buying books wanting in the library'.[18] Thus the First Folio of Shakespeare was disposed of on the receipt of a subsequent edition, fortunately in this case to be repurchased several hundred years later. But the study of the dictionaries and grammars in Bodley in 1605 clearly shows that this system with regard to duplicates or later editions was followed fairly rigorously in at least this field. Happily, later acquisitions or donations have filled these early gaps but not of

[16] J.D.A. Barnicot and J.S.G. Simmons, 'Some Unrecorded Early-printed Slavonic books in English Libraries', *Oxford Slavonic Papers*, ii, 1951, p. 98. For Hungarian material, see R.J.W. Evans, 'Hungarica in the Bodleian: a Historical Sketch', *Bodleian Library Record*, ix, 1978, pp. 333–345. I am particularly grateful to Dr Simmons for his generous help and especially for his guidance in relation to the Slavica.

[17] Lievsay *Op. cit.* (n. 11), pp. 35–42.

[18] Philip *Op. cit.* (n. 3), pp. 59, 84–86.

course, in these cases, with the original copy. Outstandingly the Selden collection, received in 1659 and especially rich in language books caused a particular decimation of earlier copies. Later benefactions brought in a number of items, some important early works were acquired by purchase in the early nineteenth century and of course this field in particular was greatly enriched in 1834 by the Douce bequest, but these later acquisitions appear to have been treated as mere additions to stock. Items of which the library seems to have lost its early copies in the seventeenth century include for example the first, 1598, edition of Florio's dictionary, which features in the 1605 and 1620 catalogues but has disappeared by 1674 – although today, happily, the library possesses the Tanner and Douce copies.

Growth and expansion have always posed problems for librarians, but growth in this field does at least show that there was local interest in it. In 1617 John Minsheu, completing his *Ductor in linguas* after a period of time first at Cambridge then at Oxford, persuaded the Vice-Chancellor at Oxford, John King, to give him a testimonial to the importance of his work. Minsheu was evidently an able promoter of his book and the subscription list, one of the first in England, lists not only the king and queen but also some fifteen Oxford and Cambridge colleges as well as eighteen private persons at Cambridge and some thirty at Oxford. Slightly later, Selden's enthusiasm for language books was not an unusual one for his day and a measure of such interest is to be found in the 1674 catalogue, which breaks out from its regular alphabetical author listing to provide a special dictionary heading under Lexicon, where no fewer than a hundred and fifty-three entries are provided, these works covering Arabic, Annamite, Bohemian, English, French, Georgian, German, Greek, Hebrew, Hungarian, Italian, Japanese, Latin, Malay, Polish, Slavonic and Spanish to name but the main ones. The popularity of foreign books of this sort, based perhaps on curiosity rather than on linguistic enthusiasm it is true, is however well illustrated by the fact that as early as 1645 the library records show that a list had to be made showing in particular the books in 'the French, Italian, Spanish and other vulgar tongues for the use of all such as desire a transcript from time to time, a worke much desired by most gentlemen and strangers that come to the Library'.[19] It would appear likely that, apart from Protestant refugees such as John Verneuil (sub-librarian 1618, died 1647), most senior members of the university, even those as learned as Robert Burton, the author of the *Anatomy of Melancholy*, had little or no knowledge of foreign languages; but that there was interest in them and a desire to provide for this, be it for academic or for practical reasons, there seems little doubt. The sixteenth century, rising with Humanism, had been a time of exploration; for some the mid-seventeenth would be one of exile in Europe, while for others the later seventeenth and the eighteenth would be the age of the Grand Tour. At all these times therefore the business of learning foreign languages was essential and while few other traces of this unprescribed and unexamined Oxford activity remain, at least this bibliographical jetsam allows the contemporary tools for the job to be known, investigated, and appreciated.

[19] *The Bodleian Library Account Book 1613–1646*, ed. G. Hampshire, Oxford, 1983, p. 151.

In Pursuit of Provenance: The George Graham Proto-Orreries

Silvio A. Bedini

Man's preoccupation with the structure of his physical world has been manifested from early in the present millenium by his efforts to reproduce it in mechanized astronomical models called planetaria. His efforts to know and understand the components of his universe have persisted to the present, resulting most recently in splendid sophisticated computerized optical planetaria. In Ptolemaic models, planetaria generally demonstrated the rotations of the sun and other planets around the earth. The adoption of the Copernican system and the diffusion of Newtonian philosophy in the seventeenth century led to a renewal of interest in operating models demonstrating the rotations of the earth and other planets around the sun. Some models were simple in form, requiring the movement of their parts by the hand of the demonstrator, while others operated by means of gearing utilizing a hand crank or clock movement.[1]

A major development prior to the emergence of optical planetaria occurred during the first decade of the eighteenth century with the invention of an instrument by the London clockmaker George Graham, which not only demonstrated the annual motion of the earth around the sun, but also exhibited the diurnal rotation of the earth upon its axis, as well as the revolution of the moon around the earth.

Graham's invention initiated the tradition of English planetaria that survived for more than one hundred fifty years. Despite the significance of this invention, however,

[1] Derek J. De Solla Price, 'On the Origin of Clockwork, Perpetual Motion Devices and the Compass', *Contributions from the Museum of History and Technology, United States National Museum Bulletin 218*, 1959, pp. 86–90; Henry C. King in collaboration with John R. Millburn, *Geared to the Stars. The Evolution of Planetariums, Orreries and Astronomical Clocks*, Toronto & Buffalo, 1978, pp. 150–57; David Layton, 'The Motion of the Planets: Newton's Effect on English Thought', *History Today*, May 1957, pp. 388–395.

From *Learning, Language and Invention: Essays presented to Francis Maddison*, ed. W.D. Hackmann and A.J. Turner. Published by Variorum, Ashgate Publishing Ltd, Gower House, Croft Road, Aldershot, Hampshire, GU11 3HR, Great Britain.

the only documented information about the two examples he produced and their subsequent history is confused and uncertain.[2]

Selecting appropriate nomenclature for the Graham instrument creates some problems. It cannot properly be identified as either a 'planetarium' or 'planetary machine', because the only planet incorporated is the planet Earth. Although the Graham instrument is generally identified as an 'orrery', the name was not in fact given to this type of instrument until a few years after Graham produced it, to describe a more elaborate form based on the Graham instrument by another London instrument maker, John Rowley. Consequently, inasmuch as the Graham instrument pre-dated the 'orrery', it is more appropriately referred to as a 'proto-orrery'.[3]

After serving an apprenticeship with the London clockmaker Henry Aske, in 1696 Graham joined the shop of the pre-eminent English clockmaker Thomas Tompion as a journeyman. This was the beginning of a long and successful association, and a close friendship developed between them. Later Graham married one of Tompion's nieces. From 1710 clocks and watches produced in the Tompion shop were signed with the names of both Tompion and Graham. During Tompion's last years, Graham was assigned full management of the shop, and after Tompion's death in 1713 he bequeathed his shop, stock and work in hand to Graham.

Graham continued to carry on the business in Tompion's house 'At the Sign of the Dial and Three Crowns' until 1720, then moved to a new location 'At the Sign of the Dial and One Crown' near Fleet Bridge. In the same year he was elected to membership of the Royal Society and two years later served on its Council. At the same time he was active in the affairs of the Clockmakers' Company, becoming master in 1722.[4]

Graham's accomplishments in the field of horology and astronomical instrumentation are among the most significant achieved by English clockmakers in the eighteenth century. He developed the form of dead-beat anchor escapement that was to come into general use, and made important experiments with the compensated pendulum, inventing that with mercury-compensation. He produced important sidereal clocks as well as instruments for the leading astronomers of his time. These ranged from small pocket telescopes and levels to the famous eight-foot mural quadrant that he made in 1725 for Edmund Halley at the Royal Observatory, Greenwich, and which introduced a new form of calibration used at Greenwich until 1750.

During the same period, Graham constructed two zenith sectors, one for Samuel Molyneux and another for James Bradley for use at the Royal Observatory. Other

[2] Francis Maddison, 'An Eighteenth-century Orrery by Thomas Heath and some Earlier Orreries', *The Connoisseur*, cxli, no. 569, April 1958, pp. 163–167; Howard C. Rice, Jr., *The Rittenhouse Orrery, Princeton's Eighteenth Century Planetarium, 1767–1954. A Commentary on an Exhibition held at Princeton University Library*, Princeton, 1954, pp. 6–10; King and Millburn *Op. cit.* (n. 1) pp. 152–154.

[3] Nomenclature proposed by John R. Millburn, communication of January 25, 1993.

[4] H. Alan Lloyd, 'George Graham, Horologist and Astronomer', *Horological Journal*, xciii, November 1951, pp. 708–717, *vide* p. 714; reprinted in *The Journal of the Royal Society of Arts*, c, no. 4861, November 30, 1951, pp. 30–49; David Landes, *Revolution in Time*, Cambridge, 1983, pp. 140–141, 221.

instruments designed and made by Graham included transit instruments, an equatorial sector and a micrometer for insertion into the plane of a telescope. He experimented extensively with magnetism and the variation of the compass. He also personally made astronomical observations and published a number of papers on the results of his work. He is properly identified in his obituary notice as 'the first general mechanician of his day'.[5]

It is not known just when and how Graham became interested in astronomy, inasmuch as it was not until Tompion's death in 1713 that he began to produce his sidereal clocks and astronomical instruments. Tompion had made astronomical instruments for Robert Hooke; and possibly inspired by his master, Graham may have undertaken study of the subject, and mastered the practical astronomy of the period. He presented papers before the Royal Society on solar and lunar eclipses, the occultation of Mars by the moon, the transit of Mercury and other related subjects.

It was undoubtedly the duality of his production of horological and astronomical instruments that inspired Graham to combine them in an effort to produce a mechanized model of the universe, perhaps for his own diversion or as an aid to his astronomical studies. A gradual progression is apparent in Graham's transition from clocks and watches to the invention of his proto-orrery, to sidereal clocks and finally the design and construction of instruments for the Royal Observatory. His experimentation leading to his proto-orrery began during the early years of his association with Tompion, certainly between 1705 and 1710, the year he became Tompion's partner. No surviving records relating to Tompion and Graham mention the proto-orrery, however, and the only tangible remains of this interest are in fact the two examples of the instrument.[6]

Still to be resolved is the question, which of the two instruments was the first to be made. Hereinafter they will be identified by the names that appear on them, 'Tompion–Graham' and 'Graham' [Figures 1 and 2]. Both instruments are contained in ebony cases of the period with decorative mouldings. Each is equipped with a carrying case, that of the Graham instrument is made of mahogany with an interior lining of red velvet, while that of the Tompion–Graham instrument is made of ebony having the interior lined with contemporary green velvet. They are equipped with two locks, and a glass panel on the upper surface and a glazed door at the side in front of the dial permit manipulation of the crank and allow viewing of the instrument in operation without removal of the carrying case.[7]

---

[5] C. Doris Hellman, 'George Graham, Maker of Horological and Astronomical Instruments', *Vassar Journal of Undergraduate Studies*, v, May 1931, pp. 221–251; John R. Millburn, 'The Fleet Street Address of Graham and his Successors', *Antiquarian Horology*, viii, no. 3, June 1973, pp. 299–301; James Oakes, 'Early English clockmakers. Thomas Tompion', *The Englishman*, November 28-December 1, 1713.

[6] Derek Howse, 'The Tompion Clocks at Greenwich, and the Dead-beat Escapement,' *Antiquarian Horology*, December 1970, pp. 18–34 and March 1971, pp. 114–133; R. W. Symonds, *Thomas Tompion, His Life and Work*, London, 1951, p. 53; communication of October 28, 1991 from Miss Pauline Snelson, Chief Editor, B. T. Batsford, Ltd, reported that the firm's correspondence files from the 1950s were no longer retained and information concerning the publication of Symonds' work was no longer available.

[7] George H. Gabb and Frank Sherwood Taylor, 'An Early Orrery by Thomas Tompion and George

Figure 1. The proto-orrery bearing the signatures of Thomas Tompion and George Graham (Courtesy of the Museum of the History of Science, Oxford, inv. no. 48–27)

Figure 2. The proto-orrery bearing the signature of George Graham. Accession No. A-156 (Courtesy of the Department of the History of Astronomy, Adler Planetarium and Astronomical Museum, Chicago)

Figure 3. Contemporary portrait of Prince Eugene of Savoy. Engraving, 18th century (Courtesy of The Library of Congress)

Figure 4. The orrery stated to be the 'original' made by John Rowley for the Earl of Cork and Orrery (Courtesy of the Science Museum, South Kensington, London)

The case of the Graham instrument is relatively simple in form, with minimum decoration, while that of the Tompion–Graham instrument is equipped with elaborate silver mounts. The Graham instrument can be manually operated by means of a hand crank as well as mechanically by clockwork, while the Tompion–Graham instrument has only a hand crank. An elaboration of the latter consists of two fixed pointers indicating both the Julian and Gregorian calendars, a feature lacking in the Graham instrument, suggesting that the latter had deliberately been made for use in Great Britain and not on the Continent. At a later time, however, a modification was introduced to convert the Graham instrument from the Julian to the Gregorian calendar, a conversion that was initiated in Great Britain at the beginning of the year 1752. This conversion would have been required prior to that date if it had been sold to a client on the Continent.[8]

The earliest recorded published mention of the Tompion–Graham instrument attributed its invention solely to Tompion, without mention of Graham's name. It is in this account that mention is made for the first time of the identity of the individual for whom the instrument was made. Recorded in the *Minutes* of the Gentlemen's Society of Spalding for the year 1713, eight months before Tompion's death:

> Monday, March 2nd, 1713. Mr. Johnston gave the Society an account of Mr. Tompion's curious machine for explaining the motion of the sun, moon, and earth, according to the Copernic system, which machine was made for Prince Eugene [Figure 3].[9]

Maurice Johnson was the founder of the Society. Originally from London, where he was associated with Joseph Addison, Sir Richard Steele, John Gay, Alexander Pope and William Stukeley in the founding of the Society of Antiquaries, Johnson was inspired in 1710 to form a similar body in south Lincolnshire. The purpose of the new society, as formulated by Johnson, was to '.... deal in all arts and sciences, and exclude nothing from our conversation but politics, which would throw us all into confusion and disorder'. The membership included such notable figures in the arts and sciences as Jean Theophilus Desaguliers, John Evelyn, John Gay, Alexander Pope, Sir Hans Sloane, and William Stukeley; the Society continues in existence to the present time. The early *Minutes* provide an interesting record of contemporary scientific concerns, ideas, and achievements, and reflect the influence of the Royal Society on the English world of learning.[10]

Graham Recently Acquired by the Museum of the History of Science, Oxford', *The Connoisseur*, September 1948, pp. 24–26, 55; Symonds *Op. cit.* (n. 6), pp. 294–295; King and Millburn *Op. cit.* (n. 1), pp. 152–154; communications from Francis R. Maddison, August 19, September 22, October 7, and October 28, 1964, March 9, 1966.

[8] Communication from William J. F. Andrewes, August 26, 1991, who first noted this feature in 1975 or 1976.

[9] *Minutes* of the Spalding Gentlemen's Society, vol. I, 1710–1719, entry for March 2, 1713; communications from W. A. James, Hon. Secretary, Spalding Gentlemen's Society, September 22 and October 2, 1964.

[10] H. J. J. Winter, 'Scientific Associations of the Spalding Gentlemen's Society During the Period 1710–1750', *Archives Internationale d'Histoire des Sciences*, January 1950, pp. 77–81. It is to be noted that

On October 20th of the same year, the invention again came to public notice in the newspaper *The Englishman,* in the writings of the London journalist and popular figure about town, Sir Richard Steele. A member of the Spalding Society, he had undoubtedly learned of the instrument from Johnson's account in March. In describing and illustrating it and its construction, Steele attributed its invention to neither Tompion nor Graham, but to John Rowley:

> I sit down therefore at present to do Justice, and consequently great Honour, to that worthy and ingenious Artificer Mr. John Rowley; who has lately distinguished himself by the Invention of a Machine which illustrates, I may say demonstrates, a System of Astronomy, as far as it relates to the Motions of the Sun, Moon and Earth, to the meanest Capacity. That which would have taken up a Year of Study to come to a familiar Apprehension of it, is communicated in an Hour.
>
> The Honest Man calls his Machine the *Orrery,* in Gratitude to the Nobleman of that Title; for whose Use and by whose Generosity and Encouragement he began and accomplished the Undertaking. But to turn a Matter of this Nature to the suitable Advantage of the Artificer, is not to be expected from any thing less than a Prince, or from a general Fund of many Encouragers. His late Royal Highness, the Prince of Denmark, honoured Mr. Rowley with his Notice; and had it pleased God to have spared the Life of that excellent Man, who had a very solid and great Capacity for these Enquiries (though he was above all sort of Ostentation) this Artist would not, at this Day, have wanted one who would have esteemed and favoured him according to his Merit.

Steele went on heaping further praise on Rowley's achievement, providing descriptions of the instrument's elements and functions. The name 'orrery' had apparently been given to the instrument by Rowley himself, although Desaguliers attributed the naming of it to Steele. E. Budgell, in his biography of Charles Boyle and the Boyle family published several decades later, went so far as to credit the conception of the instrument entirely to the 4th Earl of Cork and Orrery, who, he wrote, had hired 'an ingenious Artificer' to construct it for him.[11]

In addition to a description in Desaguliers' *Lectures of Experimental Philosophy* published in 1719, the instrument made by Rowley for the Earl of Orrery [Figure 4] was also described and illustrated in several contemporary works. It was featured in an unusual little publication by John Harris, a Fellow of the Royal Society, *Astronomical Dialogues between a Gentleman and a Lady* .... It was also described in general terms and illustrated by Edmund Stone in 1723 in his translation of Nicolas Bion's *Traité de la construction et des principaux usages des instruments de mathematique.*[12]

---

the date of the meeting, February 23, 1713 as given by Winter is incorrect. For a comprehensive history of this organization, see W. Moore, *The Gentlemen's Society of Spalding: its Origin and Progress,* London, 1851.

[11] Sir Richard Steele, *The Englishman,* No. 11, October 29, 1713, pp. 50–53; E. Budgell, *Memoirs of the Life and Character of the Late Earl of Orrery, and of the Family of the Boyles,* London, 1732, p. 236; Gabb and Taylor *Op. cit.* (n. 7), p. 26; King and Millburn *Op. cit.* (n. 1), pp. 156–157.

[12] J. T. Desaguliers, *Lectures of Experimental Philosophy wherein the principles of mechanicks, hydrostaticks, and opticks, are demonstrated and explained at large, by a great number of curious experiments ... Illustrated with several curious copper Figures suitable to each subject. To which is added, a description of Mr.*

It was not until several decades had passed after Steele's account had been published that the error of attribution was corrected, by Desaguliers. A close associate of Maurice Johnson, he may have been present at the meeting during which Johnson described the invention, and he was certainly aware of Johnson's account. This fact, taken together with his awareness of Steele's error in attributing the invention to Rowley, provides even greater emphasis to Desaguliers' attribution of the invention to Graham. In his work, *A Course in Experimental Philosophy*, published in 1734, he stated:

> Mr. George Graham (if I am rightly inform'd) was the first person in England, who made a movement to shew the Motion of the Moon round the Earth, and of the Earth and Moon round the Sun, about 25 or 30 years ago. In this Machine everything, that was shewn, was well and properly executed – as the Phaenomena of Day and Night – and their gradual Increase and decrease, according to the Seasons, the Place of the Earth where the Sun is successively vertical, and seems to describe its Parallels, the real annual Motion of the Earth, which gives the Sun an apparent annual Motion, the Rotation of the Sun about its Axis, the periodical and synodical Month, the solar and sydereal Days, the successive Illumination of all the Parts of the Moon, Etc. The Machine being in the Hands of an Instrument-maker, to be sent with some of his own Instruments to Prince *Eugene*, he copied it and made the first for the late Earl of Orrery, and then several others, with Additions of his own. *Sir Richard Steele*, who knew nothing of Mr. *Graham's* Machine, in one of his Lucubrations, thinking to do justice to the first Encourager, as well as the Inventor, of such a curious Instrument, call'd it an *Orrery*, and gave Mr. *J. Rowley* the Praise due to Mr. *Graham*.[13]

Desaguliers undoubtedly was personally familiar not only with Graham but also with the Graham proto-orrery signed only by Graham. In 1720 Philippe Vayringe, a young self-taught clockmaker and instrument maker at the court of Duke Leopold Joseph of Lorraine, arrived in London and spent thirteen months studying and working with Desaguliers. The tour of study was sponsored by the Duke, who was in the process of building an elaborate cabinet of experimental physics in his château at Luneville, which was to become Vayringe's responsibility upon his return to Lorraine.

During his stay in London, Vayringe had an opportunity to examine the Graham proto-orrery, either in Graham's shop or in the hands of someone in London who may have purchased it. Shortly after his return to Luneville in 1721, Vayringe constructed a similar clockwork-driven instrument based on Graham's design, with only a few variations, and housed it in a cylindrical glass case upon an elaborate pedestal.

Rowley's machine, called the orrery, which represents the motion of the moon about the earth, venus and mercury about the sun, according to the Copernican system; all carefully examined and corrected by Mr. Desaguliers, London, 1719; J. H. [John Harris], *Astronomical Dialogues between a Gentleman and a Lady; Wherein the Doctrine of the Spheres, use of the Globes, and the Elements of Astronomy and Geography are Explain'd in a Pleasant, Easy and Familiar Way*, London, 1719, pp. 159–180; Edmund Stone, *The Construction and Principal Uses of Mathematical Instruments. Translated from the French of M. Bion ... To which are added, the construction and uses of such instruments as are omitted by M. Bion; particularly of those invented and improved by the English*, London, 1723, pp. 189–191, pl. XIX.

13 J. T. Desaguliers, *A Course of Experimental Philosophy*, 2 vols, London, 1734–44, vol. 1, p. 430.

The instrument pleased Duke Leopold to such a degree that he sent it to Vienna by the hand of Vayringe as a gift to the then Archduchess Maria Theresia, who later became his daughter-in-law. It subsequently became part of the imperial collections in Vienna and eventually was transferred to the Naturhistorisches Museum in Vienna, where it is presently to be found.[14]

One final eighteenth-century reference to the invention occurred in Graham's obituary notice in the November 1751 issue of *The Gentlemen's Magazine*, which listed among Graham's many other accomplishments, if somewhat inaccurately, that:

> He comprized the whole planetary system within the compass of a small cabinet, from which, as a model, all the modern orreries have been constructed ...[15]

Of particular significance is the mention in the Spalding Society's *Minutes* of 1713 and in Desaguliers' account that the Tompion–Graham instrument was made for 'Prince Eugene'. An undocumented account published by R. T. Gunther in 1923 appears to confirm the identity of the recipient of the instrument as Prince Eugene:

> Before the year 1715 an instrument was contrived and made by GEORGE GRAHAM, which represented at the same time both the annual and diurnal motions of the earth, and also the synodic period of the moon; these motions had never before been exhibited together by any mechanical contrivance, at least in England. The instrument was put into the hands of a workman to be packed and sent to Prince Eugene. Whether this workman was Rowley, or some other person employed by him, does not appear; but it is said that he copied the model put into his hands, and from it Rowley made a machine for the Earl of Orrery, with additions of his own invention.

The account of the manner of the instrument's shipment was reported once more by George H. Gabb and F. Sherwood Taylor in 1948 in an article about the Tompion–Graham instrument, as well as in the biography of Tompion by R. W. Symonds. The latter work was first published in 1951, although it had been written several years earlier. To the accounts by Gabb and Taylor and by Symonds of the instrument's shipment was added a legend relating that the Tompion–Graham instrument had been commissioned from Tompion by Queen Anne, who planned to present it to the Duke of Marlborough in acknowledgement of his triumphs on the battlefield on behalf of the British Crown.[16]

John Churchill, the first Duke of Marlborough, had already received the highest honors and rewards that his country could bestow by the time of Queen Anne's succession to the throne in 1702. His brilliant defeat in 1703 of the French armies,

---

[14] P. Michel, *Biographie historique et pencalogique des hommes illustrés de Lorraine*, Nancy, 1929, pp. 509–510; LePage, 'Notes sur le mecanícien Vayringe', *Journal de la Société d'archéologie lorraine*, 1855, pp. 79–83. [Tardy], *Dictionnaire des Horlogers français*, Paris, 1971–72, vol. A–K, p. 634; King and Millburn *Op. cit.* (n. 1), p. 153. The similarity between the Vayringe instrument and the Adler proto-orrery was first reported by King and Millburn.

[15] *The Gentlemen's Magazine*, November 1751, p. 524.

[16] Desaguliers, *Op.cit.* (n. 13), vol. 1, p. 430; Gabb and Taylor *Op. cit.* (n. 7), pp. 24–26, 55; Symonds *Op. cit.* (n. 6), pp. 294–295.

and the subsequent campaigns in which he was aided by Prince Eugene of Savoy, inspired British writers to perpetuate Marlborough's name and fame in contemporary literature. He was awarded the manor of Woodstock and thereupon a palace was built for him at royal expense. He was created a prince of the empire, and the principality of Mindelheim was formed in his honour.

Very soon thereafter, however, the Duke of Marlborough's popularity at court began to fade. His subsequent military action against the French at Mons in 1708 and 1709 proved to be a less than spectacular victory, and served as a prelude to his fall from royal favour. Party intrigues at home did much to undermine his influence at court, followed by unfortunate involvement in political conflicts, and strong suspicion of his misappropriation of military funds. Meanwhile, growing personal differences and disaffection between the Duchess of Marlborough and the ailing Queen widened the breach irrevocably, and thereafter the Marlborough family was subjected to repeated humiliations.

By 1711 the Duke had finally fallen completely out of favour at court, and he left England the following year, residing on the Continent until the Queen's death in 1714. Consequently it seems most unlikely that the Queen would have commissioned the instrument to reward the Duke just at the time that he was least in favour. All efforts to identify the source of the legend reported by Gabb and Taylor have failed, and it may have been a modern fabrication.[17]

It is entirely plausible, however, that the Queen might have commissioned such an instrument from Tompion, if she had wished to have one. By this time Tompion had become the foremost clockmaker of the realm, having already produced timepieces for Charles II, William III as well as for Queen Anne. Between 1702 and 1708, she had commissioned Tompion to produce a 'swing clock' and other timepieces. Furthermore, he was the responsible for the maintenance of clocks in the royal palaces.

According to the legend, after having received the royal gift of the Tompion–Graham proto-orrery, the Duke of Marlborough presented it to his military ally, François Eugene, Prince of Savoy. That he might have done so is entirely possible, for they were great friends. The Prince had served brilliantly in the Austrian imperial army of his uncle, Emperor Leopold I, and he won numerous military campaigns against the French. In 1704 in Bavaria he fought his first campaign allied with Marlborough's English forces, and a close friendship developed between the two leaders. When Marlborough fell from grace at home, Prince Eugene immediately departed for London with the intention of attempting to help restore his friend's influence at court. He was anxious also to attach England once more to the coalition with the Austrians, but his mission was doomed to failure. Arriving in London on January 16, 1712, the Prince remained until March 17th. He was instantly popular and constantly fêted, and it is conceivable that he was frequently Marlborough's guest at his estate at Blenheim Castle before the Duke departed from England later that year.[18]

[17] Winston S. Churchill, *Marlborough, His Life and Times*, 4 vols, London, 1936, vol. 3, pp. 29–30, 312–24 and *passim*; vol. 4, pp. 64, 136, 186 *et seq*., 212 *et seq*., 363–366, 368–369, 481–482.

[18] Nicholas Henderson, *Prince Eugen of Savoy*, New York, 1965, pp. 190–92, 288–92; Symonds *Op. cit.* (n. 6), pp. 41–43; Churchill *Op. cit.* (n. 17), vol. 4, pp. 514–520.

There was much comment and gossip about the Prince at court, and his every move was closely observed and reported. Numerous references to his sojourn occur in the private papers and correspondence of the British ambassador to The Hague, Thomas Wentworth, Lord Raby, created Earl of Strafford. In one letter to her husband, Lady Strafford archly commented, 'I have taken it into my head that Prince Eugen has been some time or other wounded in the ----, for I never saw any body sit down so stiff'. Again she wrote, 'Prince Eugin was with the Queen a Sunday night, Mr. St. Johns went with him. I hear he is very much mortified at the Duke of Marlborough being out of favore'. A week later she commented in a letter to her husband, 'The Ladys here dont admire Prince Eugene, for he seemes to take very little notis of them', and in the same letter she remarked, 'that fruitfull creature Prince Eugene takes up everybody's conversation'.

In a letter written before the Prince's arrival, Lord Berkeley of Stratton suggested, 'Prince Eugene I suppose hath altered his plan, since he finds the D[uke] of Marlborough out of all his places, and they say he will stay here only a little while. There are mighty crowds in the morning to see him, and he is invited every day about to dinner and I think generally liked'. Peter Wentworth informed his brother Lord Raby, 'On her Birth day she [Queen Anne] gave P----- E----- a sword sett with diamonds, the Queen being to be carried in her chair from her dressing room to the great Drawing room, everybody but the ladies in waiting and my Lord Chamberlain was kept out of that appartment; but when the Queen goes in her chair 'tis my business to be there, so I saw my Lord Chamberlain come in with P----- Eugene alone and go into the dressing room, and after staying 2 or 3 minutes he came out with the Sword the Queen had given him in his hand, and then pull'd off his own Sword and gave it to a page of the back stairs that stood at his elbow, and put on t'other.'[19]

A search of the Marlborough family archives at Blenheim Castle and in the Record Office, as well as a survey of published biographies and accounts of court life of the period, have failed to reveal any mention of the Tompion–Graham instrument in connection with Queen Anne or Marlborough. No earlier account of the legend recounted by Symonds or Gabb and Taylor has been found.[20]

During his two months sojourn in London, Prince Eugene saw much of the city, skilfully escaping the large crowds of admirers and curiosity seekers who followed his every move. He took pleasure in wandering alone to explore the sights, especially during the long periods he was forced to await the scheduling of meetings with the British ministers. Often accompanied by Georg Wilhelm von Hohendorff, his confidant and travelling companion, he particularly enjoyed seeking out the London bookshops where he made purchases for his personal library.

---

[19] James J. Cartwright, ed., *The Wentworth Papers, 1705–1739. Selected from the Private Correspondence of Thomas Wentworth, Lord Raby, Created in 1711 Earl of Strafford of Stainborough, Co. York. With a Memoir and Notes by James J. Cartwright*, London, 1883, pp. 224–265, *vide* 247–248.

[20] Gabb and Taylor *Op. cit.* (n.7), p. 26; Henderson *Op. cit.* (n. 18), p. 202.

Prince Eugene may have met Tompion in the royal palaces, and learning of his reputation, he may during his wanderings through the city, have called at Tompion's shop 'At the Sign of the Dial and Three Crowns'. There he may have observed the Graham instrument, either as a work in progress or in completed form, and commissioned a more elaborate version to be made for himself to be shipped to him in Vienna after his departure. The science lecturer and writer, Desaguliers, reported that the Prince had in fact called at the shop of John Rowley, another London instrument maker, from whom he purchased a selection of scientific instruments to be sent to Vienna. It was Desaguliers, furthermore, who reported that the Tompion–Graham instrument had been sent to Rowley's shop to be packed and shipped to Vienna with the purchases Prince Eugene had made from Rowley.

After his departure from London and return to Vienna in the spring of 1712, Prince Eugene continued to pursue a most active and successful military career until his retirement in 1735. He died in the following year at the age of seventy-two. During his lifetime, Prince Eugene amassed considerable collections of important paintings, sculpture, manuscripts, prints and books with which he filled his splendid Stadtpalais in Vienna and the Upper and Lower Belvedere palaces in the city's outskirts. He was unmarried, and he bequeathed his immense estate and fortune to Princess Anne Victoria of Savoy, the daughter of his oldest brother. She was a singularly unattractive spinster of fifty-two at the time, and had neither fondness nor respect for her uncle.

Princess Anne Victoria immediately and energetically set out to convert her vast inheritance into currency as quickly as possible. Her uncle's estates and collections, his military decorations, including the sword he had received from Queen Anne, his household furnishings, paintings, sculpture and *objets d'art*, all were removed quickly from his three palaces and sold as rapidly as possible, often for minimal prices. No sentiment entered into her dispersal of the properties, and the disposition was accomplished as expeditiously as circumstances permitted. No records of these sales have been preserved. Now having become a lady of substantial means, she looked about for a suitable husband, and selected the Imperial Quartermaster, Prince Friedrich von Sachsen-Hildburghausen. The marriage did not prosper, however, and soon thereafter she left her husband and went to live first in France and then in Turin, where she died.[21]

Although it is extremely questionable whether there was any substance to the legend of Marlborough's gift to Prince Eugene, it seems certain that Prince Eugene acquired the Tompion–Graham instrument, by purchase if not by gift, either between 1705 and 1710 prior to his visit to England, or during the two months of 1712 that he spent in London. Undoubtedly he kept it with other scientific instruments and cherished relics such as Queen Anne's sword in one of his palaces in which he spent his final years.

[21] Frederick Shoberl, *Memoirs of Prince Eugene of Savoy; Written by Himself [C. J. Prince de Ligne]*. 2nd edn, London, 1811, pp. 145–147, 288–290; Henderson *Op. cit.* (n. 18), pp. 288–292.

The Tompion–Graham instrument was sold off by his niece with the rest of his properties, and in 1737 it was acquired from Prince Eugene's estate by the man he had served, Emperor Charles VI, as part of the Emperor's purchase of Prince Eugene's Belvedere palace and his library from the Stadtpalais. This library, consisting of fifteen thousand printed works and two hundred thirty-seven manuscripts, is still kept together in Vienna's Nationalbibliothek. Prince Eugene's other two palaces, the Stadtpalais and the Schlosshoff, were purchased by Charles VI's daughter, the Archduchess Maria Theresia. Eugene's scientific instruments probably were first installed in the second of the three rooms that contained the library, the one in which the rarities were housed. An entry in the 'General Index of the Physics-Astronomy Cabinets of the A. H. Imperial Dwellings from 1748 to 1867' states:

> Not until Emperor Charles VI did the same [collections] by means of larger purchases in 1738 gain importance, of which the most significant originated from the collection of Prince Eugene of Savoy such as a large burning glass and a Copernican planetary system.[22]

Another reference to what appears to have been the Tompion–Graham instrument occurs in an account of the career of Johann Georg Nesstfell, a master carpenter and cabinetmaker of Alsfeld. In 1745 he had constructed a wooden Copernican planetary model operated by means of an eight-day pendulum-regulated clock. Grand Duke Francis III of Tuscany, husband of Maria Theresia, was informed of Nesstfell's achievement and invited the artisan to bring his machine to the Austrian court. Grand Duke Francis was then in Vienna preparing for his forthcoming coronation as Francis I Stephan, Emperor of the Holy Roman Empire.

Francis was greatly impressed with the model and rewarded Nesstfell handsomely. He then commissioned him to produce a similar model made of metal. On the same occasion Duke Francis also showed the artisan an English planetary model in his possession, of which, he said, 'the mechanism was spoiled, but no one was found in Vienna who was capable of bringing it back into order again'. Wilhelm Hess, Nesstfell's biographer, noted that it had been built in 1723, 'by the Englishman Rowley (Rouwley, Roley) for Prince Eugene of Savoy, and stood a long time in this library, until in 1737 with the Prince's collection it came into the possession of Emperor Karl VI and thereby landed in the Hofbibliothek'. Neither Hess nor his sources appear to have been certain of the identity of the English maker, that he was in fact Rowley, for Hess reported the name phonetically and variously.

Nesstfell agreed to repair the instrument and took it away with him to Banz. He completed the commissioned metal Copernican model in 1752 and brought it to the Viennese court in the following year. At the same time he also returned the English instrument fully repaired. Emperor Francis I Stephan was delighted with both and rewarded Nesstfell once more, this time with a gold medal and a gold chain. He also

[22] Österreichisches Staatsarchiv, Vienna, *Oberst-Kämmerer-Amt*, Sonderreihe 13, seite 25, 'General-Index des physikalisch-astronomischen Cabinetes des A. H. Kaiserhauses von 1748 bis 1867' (General Index of the Physics-Astronomy Cabinets 1748–1867). Communication from Hofrat DDr. Gottfried Mraz, May 25, 1992.

appointed him *Hofmechanicus* with an annual salary. The Emperor then placed Nesstfell's metal planetarium and the English instrument together on display in the Prunksaal or exhibition gallery of the Imperial Library.[23]

In 1768 another instrument was added to the display of these two planetary models, identified as 'a third Copernican machine'. It was an astronomical clock dedicated to and presented to Empress Maria Theresia, by its designer Father Francesco Borghesi, an Italian secular priest of Mechel in the valley of the Non of Venezia Tridentina (now Trentino–Alto Adige) in the Italian Alps. The clock had been made from his drawings and under his supervision by Bartolomeo Bertolla, a local clockmaker in Mocenigo di Rumo.

Later the Emperor brought together the several collections of scientific instruments and art objects scattered between his various imperial properties, including the Hofburg, Stallburg, and Lustschloss Ebersdorf. He transferred them all to the K. K. (Imperial Mechanisch-Physikalischen Kunstkabinett) where they were repaired and restored.[24]

During the Emperor's lifetime both the Physikalischen- and the Naturalienkabinett were accommodated in the Leopold wing of the palace. During the years following his death in 1765, his extensive scientific collections including the instruments were moved from one location to another by his restless widow. In 1765 she ordered that the Kunstkabinett collection be divided and installed in several salons adjacent to the new Augustinergang that she had established the previous year in the Hofburg. The Naturalienkabinett occupied the first two rooms, the Physikalischenkabinett in the next two, and the Münzkabinett in the next five rooms. In 1791 the collection was removed from the Hofburg and temporarily stored in one of the halls of the Schweizercourt. Six years later it was brought together once more in the Tierkabinett in the left wing of the Hofbibliothekgebäudes. In 1802 the long-standing Naturalien-(Mineralien-) Kabinett also was annexed to the collections already in the same location. In 1806 the imperial scientific collections were again divided, and the Mechanisch-Physikalischenkabinett transferred to the Physikalischen–Astronomischen. In 1810 it was to be found in the Schweizerhofe under the astronomical tower in the imperial city, and there it remained until the Kabinett was disbanded in 1886. On any of these

---

23 Wilhelm Hess, *Johann Georg Nesstfell, Ein Betrag zur Geschichte des Kunsthandwerkes und der Physikalischen Technik des XVIII Jahrhunderts in den Ehemaligen Hoch stiftern Würzburg und Bamberg. Studien zur Deutschen Kunstgeschichte*, 98 Heft, Strasbourg, 1908, pp. 22–23, 62–63; [J. G. Nesstfell], *Kurzgefasste. doch gründliche Beschreibung Der --- von mir Johann Georg Nesstfell Erfunden --- und verfertigen accuraten Copernicanischen Planeten-Maschine ...* , Bamberg, 1761, Preface, pp. iii-iv; Klaus Maurice, *Die Deutsche Räderuhr*, 2 vols, Munich, 1976, vol. 1, pp. 273–274; Heinrich Kriesel, *Die Kunst des Deutsches Möbels*, 3 vols, Munich, 1970, vol. 2, p. 95; King and Millburn *Op, cit.* (n. 1), pp. 157, 229.

24 Silvio A. Bedini, *The Borghesi Astronomical Clock, Contributions from the Museum of History and Technology, Smithsonian Institution*, Paper No. 35, Washington, 1964; King and Millburn *Op. cit.* (n. 1), pp. 251–253; *Wiener Diarium*, May 18, 1768; Johanna Schönburg-Hartenstein, *Josef Anton Nagel – ein Direktor des physikalischen Kabinettes. Philosophisch-Historische Klasse Sitzungsberichte, 482 Band; Veröffentlichungen der Kommission für Geschichte der Mathematik, Naturwissenschaften und Medizin*, Band 46, Vienna, 1987, p. 100.

moves, the Nesstfell planetarium could have been added to the collection of the Naturhistorisches Museum, where it has been in storage.[25]

The 'English instrument' disappeared from the imperial collections at some time during these moves. Although described by Hess as the work of Rowley, the probability cannot be overlooked that the English Copernican model was in fact the Tompion–Graham proto-orrery, which had, as Hess confirmed, 'stood for a long time in the imperial library'. The circumstances of its disappearance from the imperial collections are a matter for speculation. Having been a prized possession of first her father and then of her husband, the Empress would not have disposed of it as a gift, and the likelihood is that it was appropriated by a member of the staff and sold.[26]

Despite the possibilities offered in the accounts of the work of Vayringe and of Nesstfell, there remains no documented record of the two Graham proto-orreries for two and one-half centuries after their production, until 1930. Then, as reported by Symonds in his biography of Tompion, 'in about 1930' the Tompion–Graham instrument was purchased by a certain Jesse Myer Botibol from a dealer in Austria. Botibol, believed to have been of Austrian origin, was a dealer in antiquities with a shop situated at various times between 1921 and 1954 at Nos 26, 28 and 30 Hanway Street in London, and a residence at Albert House, Camden Road, London. He was established as a dealer in London as early as 1917, and sold merchandise at Christie's auction sales, but little else is known about him. According to Symonds' work, Botibol stated that the Austrian dealer from whom he acquired the instrument had reported that he had purchased it directly from a monastery near Salzburg. Botibol's source indicated that the instrument had been owned by the monastery for almost two centuries, which coincides approximately with the date of the disappearance of 'the Rowley orrery' from the imperial library in Vienna.[27]

The possession and use of astronomical instruments by Austrian monasteries is not particularly unusual. The collection in the observatory of the monastery of Kremunster, for example, is well known. Part of another notable collection of scientific instruments, originally belonging to Emperor Rudolph II, was presented in about 1620 by Archduke Ferdinand III, Regent of the Steiermark and later Emperor Ferdinand II, to the Kloster Rein, a Cistercian monastery in Steiermark near Graz. There the instruments remained until about 1935, when they were sold piecemeal by the Kloster's administration to several Austrian dealers. A few of these were acquired by the Kunsthistorisches Museum and others, including instruments made by Erasmus Habermel, were

[25] Alfons Lhotsky, 'Die Sammlungen Kaiser Rudolfs II in Prag', *Festschrift der Kunsthistorisches Museums zur Feier des Fünfzigjährigen Bestandes*, Vienna, 1941–1945, part II, no. 1, pp. 237–298.

[26] *Litteratur des katholischen Deutschlands*, Coburg, 1775–1783, Vierter Band, 1783, pp. 151–52; Hess *Op. cit.* (n. 23), pp. 23–24.

[27] Symonds *Op. cit.* (n. 6), p. 295; communications from Mr. Timothy Harris, for the Head Archivist, Office of the Town Clerk, Greater London Record Office and Library, Corporation of London, September 20, 1991; from Miss Phyllida Melling, Guildhall Library, September 24, 1991; and from Mr. Jeremy P. Collins and Miss Sarah O'Brien, Christie, Manson & Woods Ltd, London, August 30, 1991.

purchased in 1936 by the late Henri Michel. These subsequently became part of the Billmeir collection in the Museum of the History of Science at Oxford.[28]

It is not surprising, therefore, that an instrument from the imperial collections in Vienna found its way into an Austrian monastery in the second half of the eighteenth century. The monastery in question was in fact the Benedictine monastery of Erzabtei St Peter in Salzburg, designated 'the principal monastery' of the region. Situated in the centre of the city, on the west side of the Kapitalplatz, the Erzabtei St Peter was founded in 696 by St Rupert and served as the residence of the cardinal archbishops until 1110. That it was the Erzabtei St Peter that had owned the Tompion–Graham instrument and had sold it in 1930 was recently confirmed in correspondence from Pater Berthold Egelseder of the nearby Benedictine monastery in Michaelbeuern.[29]

After the return of the Tompion–Graham instrument to England, three dealers in antiquities became involved with its sale, a situation causing some confusion. The instrument apparently remained in Botibol's possession for a decade or more until he consigned it to be sold to Soloman Nathan Nyburg, proprietor of the Antique Galleries at 16c Grafton Street, London, who then placed a price of £1 500 on it. Nyburg was well known to the Museum of the History of Science, Oxford, and had previously had dealings with it.[30]

The instrument next appeared in the hands of Francis Harper, a dealer in antiquities living at Wickin's Manor, Charing, Kent. Harper may have been acting as agent for Botibol or Nyburg, although it later transpired that he had purchased it from either Botibol or Nyburg. Early in 1943 Harper communicated with Frank Sherwood Taylor, curator of the Museum of the History of Science, Oxford, offering several instruments for sale, including the Tompion–Graham instrument. He commented that he was contacting Taylor on the advice of Sir Geoffrey Callender, first director of the National Maritime Museum then living in Oxford, who suggested that the Museum would be interested in photographs that Harper enclosed with his letter. 'I am sure you can tell me many most interesting things about them', he wrote, 'I have them here & could send them to you – though I am a bit frightened of trusting valuables in the post these days'. Harper had a long association with Callender, having sold him instruments in the course of the past several decades.

---

28 Communication from Erwin Neumann, April 13, 1970, May 11, 1970, August 8, 1970; communications from Henri Michel to Derek Price, March 4, 1955, May 8, 1956; communication from Henri Michel, June 15, 1962, August 23, 1936; communication from Francis R. Maddison, June 15, 1962 enclosing Michel's account of purchase, August 7, 1962; [C.H. Josten], *Scientific Instruments (13th–19th Century). The Collection of J. A. Billmeir, C.B.E.* and [F.R. Maddison], *A Supplement to a Catalogue of Scientific Instruments in the Collection of J. A. Billmeir, Esq., C.B.E.*, Oxford, 1954 and 1957); Leopold Grill, *Stift Rein*, Steiermark, 1955.

29 Communication from Pater Berthold Egelseder, OSB, Schriftleiter, Kloster Michaelbeuern, Michaelbeuern, Austria, September 2, 1992.

30 Communications from Rosemary Smith, Science Museum, London, November 11, 1991; Phyllida Melling, Guildhall Library, September 24, 1991; Sarah O'Brien and Jeremy P. Collins, Christie, Manson & Woods Ltd, August 30 and October 30, 1991; and Tim Harris, The Head Archives, Greater London Record Office, September 30, 1991.

The photographs were of several instruments, including 'a compass dial and moon dial' by Humphrey Cole dated 1590, and the Tompion–Graham proto-orrery. Taylor expressed interest in the instruments, and informed George H. Gabb, a prominent collector of scientific instruments, suggesting that he accompany him on a visit to Harper. Apparently both had been aware that the Tompion–Graham instrument had been previously offered by Nyburg, but Taylor had been unwilling or unable to meet the price asked for it. The reason that it was not sold during this period undoubtedly resulted from a combination of its elevated price and the war. It is not certain whether the instrument was still on consignment to Nyburg at this point, or had now become the property of Harper.[31]

Whether Taylor and Gabb visited Harper cannot be determined from the surviving correspondence, but several months later, when the important collection of microscopes of Reginald S. Clay was being offered for sale, Gabb wrote to Taylor, '... even at £5 000, it's a relative bargain compared with the Tellurion at Nyburg's at £1 500'.[32]

No further mention of the Tompion–Graham instrument occurred until three years later, in October 1946, when Taylor's interest in acquiring the instrument was revived, perhaps after having read the instrument's description and seen its illustration in Symonds' biography of Tompion. It is likely that he communicated with Harper at this time to determine whether the instrument still remained unsold. It was at this time, as he informed Gabb, that he found a prospective patron willing to purchase it for the Museum.

Although as anxious as Taylor was to acquire the instrument for the Museum, Gabb advised caution. He wrote:

> About that Rowley [*sic*] Tellurian. I hope you won't allow your would be benefactor to be fooled into paying £1 300 for it – an inflated price *in these days* would be £750 ... As you have *virtually* got the money, my advice is this. Write Nyburg saying 'You are now prepared to make an offer of £1 000, which is much *beyond its value*, but are *quite unable* to increase this offer, & unless he accepts it within ten days, you regret you will be *obliged* to withdraw it & so will not be further interested in it'. And if he won't accept £1 000 – well, 'Let him stew in his own juice!'

[31] Oxford University, Museum of the History of Science Archives (Hereafter M.H.S.), letter from Francis Harper to F. Sherwood Taylor, January 12, 1943; letter from George H. Gabb to Taylor, September 29, 1943. Harper, formerly at The Whitehouse, Ham, Cheltenham, had corresponded with and sold scientific instruments to Callender for several decades. George H. Gabb (1860–1948) was a prominent well-informed collector of scientific instruments. Having first become interested in the subject in 1885, he amassed a large collection of important pieces, which was acquired by the National Maritime Museum in 1947. See F. Sherwood Taylor, 'Mr. G. H. Gabb', *Nature*, no. 707, 1948. Frank Sherwood Taylor (1897–1956) was the founding editor of *Ambix*, author of *The Young Chemist* (1934); *The World of Science* (1936); *Galileo and the Freedom of Thought* (1938), and a 3-volume textbook for the schools (1939). In 1940 he succeeded R. T. Gunther as curator of the Museum of the History of Science, a position he retained until 1950 when he became director of the Science Museum, London. See A. V. Simcock, 'Alchemy and the World of Science: an Intellectual Biography of Frank Sherwood Taylor', *Ambix*, xxxiv, part 3, November 1987, pp. 121–139.

[32] M.H.S. (n. 31), letter from Gabb to Taylor, September 29, 1943.

In a postscript, Gabb added, 'I would like to know who is ready to give him "£1 000" for it? He has probably already tried America.'[33]

Accordingly, Taylor planned to visit Harper 'at his house 78 Royal Hospital Road, near Tite Street, Chelsea', to see the instrument. He sent Harper's reply to Gabb, suggesting that he join him, adding 'It speaks for itself and answers our question as to the identity of the orrery illustrated in Symonds' book'. Taylor's reference to Symonds' biography indicates that Symonds' manuscript had been completed quite some time before its publication in 1951, probably just prior to or early during the war years. Publication may have had to be delayed due to financial and/or technical considerations until after the cessation of hostilities. Meanwhile, Taylor was informed of the work and familiar with its contents. In his position with the Museum, he may have served as a referee for the publisher, as he was often asked to do by London and American publishers.

Taylor arranged to visit Harper to see the instrument, and confidently informed Gabb, ''t looks to me as if we are going to be asked a great deal less than Mr. Nyburg'. Gabb agreed to accompany Taylor on his visit to Harper, noting:

> The story of the Nyburg Tellerium is 'great stuff!' 'All things being equal', I shall be pleased to go with you, on Thursday, Oct. 31st, to that part of Austria called Chelsea, & interview 'The Count', named Harper, *& acquire it*, and the other two instruments – I regard them as being already 'in the bag!'
>
> A 'Left & right' should be enough to 'bag' the three 'birds', and we can then re-name 'Chelsea' – 'Shotover Furze!'
>
> In the meantime perhaps you will send me the photos of the 'Compass dial & Moon dial by Humphrey Cole 1590' he mentioned so I may consider them. It might be a good move if I went there *alone* as *your representative* & carried through the transaction by myself. It would make it look less important – I am willing to do this.
>
> This 'Graham Orrery' has set me thinking, since you were here, & I remember seeing a *finer one, an actual orrery*, by Graham, *with* a clock face, sold at Stevens *many years ago*, & after a search in my archives, I have found 'Stevens' catalogue, with my Notes on it. It records it was sold for the 'huge' sum of – £15!! – and I was the underbidder. – I enquired of 'Stevens', who was the buyer, & found a dealer, Simmons [*sic*], of Hanway Street, Oxford St., bought it.
>
> I called on him. He had two shops there, & was a 'Furniture dealer & in works of art'. I tried to get it from him, but he told me 'he bought it for America'.
>
> Further proof of what a changed world we live in today.'[34]

Taylor had already returned Harper's photographs and was waiting for confirmation of his appointment to see the instruments. Presumably Taylor and Gabb visited Harper at this time, although there is no mention of it in the surviving correspondence, and if so, decided that the price asked for the instrument was excessive. More

[33] M.H.S. (n. 31), letter from Gabb to Taylor, October 12, 1946.

[34] M.H.S.(n. 31), letters from Taylor to Gabb, October 22, October 26 and October 28, 1946; letter from Gabb to Taylor, October 22, 1946.

months passed without any reported further activity in the matter, and then Taylor learned that Harper had died unexpectedly. When informed of the situation, Gabb wrote rather unkindly:

> I hope Harper has gone to the place where he will be able to play his natural instrument, & not to be asked to act as a stoker! I hope you realise that when you discuss any question with his widow you will be really doing so with Nyburg! Let me know before you decide anything.[35]

Taylor promptly communicated with the widow reminding her of his interest in purchasing the 'orrery', and suggesting a price of £1 000. Mrs Harper informed him, however, that she had already consigned the instrument to Sotheby's auction galleries, to be featured in a forthcoming sale. Taylor hoped to bid the instrument in for a modest price, £1 200, as he suggested to Gabb, but the latter had other views. '*Why* do you think "you have a good chance of getting the instrument for £1 200"?,' he asked Taylor, 'it's not worth that in any case. I enquired at Sotheby's last week when it would be sold, & they said "It wasn't catalogued yet, & *might* not be sold till February".'[36]

The Tompion–Graham instrument was sold on January 27, 1948 at Sotheby & Co.'s London galleries as Lot 174 in an auction sale of 'Objects of Vertu, Snuff Boxes, Watches, Portrait Miniatures ... . Together with a rare Orrery by Tompion and Graham'. It was listed as 'The Property of a Gentleman (dcd).', confirming that it had come from the late Francis Harper's estate. It was purchased for the Museum of the History of Science with funds raised from the National Art Collection Fund, which had been organized by two of Gabb's friends and of which Gabb also was a member. In addition to the £400 subscribed by the Fund, two private donations of £100 each were received. Shortly after its acquisition, the instrument and its known history were described in an article co-authored by Taylor and Gabb in *The Connoisseur*.[37]

It is interesting to note that the proto-orrery signed only by Graham had in fact come to light many years prior to the Tompion–Graham instrument, as reported by Gabb to Taylor, but it disappeared again almost immediately thereafter and did not surface again until three more decades had passed. After having been purchased in 1914 or early 1915 from an auction sale at Stevens Galleries in London by an English dealer identified by Gabb as 'Simmons', it was shipped to New York and featured there in an auction sale of The American Art Association of properties of Henry Symons, described as an 'Expert of London'. Little is known about Symons except that he was, as Gabb described

[35] M.H.S. (n. 31), letter from Taylor to Gabb, October 30, 1946; letter from Taylor to Mrs Francis Harper, November 6, 1947; letter from Gabb to Taylor, November 7, 1947.

[36] M.H.S. (n. 31), letter from Gabb to Taylor, December 20, 1947.

[37] Sotheby & Co., Ltd, *Catalogue of Objects of Vertu, Snuff Bottles, Watches and Portraits Together with Very Fine Astronomical and Geometrical Instruments*, January 27, 1948, frontispiece and p. 20, Lot 174; *Committee of the Museum of the History of Science, Report for 1947–48*, p. 37; *A Brief Guide to the Museum of the History of Science Oxford*, Oxford, 1948, p. 23; Gabb and Taylor *Op. cit.* (n. 7), pp. 24–26, 55. The instrument is catalogued as Accession No. 48–27. Courtesy of Anthony V. Simcock, Librarian, Museum of the History of Science.

him, 'a Furniture dealer & works of art'. Between 1903 and 1920 he maintained his shop at several different addresses on Hanway Street, corner of Oxford Street. It is quite a coincidence that from 1921 to 1954 the same premises were occupied by the antiquities dealer Jesse Myer Botibol. Symons later sold properties at several earlier auction sales by Christie, Manson & Woods Ltd in New York in 1922 and 1923.[38]

In the auction sale catalogue, the Graham instrument, listed as Lot 1288, was illustrated and described as '"Honest" George Graham's Famous Orrery Clock'. The catalogue also noted that the instrument was equipped with a walnut carrying case having brass bands, and that '... the clock for many years seems to have disappeared. Only comparatively recently did it come to light again and find its way into the possession of the present owner'. It was sold in the sixth session for eight hundred twenty-five dollars, bid in by 'Otto Bernet, agent'. Bernet, a member of the staff of the American Art Association galleries, bid on the instrument on behalf of an unidentified client.[39]

Once more the Graham proto-orrery disappeared, and it was not until twenty-six years later, in November 1941, that it again came to notice, when it was offered for sale to the Adler Planetarium and Astronomical Museum in Chicago. The offer was made by Samuel E. Wilson, an importer of Chinese antiques, from whom the Planetarium had previously purchased a few instruments. Wilson was offering the 'astronomical clock' on behalf of the owner, Byron de Forest, an engineer in the Line Design Section of the Commonwealth Edison Company, who claimed to have refused an offer of ten thousand dollars for it. The instrument was stored in the vault of Commonwealth Edison Company.[40]

The Planetarium did not purchase it at this time, but five years later, in September 1946, the instrument was offered again, at a modified price, this time by de Forest, who described it as 'A Graham Astronomical Clock'. In his negotiations with F. Wagner Schlesinger, incumbent director of the Planetarium, de Forest claimed that the instrument was then owned jointly by seven members of his family, and was being offered for the price of $7 500. After further correspondence, Mr Max Adler, the founder and patron of the Planetarium was consulted, and he authorized its purchase.[41]

[38] M.H.S. (n. 31), letter from Gabb to Taylor, October 23, 1946.

[39] *Illustrated Catalogue of the Remarkable and Extensive Collection of Rare and Costly Art Property Belonging to Mr. Henry Symons, Expert of London*, New York, 1915, with 2 illustrations of the Graham proto-orrery. Communications from Mr. E. L. Cave, senior vice president Sotheby's Parke-Bernet Galleries, Inc., New York, July 25, 1976 and September 8, 1991. Harry Symons maintained an antiquities shop at several locations on the same street from 1903 to 1920, at No. 24 to No. 26, then he moved to No. 20, later No. 28 and 30, Hanway Street, W. 1, just off Oxford Street. Botibol maintained his shop in the same premises from 1921 to 1954.

[40] Adler Planetarium, Department of Astronomy Archives (Hereafter Adler Astronomy Archives), letter from Samuel E. Wilson to F. Wagner Schlesinger, November ND, 1941.

[41] Adler Astronomy Archives (n. 40), letter from de Forest to Schlesinger, September 16, 1946 offering the instrument with an undated account of the instrument's history and an undated notarized statement concerning the same signed by Mr Willys Oman, curator of the watch collections of George Packard and Paul Chamberlain of Detroit, Michigan; letters from F. Wagner Schlesinger to Max Adler, June 4, August 20, and November 19, 1947; letter from de Forest to Adler, August 21, 1947; letter from Schlesinger to de Forest, August 25, 1947; letter from Schlesinger to Loren G. McKinney, University of North Carolina, January 23, 1950. Byron de Forest, born in 1891, died at his home in Evanston, Illinois in 1954.

The provenance provided by de Forest was intriguing, to be sure, but was in fact entirely a fabrication. According to de Forest's claim, the instrument had been purchased in 1795 by the Wickett family of Holland, presumably from 'the Seat Council of London', and subsequently it was passed down through several generations to their descendants including the de Forest family, one of whom had brought it to the United States. De Forest further claimed that in about 1899 or 1900 his family had sold the instrument to Admiral George Dewey, the hero of Manila Bay. He did not explain how it was returned to the possession of his family so that it was enabled to sell it once more.

As part of the questionable provenance he had already supplied, de Forest reported an unverified tradition that when Graham moved from the shop he had shared with Tompion to his new location, he brought the astronomical clock with him, and that it was the only personal possession he retained until his death, at which time it was surrendered to the Seat Council of London, presumably in payment of outstanding debts.[42]

Although it made pleasant history, de Forest's so-called 'provenance' is fictional on all counts. The instrument had been brought to New York, not by descendants of the Wickett family, but by Henry Symons and sold at auction. The National Register of Archives and the Royal Commission on Historical Manuscripts as well as the Guildhall Library confirm that there was never at any time in London an organization known by the name of the 'Seat Council' or one resembling it. Graham died without outstanding debts; if a debt had been owing to London's civic authorities, such as the City Corporation or to Graham's livery company, the Clockmakers' Company, an account of it would have been filed. There is no mention of such in the records of the Guildhall Library nor in the Royal Commission on Historical Manuscripts. An examination of Graham's last will and testament reveals that although the document was made irregularly, it was duly proved in the Prerogative Court of Canterbury on November 18, 1751, when administration was granted to his executors. The testator's specific requests concerned his relatives and servants, and the poor of the Clockmakers' Company. There is no mention of an instrument nor anything to suggest that if it had still been in his possession at the time of death, it had not passed with the remainder of his property. Graham's obituary notice specified '... his whole property, except his stock in trade, was found in a strong box, which tho' it was less than would have been heaped by avarice, was yet more than would have been restrained to prodigality'.[43]

[42] Adler Astronomy Archives (n. 40), letter from Schlesinger to Loren G. McKinney of the University of North Carolina at Chapel Hill reporting acquisition of the instrument, January 23, 1950; Ena L. Yonge, *Early Globes Made Prior to 1850 and Conserved in the United States* (New York: American Geographical Society, 1968), p. 106; courtesy of James A. Slemp and Mrs. Roderick S. Webster.

[43] Communication from J. E. G. Bennell, Royal Commission on Historical Manuscripts, Quality House, London, September 2, 1976; *The Gentleman's Magazine*, November 1751, p. 524. The late Miss Yonge quoted the erroneous information concerning the Seat Council from the Planetarium's files without indicating the source, and it was again quoted from Miss Yonge's work by King and Millburn *Op. cit.* (n. 1), p. 152.

No publicity attended the Adler Planetarium's acquisition of the Graham instrument in 1947, and its presence remained un-noted in the collection until 1961.[44]

There is no clue to the ownership of the historical relic from the time it was in Graham's possession until it appeared in Stevens' London galleries in about 1914–1915, after two centuries, and from then until it came to light once more in Chicago a quarter of a century later. It remains for future historians of science to complete the missing gaps in the provenance of both the proto-orreries of George Graham.[45]

---

44 The Graham proto-orrery was discovered in the Planetarium's storage area by the present writer while making an inventory of the scientific instruments.

45 The assistance and suggestions of many who have contributed to this research are gratefully acknowledged. First of all must be mentioned Francis R. Maddison, curator of the Museum of the History of Science, who contributed substantially to this research in 1961, when the writer first initiated it, more than thirty years ago. Others include William J. F. Andrewes, curator, Historical Scientific Instruments Collection, Harvard University; E. L. Cave, Sotheby's Parke Bernet Galleries, New York; Jeremy P. Collins and Sarah O'Brien, Christie, Manson & Woods Ltd, London; Beresford Hutchinson of the British Museum and Kirstin Lippincott, curators, National Maritime Museum; W. A. James, hon. secretary, Spalding Gentlemen's Society; John R. Millburn, Aylesbury, Kent; James A. Slemp, formerly of the Adler Planetarium; Liba Taub, curator, Astronomical Instruments Collection, Adler Planetarium; Anthony J. Turner, Le Mesnil-le-Roi, France; and Gerard L'E. Turner, Oxford, England. Particular thanks are due to Roderick S. and Marjorie Webster of Winnetka, Illinois for their assistance with Adler Planetarium records over a long period of years, and to Anthony V. Simcock for searching the Museum of the History of Science's archival materials.

The Cambridge Legacy of Robert T. Gunther

J.A. Bennett

Fourteen volumes bound in dark blue, one in light – the ratio accurately reflects the institutional allegiance of Robert T. Gunther, who is rightly credited with the foundation of the Museum of the History of Science in Oxford. The dark blue volumes are those of his *Early Science in Oxford*, printed 'for the author' at the Oxford University Press between 1920 and 1945.[1] A fifteenth volume was a biography published by his son in 1967.[2] The light blue was his single *Early Science in Cambridge* volume of 1937.[3]

So it seems that Cambridge had Whipple and Oxford Gunther – friends through their shared interests in old instruments, but very different characters. Whipple was the technician, businessman, entrepreneur and collector; Gunther was the scholar and science historian of an antiquarian bent, personally aloof but an adroit university politician. Whipple stood outside the University, to which he presented his collection and his benefaction; Gunther worked from within, by challenging, pestering, cajoling and persuading. But the institutional separation is not so secure: Gunther's part in establishing a museum of the history of science in Cambridge, as well as in Oxford, deserves recognition.

[1] R.W.T. Gunther, *Early Science in Oxford*, 14 vols, Oxford, 1920–45. Anthony Turner has pointed out to me that volume 1 was originally published in three separate parts between 1920 and 1923 and then as a single volume with continuous pagination in 1923 both in the form that was to become standard for the series and as volume 77 in the publications of the Oxford Historical Society.

[2] A.E. Gunther, *Early Science in Oxford, Vol. XV Robert T. Gunther: a Pioneer in the History of Science 1869–1940*, Oxford, 1967. Gunther had planned and begun work on a fifteenth volume to deal with the history of the Old Ashmolean Building and that of the Museum of the History of Science; material he prepared for this is in R.T. Gunther, 'The Story of the Old Ashmolean 1683–1925', A.V. Simcock, ed., *Robert T. Gunther and the Old Ashmolean*, Oxford, 1985, pp. 1–42.

[3] R.W.T. Gunther, *Early Science in Cambridge*, Oxford, 1937.

From *Learning, Language and Invention: Essays presented to Francis Maddison*, ed. W.D. Hackmann and A.J. Turner. Published by Variorum, Ashgate Publishing Ltd, Gower House, Croft Road, Aldershot, Hampshire, GU11 3HR, Great Britain.

Although Gunther's title at Oxford was 'Reader in the History of Science', he was more astutely described by his friend A.S. Russell, writing the obituary in *Nature* in 1940, as 'essentially a seventeenth century antiquarian'.[4] He was a collector rather than a synthesizer – delighting in the individual and the particular – and this is as true of his books, which are more collections of material than they are histories, as of the collections of instruments he amassed in Oxford and, indeed, in Cambridge. His greatest work, the monumental *Astrolabes of the World*,[5] is a collection of the most prized of mathematical instruments – a paper collection of a scope and grandeur that, in relation to the literature on instrumentation, matched the grandeur of the Lewis Evans collection of actual astrolabes, whose presentation to the University of Oxford was one of Gunther's outstanding achievements.

From 1915 Gunther had been pursuing at Oxford what he called 'the Archaeology of Science'[6] – digging in the Colleges and scientific departments of the University, recording and photographing all the early scientific instruments he could find. This activity was eventually recorded, particularly in the first three volumes of *Early Science in Oxford*. Gunther brought together 123 of the instruments into an exhibition in the Bodleian Library, which ran for a month from mid-May to mid-June 1919. It was a preliminary exercise to promoting what was by then his ambition to see a museum housing these and his other finds in the Old Ashmolean Building in Broad Street.

The exhibition had a more immediate effect, for it attracted the interest of the owner of one of the country's foremost private collections of instruments – the paper manufacturer Lewis Evans, who was already known to Gunther. Evans offered his collection to the University early in 1920, but got a lukewarm response from the Vice-Chancellor. Gunther saw the Evans collection as a potential bait that might draw into his proposed museum the exhibits he had collected for the Bodleian exhibition and other instruments scattered through the Colleges and University Departments.

Evans then offered to lend Oxford his collection for two years, to encourage a decision over whether it could be accepted permanently, and it was arranged on display in the Bodleian in 1922. The need to find accommodation for the Evans collection forced attention onto Gunther's agenda and eventually – after much effort and argument on his part – the collection was accepted and displayed on the top floor of the Old Ashmolean. Thus the museum opened, or, as Gunther would have said, the museum reopened, as the Lewis Evans Collection in May 1925. Eventually, of course, it became the Museum of the History of Science and took over the entire building.

It was necessary to retell this Oxford story in outline to prepare us for what is to happen in Cambridge. An 'archaeology' of science heightens the sense of place in a subject more usually thought to transcend locality; for Gunther the site was the initial focus and a formative influence on the project. In a period when the torch of a new

---

[4] A.S. Russell, 'Dr R.T. Gunther', *Nature*, cxlv, 1940, pp. 541–542, see p. 542.

[5] R.W.T. Gunther, *The Astrolabes of the World*, Oxford, 1932.

[6] For Gunther's programme, see A.V. Simcock, 'A Dodo in the Arc', A.V. Simcock, ed., *Robert T. Gunther and the Old Ashmolean*, Oxford, 1985, pp. 45–83, on which I have drawn generally for my account of Gunther's work in Oxford.

history of science seemed to belong to grand intellectual syntheses like Burtt's *Metaphysical Foundations of Modern Physical Science* or Lovejoy's *Great Chain of Being*, it must have seemed quaint, odd, perhaps irrelevant that Gunther continued to offer volume after volume of *Early Science in Oxford*. But he pursued his archaeology, and in the decade after its position was secured in Oxford, he turned his attention, as best he could, to Cambridge. It is worth noting that the group of influential scientists Gunther had gathered together in 1928 as the 'Friends of the Old Ashmolean' adopted, as the first of their aims, an ambition to establish museums of the history of science and medicine 'at the older universities' – plural.[7]

The Cambridge work was initiated in March 1934 by Gunther writing to the Cambridge Philosophical Society to propose a survey of surviving instruments in Cambridge, leading to a public exhibition.[8] Contrary to Gunther's experience of struggling in Oxford, the immediate response was positive, indeed unrealistically so. The Secretary Hugh Hamshaw Thomas said he would see what rooms would be available for the beginning of the Michaelmas Term. The Society held a special Council meeting in May to consider Gunther's proposal, and appointed a committee with the Master of Pembroke, Arthur Hutchinson as Chairman and Hamshaw Thomas as Secretary. John Cockcroft was an active member, and was anxious that the Cavendish should be well represented, and the Committee also included the two astrophysicists H.F. Newall and F.J.M. Stratton.

At the May meeting the planned exhibition was postponed to Easter Term 1935, but Gunther set to work during the summer of 1934 on the archaeology of science in Cambridge – visiting the Colleges and Science Departments, and even the Inspector of Weights and Measures at the Guildhall.[9]

While the exhibition date was frequently postponed, Gunther had much with which to occupy himself in the intermittent periods he could spend in Cambridge. He began to draw up lists of the material he hoped to use, arranged under Colleges and Departments, but also mentioning two 'Private persons', 'Mr. E.S. Peck', who could provide 'Mortars / Pharmacy Jars' and 'Dr Whipple': 'Much very early Apparatus / Microtomes / Sectors / Architect's Scales'.[10] In fact Hamshaw Thomas had spoken to Whipple about the exhibition proposal as early as July 1934, and had found him very keen, and it is clear from the letters preserved in the Museum of the History of Science that Hamshaw Thomas did much work in Cambridge preparing the ground for Gunther.

Eventually a firm arrangement was made for two weeks in June 1936 and the accommodation settled as the East Room of the Old Schools. As the time drew nearer, Hamshaw Thomas became increasingly occupied with the arrangements. There is a letter to him from Joseph Needham in January 1936, with the postscript: 'What is

[7] From an announcement in the archives of the Museum of the History of Science, Oxford. I am grateful to A.V. Simcock for his help and guidance through the manuscripts and printed ephemera in Oxford.

[8] Museum of the History of Science, MS Gunther, 46.

[9] *Ibid.*

[10] *Ibid.*

happening about the collection of historic scientific apparatus? When and where am I to bring the few things we have in the Biochemical Laboratory?' Gunther came to know Whipple around this time, being invited to his home in Highgate to spend time with Whipple's collection.

'The Exhibition of Historic Scientific Apparatus' was opened by Lord Rutherford on the evening of Monday 8 June and the notes for Gunther's speech are preserved in Oxford, written on the stationery of the Cambridge University Arms Hotel. His efforts were recognized in the speeches of both Rutherford and Hutchinson, the latter calling him 'the perfect detonator' to release the potential energy of the scheme.

The exhibition contained some 200 items, including such memorable objects as Roger North's set of drawing instruments from Jesus College, the outstanding mathematical instruments from Trinity, astrolabes from King's and Gonville and Caius, a grand orrery from St John's together with surviving instruments from the College's eighteenth-century observatory, the celebrated physics apparatus from the Cavendish and Darwin's compound microscope from the Botany School. All concerned, at least in their letters to Gunther, judged the Exhibition a great success, and it was remounted in the Cavendish in August 1938 for the meeting of the British Association. Once again, this was Gunther's idea, having been suggested by him in September 1937.

As might be expected from the Oxford precedent, Gunther's ambitions at Cambridge did not stop at temporary exhibitions. He won a significant public expression of support at the opening of the first exhibition, when Rutherford announced:

> Now that the committee of the Cambridge Philosophical Society and Dr. Gunther have taken this enormous amount of trouble to collect this apparatus – and for that we owe a debt of gratitude – and now that we know we have this wealth of important historical apparatus in the University, steps should be taken to find a permanent suitable home for the more interesting and more historical part of it.

Hutchinson and Cockcroft expressed their agreement, the latter writing to Gunther of his hope that 'all your hard work will bear fruit in some home for the apparatus'.

The idea most commonly mentioned in Cambridge was a permanent display in the Fitzwilliam. Gunther is very unlikely to have favoured this, in the light of his antagonism to the Ashmolean in Oxford, whose very title he disputed, and he began to lobby for a dedicated museum and to carry his campaign beyond the University. In January and February 1937 he wrote letters to the Prime Minster, Stanley Baldwin, seeking his support for the proposal that the University of Cambridge should establish 'a Museum for the Conservation of Scientific Objects of Historic importance', emphasizing the priority he placed on *objects*:

> In all ancient institutions of good repute there are material *things* which are of greater interest and utility than some of the *persons* in office. This was proved by the fact that the British Museum paid a prodigious sum for the Codex, a higher figure than is paid in salaries to any member of the staff! The justification is that the Codex 'lasts for ever'. ... So it is, both at Oxford and Cambridge. Professors are very easy to replace. Type and unique specimens cannot be.

He continued to canvass Baldwin's support through the year.

By 1937 Gunther's supporters in Cambridge also were thinking in terms of a museum, rather than a gallery at the Fitzwilliam. Hamshaw Thomas told him in December that Hutchinson had died, and so 'we have now lost two good supporters of the idea of a museum of science instruments'; the other was Rutherford, who had died the same year. However, he judged that the movement was gaining ground, citing the successful continuation of the lectures in history of science and the plan for an exhibition at the British Association meeting. In the following year Gunther persuaded the General Committee of the Cambridge meeting of the British Association to express to the University their hope that 'the scheme for a permanent exhibition of historic scientific instruments at Cambridge would be brought to fruition'.

The book *Early Science in Cambridge* was part of this campaign. It was completed in 1936, the same year as the exhibition, and shows signs of having been hastily prepared, with many mistakes and examples of hurried editing. One of the most embarrassing slips was pointed out immediately by Arthur Hutchinson when he received his copy in mid-January 1937. (The printer's estimate was dated only 7 December 1937.) The portrait of Lord Rayleigh had been captioned as one of J.J. Thomson:

> It is most unfortunate that this substitution should have occurred as I fear it will rouse much unfavourable comment here. – Can anything be done about it?

Fortunately not many copies had been distributed, probably not many even bound up. One that had been sent was to Rutherford, and Gunther had to write asking him to return the offending plate.

Perhaps even more than any of his other books, *Early Science in Cambridge* is a collection of miscellaneous information – biographical facts, anecdotes, quotations – for the most part simply arranged chronologically within subject chapters. But embedded within this, running throughout the text with sequentially numbered items, is a much more valuable record – an illustrated catalogue of the instruments Gunther found in Cambridge. Since much of this material has now found its way into the Whipple Museum, this is an invaluable record of locations and sometimes conditions, together with other information Gunther picked up at the time.

Early Science in Cambridge was reviewed by the two Cambridge men most closely involved – Hutchinson in *Nature* and Hamshaw Thomas in the *Cambridge Review*.[11] They were not altogether disinterested. Hamshaw Thomas wrote to Gunther that he has secured his review copy, saying: 'I shall, of course, go "all out" for the Museum'.[12] Hutchinson in *Nature* had to admit that the book 'must be regarded more as a source book than as an orderly treatise' and that it suffered from mistakes, misprints and repetitions, and 'from a certain lack of orderly arrangement and from rather meagre documentation'. He did acknowledge the value of Gunther's efforts and the great amount of information he had crammed into his book, and ended with support for a permanent home.

---

[11] *Nature*, cxl, 1937, pp. 130–131; *The Cambridge Review*, lviii, 1936–7, pp. 434–436.

[12] Museum of the History of Science, MS Gunther, 46.

Hamshaw Thomas was a little more measured in his criticisms, and devoted more space to future hopes. He admitted that the book might 'at first sight appear to be a conglomeration of information' and that it had to be described as 'materials for a history of science in Cambridge'. His vision of a museum, however, was painted much more strongly than before. He was concerned that, though Cambridge was now aware of the wealth of her collections, it was hazardous to leave them to their fates in individual departments – 'till', as he put it, 'another King arose who knew not Joseph':

> Is not this the time for the establishment in Cambridge of a museum of the history of science? We can scarcely hope for accommodation like the Old Ashmolean at Oxford, but it ought to be possible for the University to find accommodation for our collections on some part of the museum's site. If a suitable place were provided there is little doubt that some colleges and departments would be glad to hand over much of the old apparatus which they have no room to display. A suitably exhibited collection with appropriate portraits and documents would be of permanent value to the University, a constant source of interest to students, and a centre which would be visited by scientists from all parts of the world.

As a prophetic vision, this came close to the mark.

Gunther died in 1940 and did not see the long-term consequences of his work in Cambridge. There can be little doubt that Whipple's decision to present his magnificent collection – and eventually a benefaction to extend the collection – derived from the Gunther exhibition and the consequent movement to establish a permanent museum. The collection was presented in October 1944, and the next exhibition was of Whipple's collection at the Old Schools in November. It was afterwards stored, first at the Fitzwilliam, then in a war-time factory building near the Observatories and, after a fire there, in Girton College, until in May 1951 the Museum at last opened its doors in Corn Exchange Street. It moved to its present home in Free School Lane in 1959.

Although Gunther did not see the outcome in Cambridge, the parallels between the policies he adopted in Oxford and in Cambridge are striking – the archaeology of science activity in the colleges and departments, the publication of its results in the *Early Science* series, the organization of a public exhibition to recruit interest and support for a permanent museum, and the cultivation of a significant collector in the belief that a large presentation would force the University to act in a matter that otherwise had no urgency and in the further belief that the existence of such a collection would act as a magnet to the various collections scattered throughout the University and Colleges. We began with a juxtaposition of Gunther in Oxford and Whipple in Cambridge; we now see that this needs to be redrawn. The Cambridge equivalent of Lewis Evans was Robert Whipple; the Cambridge equivalent of Gunther was Gunther.

A Musician among 'The Measurers'

John Bergsagel

In the Museum of the History of Science in Oxford there hangs a small painting, done on a wooden panel, depicting a scene of great industry and activity [Figure 1]. In a building, or group of buildings, of indeterminate character and architecture, a number of people are busily engaged in a variety of occupations; on closer examination, however, one comes to the realization that, though all are doing different things, they are nevertheless all doing the same thing – they are measuring. In the foreground a woman is measuring cloth and a man is measuring (or inscribing) a geometric drawing; in the middle ground a pair are seen filling containers with dry measure, while further back, in what to judge from the sign projecting from the wall outside may be a tavern, still others are drawing off containers of liquid measure. Looking through the house one sees that in the courtyard to the left some men are weighing large sacks, perhaps of corn or coal, whereas in a field to the right two men are taking sightings, one measuring horizontal angles with a surveyor's cross, the other holding a quadrant to measure altitude. In addition, spread out over the floor in the foreground are a globe and an assortment of instruments of measurement.[1]

[1] These have been identified for me very kindly by Anthony Turner (whom I should like to thank for valuable advice) as follows: From left to right the instruments shown in the painting are a large sack balance in use with weights scattered around in front of it; an arched pair of adjustable scribers, a (probably) celestial globe, a pair of one-handed dividers being used to scribe a circle on the board held by the craftsman who also has a rule on his lap; beside his right foot are a plumb-line on a board, a pair of flat-leg dividers, a plumb-line level or *archipendolus*; by his left foot a geometrical square or plotting board with an adjustable rule pivoted from one corner and a pair of dividers lying on it; in front of this is a geometric quadrant for taking altitudes and an indeterminable object; a polyhedral sundial with in front a straight rule and a set square; an apothecary's balance; the woman is measuring cloth using a tailor's rule (perhaps an aune) and a second is lying on the floor in front of her beside a pair of scissors. Behind a dry capacity measure is being used, while in the distance the measurers in the fields are using an altitude quadrant and a surveyor's cross of the open-sights crown type.

From *Learning, Language and Invention: Essays presented to Francis Maddison*, ed. W.D. Hackmann and A.J. Turner. Published by Variorum, Ashgate Publishing Ltd, Gower House, Croft Road, Aldershot, Hampshire, GU11 3HR, Great Britain.

Figure 1. *'The Measurers', formerly attributed to Hendrik van Balen, c. 1600. Museum of the History of Science (inv. no. 33–73)*

Two groups remain whose relevance to this uniform activity is not clear. One is the intimate couple that can just be glimpsed in a room at the very back of the premises; they have apparently dined together and, if indeed the scene has any significance in the context at all and is not just confirmation of the fact that we are looking into a tavern, perhaps it is that the man is now taking his sweetheart's measure in a fond embrace, or perhaps he is proclaiming his love in the measured feet of verse – 'lisping in numbers', as Pope might say. If the intention of this vignette is obscure, the activity of the other group, displayed prominently in the left foreground, is clearly evident: here a musician is guiding a pupil through a piece of music contained in a book lying across his lap. But neither here is the significance of the scene in the context of the apparent theme of the picture immediately obvious: if this is an illustration of yet another act of measuring, what is being measured?

The Museum has obliged public interest in the picture by making it available as a postcard under the title 'The Measurers', naming Hendrik van Balen (1560–1632) as the artist, though the authority for both title and attribution is not known.[2] Whether original or not, the title is obviously appropriate, and though Hendrik van Balen's authorship has recently been denied,[3] the origin of the painting remains assigned to Antwerp (where Hendrik van Balen worked) a little before or about the year 1600. It is an amusing picture and it may be that for the general public, as it is for me, the presence in it of the musician is its principal point of interest. For many, whose concept of music has been shaped by nineteenth-century Romanticism, the inclusion of an art amongst the sciences must indeed seem incongruous. After all, though most, if not all, of the numerous instruments and tools depicted in the painting can also be seen on display in the Museum, no music is there exhibited, nor is there anything to suggest that the Museum itself considers music as coming within its province.

The historian will of course see in the painting a reflection of the medieval systematization of the *artes liberales*, which included music among the branches of learning assigned to the *quadrivium*, the mathematical sciences: Arithmetic, Geometry, Astronomy and Music, whereas the *trivium* comprised the eloquent arts of Grammar, Rhetoric and Dialectic. This association, based on a view of music attributed to Pythagoras, was propagated in the West representatively by Boethius (*c.* 480–524) and Cassiodorus (*c.* 485–*c.* 580) and remained influential well into the seventeenth century.[4] Throughout the Middle Ages it provided the basis for the teaching of music

[2] I am informed that a new printing of the card takes account of this uncertainty.

[3] By Bettina Werche (Bonn) in a letter to the Museum, 20 August 1991. The relationship of this painting to the drawing of the same scene (brought to my attention by Anthony Turner), reproduced in an engraving by C. Galle after a member of the school of Frans Floris (illustrated in Jean Bonnaveau, *Étude des instruments contenus dans les gravures de Sebastian Leclerc, 'L'Académie des Sciences et des Beaux-Arts', 'L'Académie de Physique'*, privately published, [Cannes], n.d. [*c.* 1992], p. 6), is uncertain, but obviously has a bearing on the question of the attribution of our painting, as well as on the possibility of finding in it traces of the Italian influence of which Frans Floris (*c.* 1517–1570) was a prominent representative in Antwerp.

[4] The idea of music being closely related to mathematics was still current in the eighteenth century, as I discussed with reference to Sir William Herschel in 'Music, a Science and an Art: the 18th-Century Parting of the Ways' in *Dansk Årbog for Musikforskning*, xii, 1981, pp. 5–18.

in the universities, but increasingly in the fifteenth and sixteenth centuries music's mathematical properties were felt to be inadequate to explain either its aesthetic qualities or its affective power. Boethius had already in his *De institutione musica* recognized music's ambivalent character when he wrote, '... of the four mathematical disciplines, the others are concerned with the pursuit of truth, but music is related not only to speculation but to morality as well',[5] thereby seeming to imply that music in some way constituted a bridge between the *quadrivium* and the *trivium.* Heinrich Hüschen saw the development throughout the Middle Ages thus: 'Für das Mittelalter ist es nun höchst bezeichnend, daß sich der Schwerpunkt des Musikdenken in einem kontinuierlichen Entwicklungsprozeß von der mathematisch-spekulativen Anschauung zur klanglich-empirischen Auffassung hin verlagert.'[6] And, continuing the story into the Renaissance, Claude V. Palisca concluded that 'To link music with the verbal arts, with rhetoric as well as poetry, was as characteristic of the Renaissance as it was typical of the Middle Ages to ally music with the mathematical sciences.'[7]

It is Palisca's declared aim, in contradiction of the view that is generally held, 'to show that with music, as with the other arts and letters and learning in general, the movement we call the Renaissance *began in Italy,* and that its chief source of inspiration was the revival of antiquity'.[8] Yet it is in a treatise written by a Netherlander and published in Nürnberg in 1552 that music is most uncompromisingly removed from the *musici mathematici*, whom the author regarded as being too concerned with the teaching of rules and with speculation and whose compositions deserved only to be cast aside, to be relocated ultimately amongst the *poetici*, a group which constitutes a projection of the class of musicians he calls 'the most excellent', who know how to turn theoretical rules into compositions capable of expressing all the emotions in elegantly ornamented melodies and songs. The 'poets', who are supreme in this hierarchy, are 'Belgians, Picards and French' (Italians are not mentioned); they are 'the only ones who serve in the chapels of the Pope, the Emperor and the King of France'.[9] The author, Adrian Petit Coclico (*c.* 1500 – after 1562), cannot be said to rank high as a theorist, but he expresses

---

[5] Book I, 1, quoted from O. Strunk, ed., *Source Readings in Music History,* New York, 1950, p. 80.

[6] H. Hüschen, 'Die Musik im Kreise der artes liberales', in *Gesellschaft für Musikforschung: Kongress-Bericht, Hamburg*, 1956, Kassel, 1957, p. 120.

[7] C. V. Palisca, *Humanism in Italian Renaissance Musical Thought,* New Haven, Conn., 1985, p. 332.

[8] *Ibid.*, p. 22 (the italics are mine). On p. 1, fns 1 and 2, Palisca specifies that the view of a Northern origin of the Renaissance in music which he wishes to revise is that expressed, for example, by Gustave Reese in his *Music in the Renaissance*, New York, 1954; Howard Mayer Brown in his *Music in the Renaissance,* Englewood Cliffs, 1976; and Leo Schrade, 'Renaissance: the Historical Conception of an Epoch', in *Kongress-Bericht der Internationale Gesellschaft für Musikwissenschaft, Utrecht 1952,* Amsterdam, 1953, pp. 19–32. To these references might be added Schrade's *Monteverdi, Creator of Modern Music,* New York, 1950, part 1: 'The "Perfect Art": A Prelude to Monteverdi', pp. 17–74, which emphasizes the rôle played by the revival of interest in antiquity, not in shaping the Renaissance, but in contributing to the Italian reaction against the dominant Northern Renaissance style that gives rise at the end of the sixteenth century to the ensuing Baroque period in music.

[9] Adrian Petit Coclico, *Compendium musices,* Nürnberg, 1552. Facsimile edition, ed. M. Bukofzer, *Documenta musicologica,* Erste Reihe IX, Kassel, 1954, pp. 15–17.

at least the views of a practising musician who claimed to have been a pupil of the great Josquin des Prez.

Indeed, this change of attitude with regard to the mathematical basis of music is generally considered to be one of the most characteristic signs distinguishing the Renaissance from the Middle Ages in music. At the same time, the humanistic revival of the learning of antiquity, so comprehensively documented by Professor Palisca, gave rise during the Renaissance to renewed interest in Pythagorean, Platonic and neo-Platonic ideas, of which Johannes Kepler's championing of the 'Harmony of the Spheres' as late as 1619[10] must represent the culmination of speculative thinking in music. The conception of music and of music's place in the Divine Order is thus made complicated and the simple transition from a medieval – speculative to a Renaissance – practical view is disturbed. Furthermore, the competing claims as to the origin, nature and course of development of the Renaissance in music adds interest to the geographical distribution of even the smallest bits of evidence. One asks therefore if it is possible that the little painting in Oxford can tell anything significant about the way in which music was regarded *c.* 1600 in Antwerp?

I have called 'The Measurers' a reflection of the medieval teaching of the *quadrivium*, but it is not a direct illustration of it in which, as was common in medieval books, the four mathematical sciences are represented by personifications (successors perhaps to the Muses of antiquity) identified by labels or by means of their accompanying attributes.[11] Our picture is a stage removed, so to speak, from that and shows, instead of representations of the sciences, the scientists themselves at work – or at least practitioners of the sciences, which is not the same thing, as Boethius made clear when he wrote 'It is far greater and nobler to know what someone does than to accomplish oneself what someone else knows, for physical skill obeys like a handmaid while reason rules like a mistress'.[12] The people in our picture are measuring, which we may define as 'describing in terms of numbers' or 'quantifying'; what interpretation of music does the artist thus reveal in his treatment of the musician in this picture? Perhaps it will help to put it into perspective if we look first at some other treatments of the theme.

A representation of music in a context which anticipates 'The Measurers' by two hundred years, inasmuch as it also combines music and practical measuring, is Jehan de Nizières' illustration to Chapter XIX of Jean Corbichon's *Livre des proprietés des*

[10] *Harmonice mundi libri V*, Linz, 1619. See, *inter alia*, Eric Werner, 'The Last Pythagorean Musician: Johannes Kepler', in *Aspects of Medieval and Renaissance Music*, ed. J. LaRue, London, 1967, pp. 867–882.

[11] See, for example, Werner Bachmann, 'Bilddarstellungen der Musik im Rahmen der artes liberales', in *Gesellschaft für Musikforschung: Kongress-Bericht, Hamburg*, 1956, Kassel, 1957, pp. 46–55.

[12] *De institutione musica* I, 33, quoted from *Source Readings in Music History*, *Op. cit.* (n. 5), p.85. This distinction was expressed throughout the Middle Ages in the oft-quoted lines:

> Musicorum et cantorum magna est distantia,
> Isti dicunt, illi sciunt, quae componit Musica.
> Nam qui facit, quod non sapit, diffinitur bestia.

Figure 2. Jehan de Nizières' illustration from Jean Corbichon's Livre des proprietés des choses, *1372. Paris, Bibliothèque Nationale, MS fr. no. 22532, f. 336*

Figure 3. *Peter Bruegel the Elder, 'Temperantia', 1560, engraved by Philippe Galle*

Figure 4. Hans Holbein the Younger, 'The Ambassadors', 1533. National Gallery

Figure 5. *Philippe Galle, after Johannes Stradanus* 'Encomium musices', c. *1590*

choses, a French translation (1372) of Bartholomæus Anglicus' *Liber de proprietatibus rerum* (before *c.* 1250), in the copy in Paris, Bibl. Nat. MS fr. no. 22532, f. 336 [Figure 2]. In the foreground two figures are engaged in activities similar to those depicted in the Oxford painting: the man to the left is measuring a length of cloth, while the man to the right is holding up a pair of scales, the weights for which lie under it on the floor. As a token of other kinds of measurement, perhaps, a triangle is seen between the two measurers. The middle ground of the picture is occupied by a display of scrolls bearing the roman numerals I to X, obviously the common property of the mathematical sciences, but also the property common to the 'things' of creation, as expressed by Plotinus: 'Number exists before objects which are described by number. The variety of sense objects merely recalls to the soul the notion of number.'[13] Arranged in a semicircle around these groups and providing the background of the picture are twelve instrumentalists and, in the centre, five singers standing before a choirbook on a music stand. Though it might have the appearance of a concerted performance, a kind of 'oratorio' *anno* 1400, the scene can hardly be taken as showing an actual performing situation. It is rather an illustration of the contents of Bartholomæus' Chapter XIX, *De musica*, which describes at length the voices and instruments that create the sounds of music.[14]

An even larger and equally unlikely assembly of singers and instrumentalists is depicted in Peter Bruegel the Elder's fantastic drawing of 'Temperantia' (dated 1560), in his series of the 'Seven Virtues', engraved by Philippe Galle and published by Cock in Antwerp [Figure 3]. Music is here in company not only with the other sciences of the *quadrivium,* but with the *trivium* as well. 'Temperantia', outfitted with symbols of moderation (a clock on her head, a bit and reins in her mouth, a pair of spectacles in her hand, standing on the sail of a windmill, etc.) is surrounded by exponents of the seven liberal arts, shown (as in the Oxford picture) in the active practice of their disciplines and obviously badly in need of her controlling influence. Thus the representation of music, with its juxtaposition of, for the period, incompatible elements, can be construed as music in disharmony for lack of order and proportion. As Konrad Renger observes, 'Alle Künste haben mit Maß und Messen zu tun, was auch Bruegel

[13] I quote the English translation used in John Stevens' *Words and Music in the Middle Ages*, Cambridge, 1986, p. 13.

[14] The Latin text of *De musica* is included in Hermann Müller's 'Der Musiktraktat in dem Werke des Bartholomæus Anglicus *De proprietatis rerum*', in *Riemann-Festschrift*, Leipzig, 1909, pp. 241–255. Extracts from Bartholomæus' treatise, including virtually the whole of the *De musica* chapter, were printed by Sir John Hawkins in his *A General History of the Science and Practice of Music*, London, 1776; 2nd edn, 1853, vol. 1, pp. 212–214, 266–271, in the English translation made by John Trevisa for the Earl of Berkeley in 1398, which was printed by Caxton in 1495. Hawkins recommends the work (p. 266) in terms which I think will appeal to the dedicatee of this article:

> ... the language, style and sentiment are such, as render it to a very great degree instructive and entertaining. Numberless words and phrases, not taken notice of by any of our lexicographers, and which are now either become totally obsolete, or are retained only in particular parts of this kingdom, are here to be met with... In short, to speak of the translation of Bartholomæus by Trevisa, it is a work that merits the attention of every lover of antiquity, every proficient in English literature.

bei den einzelnen Szenen in unterschiedlicher Deutlichkeit darstellt. Hierin dürfte die Gemeinsamkeit mit Temperantia – der Mäßigkeit – liegen...'[15]

Instruments of measurement and music are also displayed together in the magnificent and fascinating portrait, now called 'The Ambassadors', by Hans Holbein the Younger, dated 1533 [Figure 4]. Obviously, in this formal scene from the milieu of the court of Henry VIII there can be no question of seeing the instruments in practical use; the studied ostentation of their display must be in order to use their symbolic significance to indicate the intellectual interests of the two aristocrats portrayed. Thus Music is represented by a lute placed together with an open music book (the ends of some pipes can also be seen protruding from the shadows). The simple neutrality of this symbolism is drawn in question however when we observe that one of the strings of the lute is broken, in which condition it cannot stand as a quadrivial symbol of order and harmony but is rather a traditional symbol of discord and disharmony. Similarly, closer examination reveals that the music book has been so meticulously rendered that the music and even the edition of the print can be identified: rather surprisingly it is seen to be one of the early monuments of Lutheran church music, the Tenor partbook of the second edition of Johann Walter's *Geystliche gesangk Bucklein* (Worms 1525), lying open to expose two of Martin Luther's own hymns: 'Kom heiliger geyst' on the left page and 'Mensch wiltu leben seliglich' on the right. To be sure, in 1533 King Henry was himself challenging the Pope's authority, but all negotiations had not yet been abandoned and it was not until the following year that the Act of Supremacy was promulgated. Furthermore, after the unedifyingly abusive exchange of writings which had taken place between them some ten years earlier and following Luther's negative reply to Henry's appeal to the theologians of various universities for a judgement in the matter of his divorce, Henry was far from being disposed to regard Martin Luther benignly. Accordingly the introduction of Lutheran elements into a painting done in London at just that time must have risked being considered provocative and must therefore have had a well-considered significance. Whatever it was had been forgotten, however, together with the identity of the subjects, when the painting was brought back to England in 1792 and eventually acquired by the National Gallery in 1890. It was not appreciated again until Miss Mary F. S. Hervey published her study *Holbein's 'Ambassadors': the Picture and the Men,* London, 1900, in which she established the identities of Holbein's subjects as (left) Jean de Dinteville, seigneur de Polisy, who was King François I's Ambassador to England in 1533, and (right) Georges de Selve, Bishop of Lavour, who had secret permission to join his friend in London for some weeks in the summer of that year. In addition to being a scholar and a music lover, De Selve was fluent in German and was known to be a churchman with views sympathetic to the Reformation who hoped for, and was no doubt in England to help bring about, a healing of the religious discord which threatened Europe.[16] Thus, though

[15] *Peter Bruegel d. Ä. als Zeichner. Eine Ausstellung der Kupferstichkabinetts Berlin 19. september bis 16. november 1975*, p. 70.

[16] See, *inter alia*, John Rowlands, *Holbein. The Paintings of Hans Holbein the Younger*, Oxford, 1985, pp. 85–87, 139–140; and Paul Ganz, *The Paintings of Hans Holbein. First Complete Collection*, London, 1950, pp. 241–243.

appearing under the guise of one of the quadrivial sciences, Music's presence in the painting conveys a message which has nothing to do with mathematics.

Towards the end of the sixteenth century actual pieces of music were incorporated into a series of ten engravings by Jan Sadeler, most of them published in Antwerp. Though they are clearly and carefully copied, the engravings are not to be seen as editions of music, but as works of art, in which music is utilized as an element in the design. Of a similar kind is the title page of a set of eighteen engravings devoted to Biblical subjects having reference to music under the title 'Encomium musices', designed by Johannes Stradanus (Jan van der Straat) and engraved and published *c.* 1590 by Philippe Galle in Antwerp (who also engraved Bruegel's 'Temperantia', Figure 3). In this picture [Figure 5], a six-part motet *Nata et grata polo vocum discordia* by Andreas Pevernage (1543–91) is shown as part of a design illustrating Psalm 150. The large music book is supported by the central figure of *Musica*, assisted on either side by *Harmonia* and *Mensura*, and around them is worked an elaborate network of all the instruments with which the psalmist urges that the praises of the Lord be sounded. Here music appears alone, independent of the other mathematical sciences of the *quadrivium*, and the scene is one in which the emphasis is on *musica practica* rather than on *musica speculativa*, in keeping with the psalmist's admonition to make music. The three ladies personify concepts which over the centuries have been interpreted in a variety of ways – *Harmonia*, for example, was used by Martianus Capella in the early fifth century as a term equivalent to *Musica*, and in the middle of the thirteenth century Franco of Cologne used the designation *musica mensuralis* to distinguish polyphony from plainsong, *musica plana*. It is evident from their juxtaposition here, however, that their symbolism is meant to be understood in the most obvious way: *Musica*, standing in the centre and above the others, is the art of (practical) music, 'Frau Musica', and *Harmonia* and *Mensura*, placed to left and right on a level below her, are her 'daughters', music's two aspects of space and time – that is, the relationships of pitches to each other, on the one hand, and of their durational values, on the other. Thus, *Harmonia* is represented holding as her attribute a winged heart surmounted by a pair of ears, while *Mensura* holds the verge escapement of a clock. Standing as she does in apposition to Harmony, then, Measure here represents the quantifying of time, which reminds us that we are after all still in the realm of mathematics. Just as pitch intervals since Pythagoras had been expressed as proportions, so also from the beginning of the fourteenth century a system of 'mensurations' had made it possible to express durational values in terms of proportions. In fact, only the latter are treated in the *Proportionale musices* by Johannes Tinctoris (*c.* 1435–?1511), a Franco–Flemish composer and theorist, who, while in the employ of the King of Naples (from *c.* 1472), wrote a series of treatises on the theory and practice of music that constitute the most comprehensive treatment of the subject in the fifteenth century. It is in the *Proemium* to this treatise (and again in similar terms in the *Liber de arte contrapuncti*) that Tinctoris bears weighty witness to what has been regarded as the Northern origin of the Renaissance in music:

> At this time, consequently, the possibilities of our music have been so marvelously increased that there appears to be a new art, if I may so call it, whose fount and origin is held to be among the English, of whom Dunstable stood forth as chief. Contemporary with him in France were Dufay and Binchoys, to whom directly succeeded the moderns Ockeghem, Busnoys, Regis and Caron, who are the most excellent of all the composers I have ever heard.[17]

If we now look again at the little painting in the History of Science Museum in Oxford in the light of the various other representations of music we have seen, we can perhaps better appreciate its distinctive treatment of a familiar subject. Also here music is depicted in the context of the mathematical sciences of the *quadrivium*, but no traces of speculative thinking remain in this scene of mathematics at work in the daily life of ordinary people. Thus it is not Apollo, Pythagoras nor Boethius, but an anonymous music master who is shown instructing a pupil in the music that can be glimpsed in the book that lies across his lap. It would be very interesting to know what this music is, but a photographic enlargement of the book [Figure 6] shows that the unknown artist has not been as minutely meticulous about drawing the notes on the pages of music as were Holbein [Figure 4] and Galle [Figure 5] – at least I have not been able to construe them. It does not appear that he has made any attempt to suggest the presence of a text under the lines of music nor are the musicians' mouths open as if they are singing; furthermore, though a drum lies by the teacher's foot and the object behind the globe by his left hand could perhaps be a bass shawm, there is no implication that they have any part in the musical activity. Whether the music is vocal or instrumental is therefore not specified; the pupil is studying the music under the supervision of the teacher, working his way along the line, following the notes with his finger, hence performing it, even if silently.[18] This is no doubt the act of measuring that relates what the musicians are doing to the other activities seen in the picture – but *what* is being measured?

The book has the format of a partbook, that is, a book containing a single voice part of a polyphonic composition, in which case the relationships of the pitches under consideration are successive, not simultaneous intervals. The traditional instrument for demonstrating pitches and measuring the intervals between them was the monochord; its absence here emphasizes the unscientific, perhaps even domestic character of the scene, but it also suggests that it is not the pitches of the notes that are under examination. That the pupil's problem with the music lies in the rhythm is further suggested by the fact that the master is clearly marking time, the *tactus*, with his left hand, probably by tapping the shoulder of the pupil in time-honoured fashion. What is being measured then is Time, but here only the artificially articulated segment of it essential to a work of art, not that of the vast vision of Henry Vaughan, who:

---

[17] *Source Readings in Music History, Op. cit.* (n. 5), p. 195.

[18] As Carl Dahlhaus points out, 'With music...silent reading, insofar as it is not to collapse into thin abstraction, always represents an inner hearing, translating signs into sound. Musical meaning, in contrast to linguistic meaning, is only to a slight extent, if at all, detachable from the sounding phenomena.' (*Esthetics of Music*, translated by W. W. Austin, Cambridge, 1983, p. 12.

Figure 6. Detail of music book in 'The Measurers'

... saw Eternity the other night
Like a great Ring of pure and endless light,
All calm, as it was bright,
And round beneath it, Time in hours, days, years,
Driv'n by the spheres
Like a vast shadow mov'd, In which the world
And all her train were hurl'd;

for the measurement of which the artist has provided other instruments, such as a sundial.[19]

'The Measurers' retains the association with the mathematical sciences seen in all our examples, with the exception of Figure 5, which is devoted to Music alone. That picture, however, calls attention to the fact that music must be measured in time, which appears also to be the rôle of the musicians in 'The Measurers'. Both of these pictures emanate from Antwerp at the end of the sixteenth century, as does also Bruegel's 'Temperantia', in which the *artes liberales* are viewed satirically for their lack of moderation, a virtue that may perhaps be equated with common sense, and thus common to them all is an emphasis on the practical rather than the speculative aspect of music.[20] None of these pictures by Northern artists, least of all 'The Measurers', gives any sign of taking part in the revival of the learning of antiquity, the debate concerning problems of theory and aesthetics, tunings and temperaments and the music of the spheres, which Palisca so masterfully demonstrates occupied the thinking of many Italians in the fifteenth and sixteenth centuries and to which he attributes the origin of the Renaissance in music.

This notwithstanding, we are in no doubt that the view of music with which we are presented in our pictures is a Renaissance view, for, as Lowinsky observes:

> [In the Renaissance] the stage moves from the quiet monastic cell, in which most medieval works on music were conceived, to the noisy places of musical performance, the choirloft, the rehearsal rooms of town musicians, the house of the boy choristers, the humanistic gymnasium, the private concert halls of nobles and patricians.[21]

A detail of 'The Measurers' which has recently been uncovered provides further assistance in removing the musicians therein depicted from the speculative world of the Middle Ages and placing them within the humanist context of the Renaissance. In a letter of 10 April 1990, written to assist the art historian Bettina Werche (see n. 3) in her investigation of the painting, the curator of the Museum of the History of

[19] 'There is an objective time that pertains to things outside as well as inside the human world and includes man himself insofar as he qualifies as a thing by constituting a part of the universe. There are also many subjective and intersubjective ways of significantly modifying objective time, merely by perceiving it and reacting to its comprehensive hold over existence.

François le Lionnais, *Time*, London, 1962, pp. 78–81.

[20] Many more examples could be given, of course, such as Jan Bruegel's 'Hearing', now in the Prado in Madrid.

[21] Edward E. Lowinsky, 'Music of the Renaissance as Viewed by Renaissance Musicians', in Bernard O'Kelly, ed., *The Renaissance Image of Man and the World*, Columbus, Ohio, 1966, p. 130.

Science, Francis Maddison, wrote that it had been removed from its frame and, though no signature had been revealed, the word 'Horatius' was discovered 'written in gold script on a black band at the bottom of the picture'. The significance of this ascription was not at first realized, but when in 1993 the correspondence was made available to me I was able at once to associate it with a quotation in a famous music-historical document from which its meaning becomes obvious.[22]

The mathematical ordering of the universe, the discovery of which was attributed to Pythagoras and developed by Greek philosophers, was justified for adoption into Christian theology by having reference to such scriptural passages as 'Omnia in mensura, pondere et numero disposuit' [Thou hast ordered all things by measure, weight and number] (*Liber sapientiae* XI, 21). This is illustrated, for example, by Jehan de Nizières [Figure 7] at the beginning of the manuscript (Paris, Bibl. Nat. MS fr. no. 22532) containing the French translation of Bartholomæus Anglicus' *Liber de proprietatibus rerum*, from which also Figure 2 is taken. Against this background, a theme such as that illustrated by 'The Measurers' might appear to be a demonstration of the persistence in northern Europe of medieval habits of thought during the Renaissance, in this case as late even as *c.* 1600. However, the name 'Horatius', revealed when the frame was removed, places 'The Measurers' within quite a different frame of reference, for it shows that the unknown artist found the authority for his subject not in the Bible, but in the writings of a Roman classical author. And so did the Renaissance theorist Giovanni Maria Artusi (*c.* 1540–1613) in *L'Artusi, ovvero, Delle imperfezioni della moderna musica*, Venice, 1600, when, in the Second Discourse, he criticized some madrigals by an unnamed composer for their disregard of the accepted rules of composition. The madrigals were by Claudio Monteverdi and were subsequently published, one (in two parts) in his *Il quarto libro de madrigali* (1603), and two in the famous *Il quinto libro* (1605), in which the composer included an important letter (extensively glossed by his brother Giulio Cesare at the end of his next publication, *Scherzi musicali*, 1607) in which Artusi's criticisms were rebutted by claiming that the offending pieces represent a new style in music, a *seconda prattica*.[23] Artusi couched his criticism in the following terms:

> Do you not know that all the arts and sciences have been brought under rules by scholars of the past and that the first elements, rules, and precepts on which they are founded have been handed down to us in order that, so long as there is no deviation from them, one person shall be able to understand what another says or does? And just as, to avoid confusion, it is not permitted to every schoolmaster to change the rules bequeathed by Guarino, nor to every poet to put a long syllable in verse in place of a short one, nor to every arithmetician to corrupt the processes and proofs which are proper to that art, so it is not permitted to everyone who strings notes together to deprave and corrupt music, introducing new modes of composing with new principles founded on sand. Horace says:

[22] In an exchange of letters Francis Maddison and I discovered that, following different routes, we had independently arrived at the same realization of *Horatius'* significance.

[23] See Claude V. Palisca, 'The Artusi-Monteverdi Controversy', in Denis Arnold and Nigel Fortune, eds, *The Monteverdi Companion*, London, 1968, pp. 133–166.

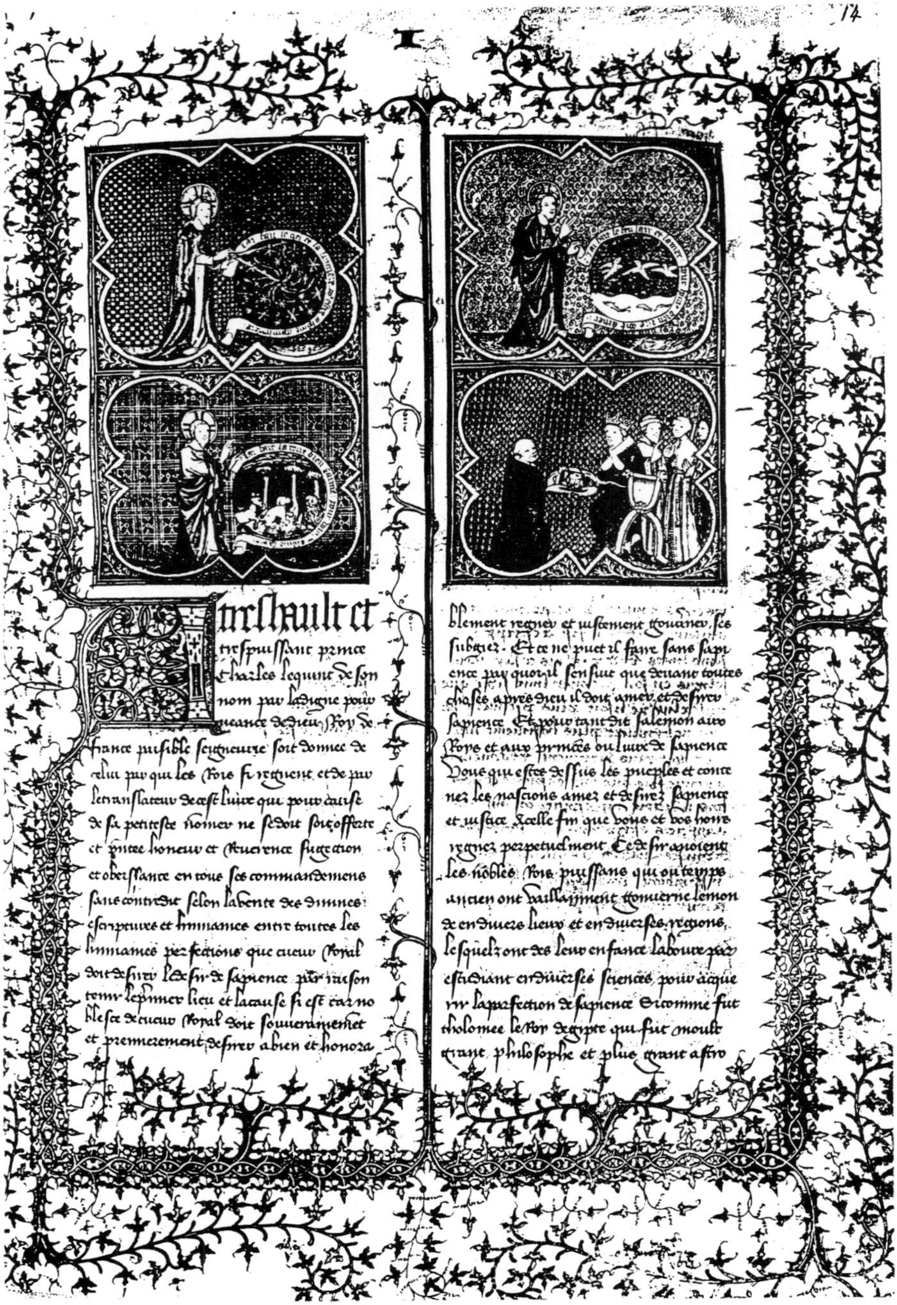

Figure 7. *Jehan de Nizières' (same source as Figure 2). Paris, Bibliothèque Nationale, MS fr. no. 22532*

> *Est modus in rebus, sunt certi denique fines*
> *Quos ultra citraque nequit consistere rectum.*
>
> (There is measure in all things. There are, in short, fixed bounds, beyond and short of which right can find no place.)[24]

Whereas *Liber Sapientiae* declared the order imposed on all things by the Divine Creator, Artusi, though a priest, interprets the passage from Horace in such as way as to attribute to the scholars of the past the rules that bring order to all arts and sciences.

Thus our anonymous artist, working presumably in Antwerp *c.* 1600, asserts a view of the world in which 'there is measure in all things', including music, cryptically giving as his authority the same secular, pre-Christian author as does G. M. Artusi in Bologna in 1600. Despite this modest coincidence there is no reason to suppose that these two events are in any way related. Yet it is an interesting fact that the two worlds represented by our principals, who, though separated by such a distance, simultaneously confirmed a similar view of natural order, actually converged at the relevant time. In the summer of 1599 Claudio Monteverdi, the subject of Artusi's criticism, was in the retinue of Duke Vincenzo Gonzaga of Mantua on a trip to Spa in Flanders. On August 21 they arrived in Antwerp and somewhat later they stayed in Brussels. They were undoubtedly in touch with Flemish artists, for in one or the other of these places the Duke apparently engaged an artist[25] to come to Mantua, and they equally certainly made the acquaintance of Franco–Flemish music (Antwerp was an important centre for music publishing) and musicians. In his *Dichiaratione della Lettera Stampata nel Quinto Libro de suoi Madrigali*, published in 1607 but written only 'some months after' his brother's letter had been printed in his Fifth Book of Madrigals (1605), Giulio Cesare Monteverdi says that Claudio introduced the '*canto alla francese* in this modern manner' into Italy on his return from Spa in 1599.[26] What this actually means is uncertain: since the *secondo prattica* proclaimed by Claudio Monteverdi is explained by Giulio Cesare as one that 'makes words the mistress of the harmony', it is tempting to associate it with the *musique mesurée à l'antique*, musical settings of verse in the quantitative metres of classical poetry, practised by the *Académie de poésie et de musique* founded by Baif and Courville in 1570, of which the leading composers were Claude Le Jeune (*c.* 1528–1600) and Jacques Mauduit (1557–1627).[27] However, it is not for his treatment of words that Artusi criticizes Monteverdi – indeed, the objectionable passages from Monteverdi's madrigals are printed by Artusi without their texts as examples of music in which the laws of the *Ars perfecta* have been disregarded. Since, as Giulio Cesare Monteverdi points out, a number of other prominent composers had committed similar

[24] Quoted from *Source Readings in Music History, Op. cit.* (n. 5), p. 401. The passage from Horace is from *Satires*, I, 1, 106–107, in the edition of H. R. Fairclough (1926).

[25] Apparently Frans Pourbus the Younger, though also Peter Paul Rubens entered the employ of Vincenzo Gonzaga in Mantua in 1600.

[26] Claudio Monteverdi's 'Letter' and Giulio Cesare Monteverdi's 'Declaration' are printed together in English translation in *Source Readings in Music History, Op. cit.* (n. 5), pp. 405–412.

[27] Schrade argues cogently against this interpretation in *Monteverdi, Creator of Modern Music, Op. cit.* (n. 8), pp. 170–178.

offences, it is perhaps an indication of the leading rôle that was being attributed to Monteverdi at the turn of the century that he should be singled out as the chief threat to the Renaissance polyphonic style, developed by Northern composers and brought to a state of perfection during a century and a half, that Artusi was determined to fight, under the banner of Horace, to preserve.

As we now know, Horace has also provided the motto under which 'The Measurers' in the Museum of the History of Science in Oxford was painted. Though we have observed that here too the music that is being studied appears to be without text, the observation carries little weight and the musical evidence that can be derived from the picture is admittedly limited. Nevertheless, if we assume that the reference to Horace carries the same implications when used here as when Artusi invoked him against Monteverdi, we may interpret the picture as representing a teacher, a Northern colleague of Artusi, who is instructing his pupil in music written in the style developed by his forefathers, the style with which music was reborn in the fifteenth century and which reached perfection in the sixteenth, the style in which the rules governing music's correct measurement must be learned in order that 'the imperfections of modern music' may be avoided. It is a pity that all, the music, the musician and the painter, are anonymous, but at least Horace is named and with him the loyalties of the artist are made clear.

Reconstructing the Angle-measuring Instruments of Pierre Gassendi

Allan Chapman

In 1992, the Provençal City of Digne-les-Bains celebrated the four-hundredth anniversary of the birth of its most illustrious figure, Pierre Gassendi. As part of these scholarly celebrations, it was decided that two of Gassendi's astronomical instruments might be reconstructed, to augment the visual exhibition, and I was asked to undertake the work. These instruments have subsequently been placed on display in the Musée de Digne, Boulevard Gassendi. None of Gassendi's original instruments survive and unlike his acknowledged technical master, Tycho Brahe, he left no clear descriptions of them, but from the numerous recorded observations that were published in his *Opera omnia* (1658) it was possible to reconstruct two of Gassendi's principal pieces. They were his five [Paris] foot quadrant, and his seven-foot *radius astronomicus.*

Gassendi made astronomical observations throughout most of his adult life, and discussion of their implications formed a regular feature of his correspondence with scientists across Europe.[1] These include telescopic observations of the sun, moon, and planets, the most famous of which was his Parisian observation of the transit of Mercury across the solar disk in 1631. Virtually all of Gassendi's observations involve the measurement of angles, however, for it was from such angular measurements that the planetary and dynamic theories of Copernicus, Tycho, Kepler, and Galileo could best be evaluated within the astronomical debate of the early seventeenth century.

Renaissance astronomical instruments displayed remarkable ingenuity in their design, as scientists and craftsmen attempted to obtain the maximum practical accuracy from a set of Euclidean principles that were embodied in simple metal and wooden

[1] *Petri Gassendi Diniensis, Opera omnia,* 6 vols, Paris, 1658. Volumes 4 & 6 contain the main body of astronomical observations and letters. Though only the 5-foot quadrant and *radius astronomicus* were reconstructed, a shadow square, a 2-foot quadrant, and other instruments are mentioned in the observations.

From *Learning, Language and Invention: Essays presented to Francis Maddison,* ed. W.D. Hackmann and A.J. Turner. Published by Variorum, Ashgate Publishing Ltd, Gower House, Croft Road, Aldershot, Hampshire, GU11 3HR, Great Britain.

structures. These designs all hinged upon the accuracy with which the circle could be divided, and tended to employ one of two principles, both of which were evident in the instruments of Gassendi. The first of these depended upon the direct division of a circular arc using beam compasses, and the further subdivision of its degrees by means of transversals, Verniers, or micrometers. But the direct division of a quarter-circle into 90 perfectly homogeneous units is geometrically impossible, as early-seventeenth-century craftsmen well knew, so that the ensuing quadrants and sextants were recognized as carrying graduations that were not perfect.

The second graduation principle employed exploited the geometrical relationship that exists between the radius, tangent, and chord of a circle, to make measurements against an equally divided straight rod. The parallactic rulers of Ptolemy had been the earliest instruments to use these principles for astronomical measuring purposes in the ancient world, and Regiomontanus, Copernicus, and Tycho Brahe had subsequently used such rulers.[2] Tycho had also experimented with the other straight-rod variant for measuring angles, the *radius astronomicus*, and as Gassendi was later to find, commented upon its convenience as a travelling instrument.[3]

Many astronomers had used the *radius astronomicus* in its various permutations of graduation in the 150 years that led up to its valedictory employment in fundamental research in the hands of Jeremiah Horrocks in 1640. Wilhelm Schickard had described a form of the instrument that was to be influential across Europe, and which was similar to Gassendi's instrument.[4] The advantage of the *radius astronomicus* against the quadrant lay in its simplicity, and the absence of any theoretical limits upon its division such as those that applied to the arcs of quadrants. Its straight rods could be several feet in length if desired, and divided into any pre-computed number of equal divisions by continuous bisection.

When one attempts the scholarly reconstruction of an instrument from the early modern period it is always with several agendas in mind. The first arises during the process of construction, when it is possible to test the practicality of various modes of graduation from the standpoints of geometry and draughtsmanship. And when the instrument is complete and mounted, it is possible to test its practical handling qualities in the act of making angular measurements against the heavens. While a modern observer who lacks the nightly familiarity in using the *radius* and similar instruments cannot hope to reproduce the practical accuracies of Gassendi and his contemporaries,

---

[2] Tycho Brahe, *Astronomiæ instauratae mechanica*, Wandesburgi, 1598, transl. H. Raeder, E. & R. Strömgren, Copenhagen, 1946, pp. 44–51. Tycho acquired the 4-cubit rulers of Copernicus from a Canon of Ermeland in 1584, and proceeded to build a set of his own. But he found the design of the instrument to contain too many faults to be capable of critical results.

[3] Tycho *Op. cit.* (n. 2), pp. 96–97.

[4] A. Chapman, *Dividing the Circle. The Development of Critical Angular Measurement in Astronomy, 1500–1850,* Chichester: Ellis Horwood, 1990, pp. 23–31. Gemma Frisius's *De radio astronomico,* Antwerp, 1545, was a very influential work in popularizing the *radius* amongst astronomers. The most complete modern study of the instrument is John Roche, 'The Radius Astronomicus in England', *Annals of Science,* xxxviii, 1981, pp. 1–32.

he can none the less make a good assessment of the stability and practical handling qualities of a particular design. Sighting systems can be tested, along with the ease and reliability with which a scale can be read against a fiducial edge in the dark, as a way of getting the 'measure' of early modern scientific skills.

Knowing the type and dimensions of a specific Gassendi instrument (usually given in the Paris Foot of 324.8 mm) one can extract its general mathematical features. Then, using data from specific observations and their scale digit sequences recorded in the *Opera omnia*, one can determine how the scales were divided and read, to make an accurate physical reconstruction.

The Quadrant

In August 1625, in a letter to the Hollander Willebrord Snell, Gassendi described making astronomical observations with his 5-foot (1624 mm) quadrant.[5] This large, flat instrument was used to measure angles in the vertical plane, probably hanging from a wall or stand. Most likely, its main use would have been to measure the meridian altitudes of the stars and planets for incorporation into a table or celestial map.

The frame of the replica [Figure 1] is made of walnut wood, using an arrangement of seven radial and chordal beams which was characteristic of quadrants of the period. The 'limb' is also made of wood, to provide a rigid support for the brass scale.

Figure 2

At the geometrical centre of the instrument (top left corner) is a brass peg around which rotates the measuring rule, or alidade. At the opposite, or scale-end of the alidade is a backsight with two fine horizontal slits cut into it. The separation between these two slits in millimetres is exactly the same as the diameter of the central brass peg.

To make an observation, Gassendi would have first looked through the top slit and adjusted the angle of the measuring rod until the desired star was visible *exactly* on the top edge of the central peg. He would then have repeated the procedure for the lower slit against the lower edge of the peg.

Figure 3

Because the star is at infinity, it is possible to see its light across both edges of the narrow peg. When the star appears equally bright in both positions, the astronomer can be sure that no parallax, or off-centre error has entered into the observation. The arrangement was first used by Tycho Brahe about 1575 and was standard across Europe by Gassendi's time.

The scale was divided into 90 degrees with beam compasses, in accordance with Euclid's geometrical proportions for the radius and circumference of the circle. The radius of a circle will always fit six times into its circumference. The compass used to draw the quadrant's quarter circle on the brass scale can, therefore, be used to strike

[5] Gassendi, *Op. cit.* (n. 1), vol. 6, Gassendi to Snell, 'Postridie Idus Augusti 1625', pp. 6–10.

Figure 1. Replica of Gassendi's 5-foot quadrant as exhibited at the Musée de Digne in 1992

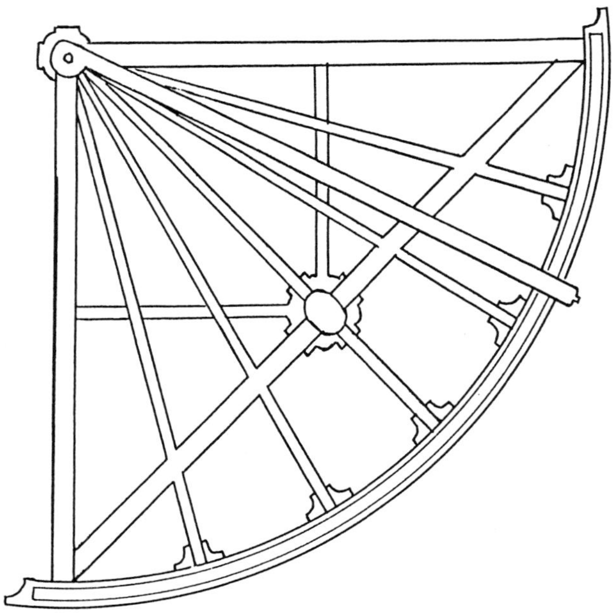

Figure 2. Replica of Gassendi's 5-foot quadrant (NB All drawings by Allan Chapman)

off its 60° point from the bottom of the brass limb. When this 60° angle is bisected to form 30°, the compass opening can be added on, to give an exact 90°. Using further bisections, trisections (threefold) and quinquesections (fivefold), the individual degrees can be drawn.

Each degree is then broken down into sixtieth parts, or minutes, by means of transversal lines. Firstly, each degree space of about 29 mm is divided into six equal parts of ten minutes. Ten equidistant lines that are concentric to it are next laid off, about 3.5 mm apart, and these are crossed by the diagonal 'transversals', six to one degree.

Figure 4

Reading an angle to a single minute then becomes simple. When Gassendi had sighted a star with the peg and slit, he would:

a) Read off the nearest full degree against the *underside* of the alidade on the scale: 36°.
b) Read off the full ten-minute spaces: 36° 20'.
c) Run his eye across the scale to see where the alidade cuts *both* a transversal line and a concentric line together. By counting which of the ten concentric lines has been cut with a transversal, he could read off the individual minute which would be added onto the existing angle: 36° 20' + 4' = 36° 24'.
d) If the alidade fell between *two* concentric lines on the transversal, then Gassendi would estimate to a half-minute, or 30 seconds: 36° 24' 30".

Though Gassendi's quadrant scale does not survive, it is clear from observations made with the instrument that it was divided into single arc-minutes, as one would expect for an instrument of large radius using Tycho's transversal scales. I have examined several transversal scales surviving from the seventeenth century, mainly in English collections, and have made micrometric measurements of six of them in an attempt to ascertain the reasonable accuracy parameters of the period.[6] The Museum of the History of Science, Oxford, has three beautifully engraved instruments from the collection of the early Savilian Professors of Astronomy. One of these instruments, a quadrant of six-foot radius, signed 'Elias Allen, Fecit Londini 1637', was used as a guide to the scale design of the Digne instrument.[7]

The five-foot Gassendi quadrant was constructed in walnut in a workshop in Digne by local carpenters, and the divisions were drawn by an engraver from Paris. The French craftsmen undertaking the manufacture worked from a detailed set of full-sized drawings which I had supplied, so that my paper templates and instructions

[6] A. Chapman, 'The Design and Accuracy of some Observatory Instruments of the Seventeenth Century', *Annals of Science*, xl, 1983, pp. 457–471.

[7] They include (1) Mural quadrant of 78.05 English inches radius, signed and dated 1637 by Allen, Museum of the History of Science, Accession no. 36.1; (2) Sextant, 73.26 inches radius, unsigned and undated but possibly by Allen, M. H. S. no. 36.2; (3) Iron quadrant, 25 inches radius, unsigned, possibly Allen, M. H. S. no. 36.3. See Chapman *Op. cit.* (n. 4), pp. 153–157, 197.

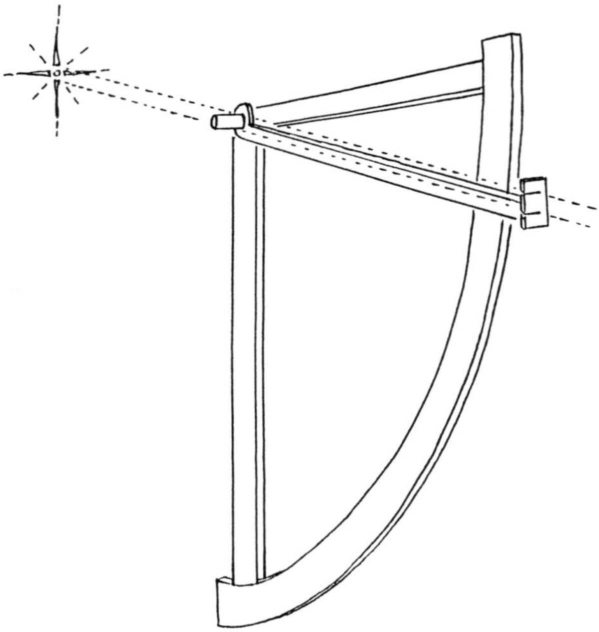

Figure 3. Sighting the star across the central peg of the quadrant

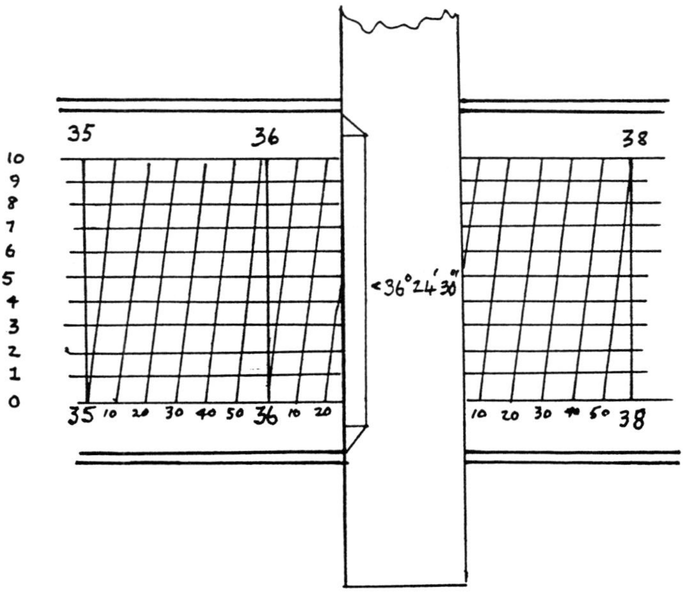

Figure 4. Reading the transversal graduations of the quadrant (full-scale)

Figure 5. Replica of Gassendi's radius astronomicus *as exhibited at the Musée de Digne in 1992*

could be used to assemble the framework of the quadrant, including joints and decorative features appropriate for an instrument of 1625. The scales likewise were copied from a fully-graduated, full-size paper scale that was part of the five-foot quadrant drawing supplied to the frame constructors. The engravers were also given precise instructions for laying off the cardinal degree points onto the scale with beam compasses, along with how to construct a seventeenth-century transversal scale reading to a single arc-minute. I fully commend the Digne craftsmen for the skill and thoroughness with which they turned my drawings into a beautiful instrument. It was decided to supply drawings, for local manufacture in Digne, to ease transportation problems.

The *Radius Astronomicus*, or Astronomical Radius

In 1619, Gassendi was making observations with an astronomical radius of 7 Paris Feet (2273 mm). This instrument was made in wood (probably beech) and was similar to that of his friend Joseph Gaultier, as described in *Opera omnia*.[8] For ease of subdivision, however, the present reconstructed instrument was slightly increased in length to the round number of 2300 mm. Unlike the quadrant, the *radius astronomicus* was made by me in Oxford and its two rods glued and jointed in Digne. It was mounted on a stand which I had designed based on a Tychonic prototype [Figure 5].

The astronomical radius was used extensively by Renaissance astronomers, being a relatively simple instrument to construct and easily dismantled for transportation. The great advantage, however, resided in its use of equidistant scales drawn upon a straight rod, which dispensed with the complexities of direct circular division.

The instrument exploited the natural trigonometrical relationship existing between the radius and tangent of a circle. All that it required was a long rod, to act as the circle's radius, and a shorter rod attached at right angles to form a 'T', and producing a tangent. Two fixed and two moveable brass sights could then be used to measure a variety of angles in the heavens once the radius was mounted onto an adjustable stand. As the quadrant was used to measure vertical (declination) angles, so the astronomical radius measured horizontal (right ascension) angles.

Figure 6

Unfortunately, Gassendi says little about his instrument beyond its rough dimensions, and there were a variety of graduation systems used by contemporary astronomers for the construction of *radii*. It has been possible to extract the characteristics of his division system, however, from the mathematical proportions used for the derivation of angles, as recorded in his observations.

The principle behind Gassendi's *radius* was the *theoretical* division of the 2300 mm rod into the 10 000 equal parts of the table of natural tangents. In reality, this rod carried no divisions at all, being a blank piece of wood, but its length was used to determine the size of the *actual* divisions drawn onto the tangent beam, and forming the 'T'.

[8] Gassendi *Op. cit.* (n. 1), vol. 4, pp. 75–76, 80 etc.

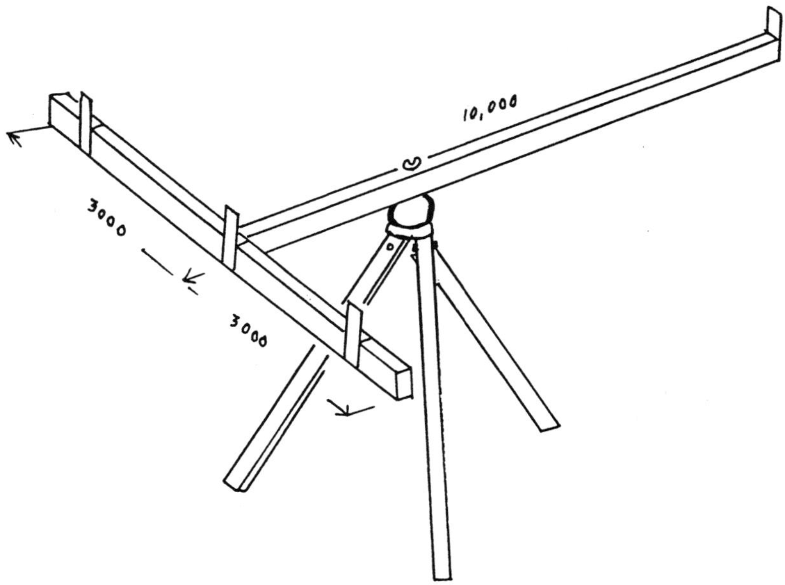

Figure 6. The radius astronomicus *on its stand showing the transom and its proportions of divisions. The central axis brass sights are fixed, while the transom sights can slide*

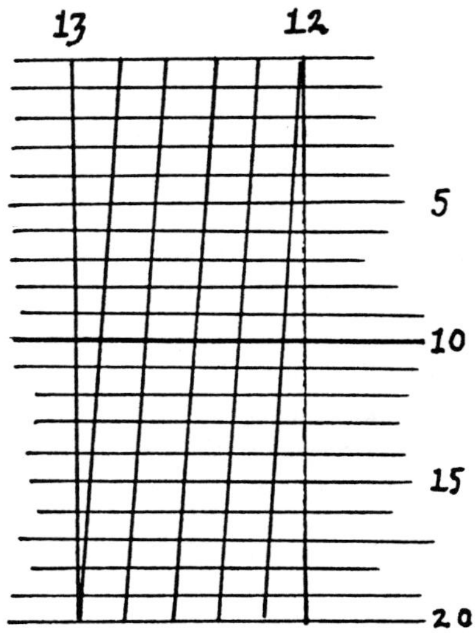

Figure 7. The transversal graduations of the radius astronomicus *(full-scale)*

If the radial rod is 2300 mm long, each 1/100th division will be 23 mm. This will give the linear amplitude of each main division drawn onto the tangent, and when each space is subdivided into 100 equal parts by means of transversal lines, each single digit will, if perfectly drawn, correspond to a 0.23 mm space.

As Gassendi recorded observations on his transom amounting to 5 800 of these single, 0.23 mm equivalent divisions, his entire transom probably carried 60 main divisions, or 6000 individual points (3 000 on each side of the central radial rod).

Though Gassendi does not indicate how the 100 individual divisions within each main space of 23 mm were laid-off, the following configuration would have been plausible, considering the dimensions of the rods.

Figure 7

Each 23 mm graduation is divided along its length into *five* equal parts, while the breadth of the tangent beam (55 mm) is divided in 20 equal parts, to produce 100 intersections by means of transversal lines. The angle would then be read in a similar manner to the transversal scale of the quadrant. The *inner* reading edge of the moving brass sight (that edge nearest to the centre of the transom) was used to locate which of the 100 intersections was crossed by a transversal line after an observation had been made. This was done by counting the whole five-digit subdivisions (20s), and adding them to the appropriate individual twentieth divisions (0.23 mm) drawn across the breadth of the transom, to obtain the total four-part figure. Thus, in the illustration, the inner edge of the sight reads:

Figure 8

– whole main divisions	=	27
– *four* whole 20ths	=	80
– single 20th divisions	=	3
	TOTAL	2783

Or, for another example:

Figure 9

– whole main divisions	=	28
– *one* whole 20th	=	20
– single 20th divisions	=	13
	TOTAL	2833

To make an observation of the angle between, let us say Mars and the bright star Arcturus (as Gassendi did in March 1619),[9] one would sight each object in turn by aligning it across the respective edges of the moving brass sights against the foresight

[9] *Ibid.*, p. 80.

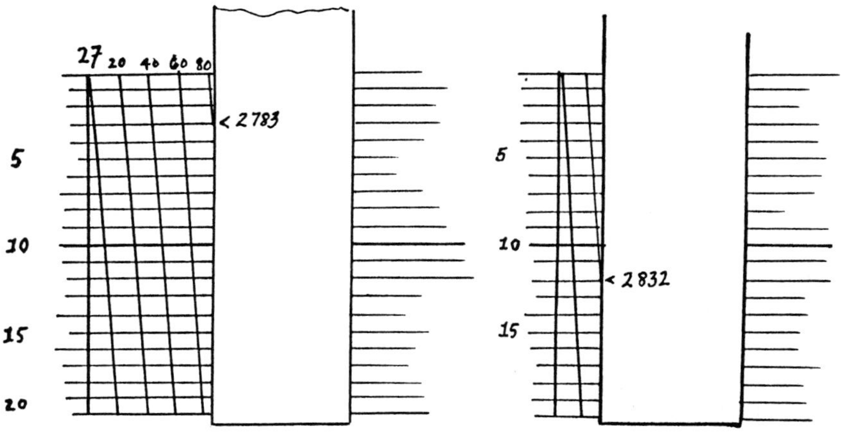

Figures 8 and 9. *Reading the transversal intersections against the edge of the brass sights on the* radius *transom (full-scale)*

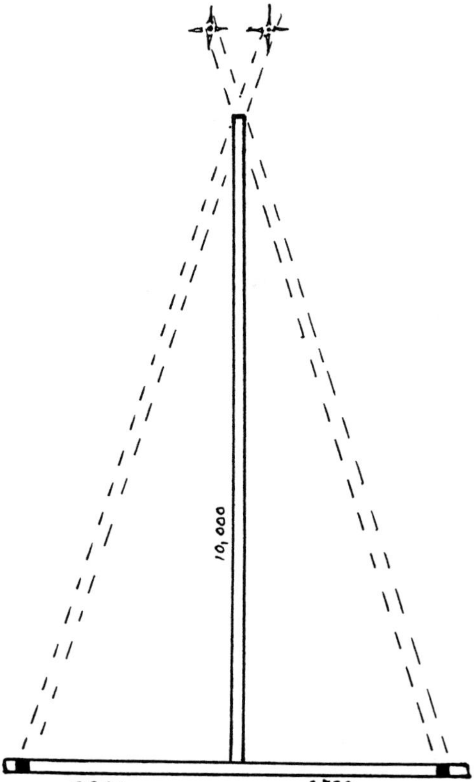

Figure 10. Radius astronomicus *seen in plan. Two observers at the extreme ends of the transom use the brass sights to sight a pair of stars across the same tip sight*

at the far end of the radial rod. The technique would have been geometrically identical to the 'peg and slit' sight of the quadrant, and would have been parallax-free, upon compensating for the thickness of the sight on the scale.

Figure 10

Left	Right
2800 divisions	2783 divisions

$$\textit{tangent}\ \frac{2800}{10000} = 0.28, \surd\ 15^\circ\ 39' \qquad \textit{tangent}\ \frac{2783}{10000} = 0.27, \surd\ 15^\circ\ 33'$$

For any angle that was beneath 16°, Gassendi could, if he desired, sight one object in the pair with the fixed central sights on the radius, and the other with either the left- or right-hand moving sight. Sometimes, he seems to have observed very small angles of one or two degrees by putting both moving sights onto the *same* transom arm, where he would have been easily able to switch his eye between the two sights.

To obtain the resulting angle, he simply took the two transom measurements and subtracted one from the other, as when in March 1619 he observed the small angle between Castor and Pollux, the two bright stars in Gemini.[10]

Figure 11

2400 divisions *minus* 1574 = 826, or tangent equivalent, 4° 32'.

It is possible (but by no means certain from his observations) that Gassendi's *radius* had a second cross-piece, or transom, at its far end, as did the instrument of his contemporary, Schickard.[11]

Figure 12

Such an instrument would have allowed angles of a greater amplitude to have been measured, though Gassendi does not seem to have recorded angles that were sufficiently large to suggest such an arrangement.

The Construction and Practical Use of Gassendi's Instruments

We possess no information about the craftsmen who constructed Gassendi's instruments. A skilled carpenter, however, could have built the wooden parts, while Gassendi himself may have drawn the divisions. Tycho Brahe and the contemporary English astronomer, Jeremiah Horrocks, were skilled graduators of their own mathematical scales. In the early seventeenth century, specialized astronomical instruments tended to be manufactured locally to suit an individual customer's specification, while the practical nature of astronomical observation itself suggests that the men who did it regularly, such as Gassendi, were probably good with their hands.

[10] *Ibid.*, for table of radius tangent scale numbers and their corresponding astronomical angles.

[11] Roche *Op. cit.* (n. 4), p. 29.

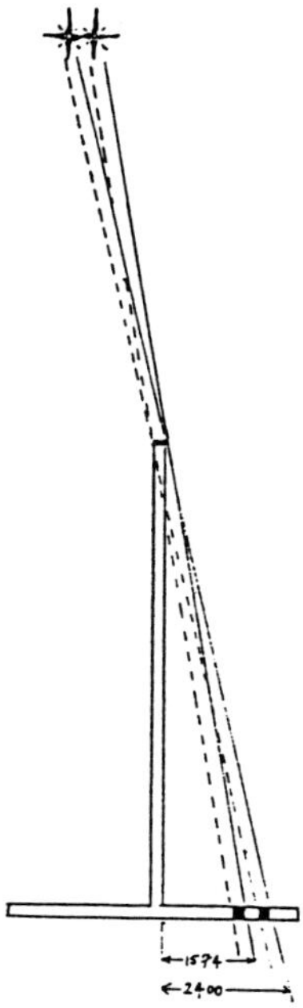

Figure 11. Measuring a small angle with the radius astronomicus, *with both brass sights at the same transom*

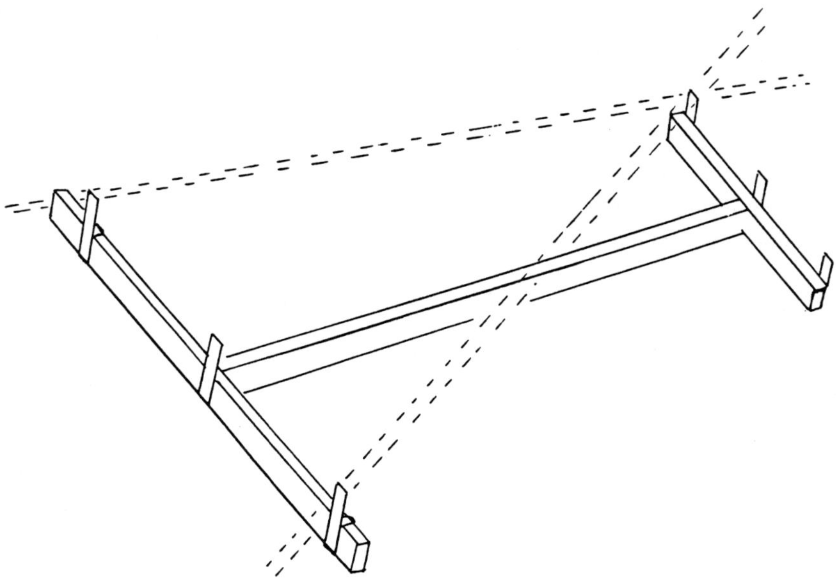

Figure 12. Variant radius *design. The addition of a second transom makes it possible to read a wider range of angles*

As Gassendi frequently travelled between Digne, Marseilles, Paris and other cities (and recorded making observations in many of them) he probably had a 'travelling kit'. In this capacity, the astronomical radius would have been ideal, because its rods could perhaps have been unpegged or unscrewed for transportation and a stand improvised. He seems, after all, to have been fond of observing the heavens from church towers. It is also possible that, having an accurately graduated transom bar and sights, (about 1400 mm long) to take on his travels, he improvised a precisely measured 7-foot (2273 mm) rod and front sight from local sources wherever he happened to be staying.

Because of the persistent cloudy weather that followed the completion of the reconstruction of Gassendi's *radius astronomicus*, and the necessity of getting it to France in good time for the opening of the Exhibition, it was not possible to test the instrument in actual astronomical observation. I did try out its scales against carefully worked-out terrestrial angles, however, and found that I could get accuracies to within a few arc-minutes. I have made astronomical observations with a replica *radius* on previous occasions and in particular, with the 36-inch instrument originally described by Jeremiah Horrocks in 1636.[12] In addition to finding the *radius* a very handy and surprisingly stable instrument to use, I found Horrocks' smaller instrument to be accurate to within well under ten arc-minutes when cross-checked against a star chart or sextant reading. This accuracy threshold could be improved, no doubt, with regular practice.[13]

The attempted reconstruction of early scientific instruments must always be undertaken with great caution and it must never be allowed to become a species of retrospective model making. This danger is always at its greatest when attempting to reconstruct instruments of conjectural design from epochs when linguistic and technical usages were different from what they are now. But when dealing with instruments of established usage, such as the quadrant and astronomical radius for which clear descriptions and sets of observations survive, one can learn a great deal from a scholarly reconstruction that is undertaken with the aim of ascertaining something about construction processes and practical handling qualities.

Acknowledgements

I wish to thank Roger Mason, FRGS, Oxford, for his assistance with seventeenth-century French linear measures, and Dr John Roche, Linacre College, Oxford, for helping me to establish a coherent system for the interpretation of Gassendi's astronomical radius values. I also thank A.V. Simcock for assistance with materials in the Museum of the History of Science, Oxford. But I take full responsibility for any mistakes.

---

[12] Jeremiah Horrocks, *Opera posthuma*, ed. John Wallis, London, 1673, 30 Aug. 1636, p. 250.

[13] A. Chapman, 'The Astronomical Art. The Reconstruction and Use of some Renaissance Astronomical Instruments', *Journal of the Br. Astron. Assoc.*, xcvi, 6, 1986, pp. 353–357. Also A. Chapman, 'Gauging Angles in the Seventeenth Century', *Sky and Telescope*, April, 1987, pp. 362–364.

Two Greek Gadgets in Wonderland?

Charles Dowsett

Charles Lutwidge Dodgson, Student of Christ Church, was interested in gadgets, and there can be little doubt that the Curator of the Museum of the History of Science could have found a place on his committee for one who had devised a *memoria technica* for the logarithms of all primes under 100, a rule for finding the day of the week for any date in the month, a substitute for glue, an improved steering gear for a tricycle, an apparatus for making notes in the dark, and, anticipating by a century one of the *grandes misères* of Oxford, a method of controlling traffic congestion. Moreover his pamphlet entitled 'Twelve months in a Curatorship' must contain many a useful hint and wrinkle. His interest in photography, in which field he counts as a pioneer, made him familiar with the world upside down, as did a question from his little cousin, Alice Raikes, standing in front of a looking-glass, as to whether the orange she held in her left hand would be in her right hand if she were 'on the other side of the glass' with the world back-to-front. If his mind needed any more contortions, one can be sure that the Oxford of his day supplied the rest. Don-like creatures abound in his work. We have it on his own authority that the Mad Hatter was once a don, as was the Duck (Robinson Duckworth of Trinity), and the Dodo (Dodgson himself who stuttered). An outside observer, Florence Becker Lennon,[1] has noted that 'the Tweedle brothers are a special university type – querulous, meticulous, infantile, quarrelsome', though it will not do for us to enquire which university she had in mind, mainly for the reasons which induce Americans to plead the 18th Amendment.

That Dodgson was a keen observer of his fellow Students is evident from the numerous photographs he took of them. One, Sir Frederick Gore Ouseley, whose musical talents were such that he played a piano duet with Mendelssohn at the age

[1] F.B. Lennon, *Victoria Through the Looking-glass*, New York, 1945; *cit.* R. Phillips, ed., *Aspects of Alice*, London, 1972, p. 76.

From *Learning, Language and Invention: Essays presented to Francis Maddison*, ed. W.D. Hackmann and A.J. Turner. Published by Variorum, Ashgate Publishing Ltd, Gower House, Croft Road, Aldershot, Hampshire, GU11 3HR, Great Britain.

of six, is credited with a remark at the age of five that belongs to the annals of Wonderland: 'Only think! Papa blows his nose in G!'[2] It is not inconceivable – indeed the theses to be unfolded below depend upon some such conception – that some of the notions in Dodgson's work might be traced to conversations at High Table. The Tutor in Mathematics was a retiring, reserved Fellow at his best with children, possibly because of his stammer, but however strong one's urge to follow Epicurus' advice (λάθε βιώσας – 'keep a low profile'), no man whom Fate has placed in a Senior Common Room is an island: the waves may not swamp but they will lap. All the ideas abroad in Oxford in his time were not confined to Christ Church of course. A photograph taken by Dodgson in 1856 shows that he was acquainted with Friedrich Max Müller, imported from Germany (as was much which had to do with philology in those days) as Taylor Professor of Modern Languages – an admirably economical chair. Much in Dodgson's work displays a preoccupation with the strophe-like qualities of human speech as if he were intent upon demonstrating the truth of Æneas' remark to Achilles that:

> στρεπτὴ δὲ γλῶσσ' ἐστὶ βροτῶν, πολέες δ' ἔνι μῦθοι
> παντοῖοι, ἐπέων δὲ πολὺς νομὸς ἔνθα καὶ ἔνθα.
>
> (Tortuous is the tongue of Man, with manifold inflections
> And words whose range of meanings extends in all directions)[3]

Semantics were of special interest for Dodgson, in particular those of the contrary kind, William Empson's seventh type of ambiguity, whereby words mean the opposite of themselves. Had Müller or the Laudian Professor told Dodgson of Arabic, in which, according to the best dictionaries, a word can mean itself, a divine attribute, a kind of camel, something obscene, and its own opposite?

> 'When you say "hill",' the Queen interrupted, '*I* could show you hills, in comparison with which you'd call that a valley.'
> 'No, I shouldn't' said Alice, surprised into contradicting her at last. 'That would be nonsense –'
> The Red Queen shook her head.
> 'You may call it "nonsense" if you like,' she said, 'but *I*'ve heard nonsense, compared with which that would be as sensible as a dictionary!'.[4]

It may be that Dodgson was as aware as his contemporary Matthew Arnold of the existence of that village situated on a high hill three miles from Oxford, whose name, Cumnor, is composed of a Celtic word for 'valley' (Welsh *cwmm*) and a Germanic word for 'rock' (Westfalian *nörre*). Max Müller could have told him that *cwmm* was cognate with English *hump* 'i.e. a hillock', and with Greek κύμβος 'cup', but it would not take any time for Dodgson to grasp that whether a cup was an image for a hill or a valley depended upon which way up it was placed, or whether one looked

---

[2] 'Through the Looking Lens', *Evening Standard*, 11.1.1974, p. 8, from *DNB*.
[3] *Iliad*, xx, 248–249.
[4] Lewis Carroll, *Through the Looking-glass*, ch. 2.

at it with the naked eye, or through a camera. Though the Cheshire Cat's remark that 'we're all mad here' was somewhat alarming, the Queen had a point. But what about her 'sensible as a dictionary'?

One sensible dictionary of which Dodgson must have been physically aware for the whole of his career at the 'House' (1852–1881), was the Greek–English Lexicon first published by the Clarendon Press in 1843, of which a third revised edition was issued in 1849.[5] Made in conjunction with Robert Scott, Master of Balliol, this was the life-work of Henry George Liddell, Dean of Christ Church at the time of Dodgson's election as Student. Though (inevitably?) based on the work of continental scholars, principally 'the German work of Franz Passow' whose name remained on the title-page until the fourth edition of 1855, the *Lexicon* was the most impressive work of Oxford scholarship in the first half of the nineteenth century, bristling with deep and extensive erudition with respect to all aspects of Greek culture, including etymological comparisons with Sanskrit, Gothic, Lithuanian, and the like, derived from the work of Pott, and later Georg Curtius. Given his interest in language, such a work was obviously up Dodgson's street. However, despite this, and despite his friendship with the Dean's daughters, Dodgson's relations with the Dean were not cordial (with the ludicrously snobbish Mrs Liddell they were far worse, to the extent that Dodgson made sure that she was not in the Deanery when he collected the children for their excursions on the river). In 1854, the year in which he presented his own illustrated copy of *Alice in Wonderland* to the daughter, Dodgson vehemently oppposed the Dean over the question of the appointment of Junior Students. In 1872 he publicly ridiculed the alterations to the fabric of the House in *The New Belfry*, and, in 1873, in *The Vision of the Three T's*. In 1874 he thwarted the Dean's plan to build cloisters round Tom Quad by open letters to the *Pall Mall Gazette*.[6] Such appeals, laced with sarcasm, to the general public do not lead one to suppose that Dodgson was an easy colleague or one with much respect for the confidentiality of College business. There is no actual statutory obligation on the part of a Fellow to see eye to eye with the Master, but it is mainly at Cambridge, judging by Dr Bentley and the Fellows of King's and the little affair at Christ's written up by C.P. Snow, that one may seek glory in outright antagonism. One need not consider it to have been a very earnest thought that prompted Dodgson to write to one of his sisters about a supposed minor misdemeanour on his part, 'if I had shot the Dean I could not have had more said about it'. Certainly it is a good thing that no gun went off accidentally, for Dodgson's lawyer would have had a difficult time pleading non-premeditation with such a document in the hands of the prosecution. One can well imagine that the Dean came to dread the Tutor in Mathematics' stammered interventions in Governing Body meetings, and not only because, as we know from the example of Le Père Grandet, stammerers often get things their own way by attrition.

[5] Henry George Liddell and Robert Scott, *A Greek–English Lexicon*. Further editions published in Dodgson's time were the 4th (1855) to the 6th (1869). Liddell worked on until the 7th edition in 1882, and for part of the 8th of 1897, dying aged 87 in 1898.

[6] R.L. Green, *Lewis Carroll*, London, 1960; Phillips *Op. cit.* (n. 1), p. 37.

Geniuses (and there can be no doubt that the versatile Dodgson comes into that category) make bad bed-fellows for the merely learned. Dodgson had shown inventiveness, albeit not on the scale of Leonardo, from an early age, and though a paragon of modesty in the company of children, he knew he was clever. In describing in 1887[7] the process of the creation of the *Alice* books, he demands to know, with respect to original writing, 'what other writing is worth preserving?' – not, one might think, the most tactful question to put to an exponent of Dr Johnson's already self-denigrating profession.

None the less, of any of the dictionaries the Queen of Hearts might have known, Liddell and Scott's was among the most eminently sensible. Moreover, since Henry George Liddell was the Dean, and presided at Governing Body meetings and at High Table, and Masters and Fellows of Oxford Colleges are what they are, it is inconceivable that the Dean did not at times hold forth upon his main academic activity, or that the Students and Canons did not, even if only out of politeness or with nothing better to do at the time, occasionally listen.

The Gryphon, one recalls, though untrained in the arts of Drawling, Stretching and Fainting in Coils, had, unlike the Mock Turtle, been a pupil of the Classical Master – that he was an old crab may or may not be relevant – for Laughing and Grief. The illustrations that he made for his manuscript *Alice in Wonderland* show that Dodgson had affinities with the Mock Turtle. Given the usual curriculum of the schools of his day, he could not but have been part Gryphon too. At the beginning of the nineteenth century 'Greek learning was perhaps at its lowest level of degradation'.[8] Yet, 'the classics' had always been the principal study in the grammar schools, and an obligation to teach Latin and Greek was included in the foundation statutes of many of them.[9] By the 1840s, when Dodgson was at school, owing partly to the efforts of headmasters like Arnold and Butler, 'the emphasis in school curricula had switched from Latin to Greek'.[10] One must certainly not deduce from the fact that Dodgson's subject as a tutor was Mathematics, that he was not educated in the Classics. Indeed, when the Classical Tripos was at last instituted at Cambridge in 1822–24, it was 'open only to those who had mathematical honours'.[11] Exactly why it is difficult now to imagine, unless it be that somebody knew that Euclid and Pythagoras were Greeks. Dodgson himself exemplified this connection, at Oxford also reading for Greats, achieving a Second in Honour Moderation and a Third in the Final Honour School, a result which may not seem what one might expect of a Student of the House: but how many primary Hellenists could achieve a First in Mathematics?

How much of Homer a schoolboy and an Oxford classics graduate would have known at that time appears to have been less than would be required now. Ogilvie records[12]

[7] Phillips *Op. cit.* (n. 1), p. 28.

[8] R.M. Ogilvie, *Latin and Greek*, pp. 82–86 quoting H.F. Clinton.

[9] *Ibid.*, p. 97.

[10] *Idem*, p. 100.

[11] *Idem*, p. 84.

[12] *Idem*, p. 138.

that although Shelley read and re-read him year after year, Homer was ignored at Winchester as late as the 1870s, neglected at Shrewsbury, and little read at Oxford. But whatever its place in 'serious' curricula, stories from the *Odyssey* are such that not even an old crab of a Classics master would not find them useful material for the lower forms. We can safely assume that Dodgson did not have to rely upon Dr Liddell to inform him of the contents of the *Odyssey*, into which he certainly dipped his finger to pull out a plum. When the White King, on being told by Alice that she can see nobody on the road, fretfully wishes that he had such eyes as 'to be able to see nobody', this may well reflect Odysseus' telling Polyphemus that his name, pehaps punningly adapted to Cyclopean phonetics, is Οὔτις 'Nobody'.[13]

Another example of Greek-inspired 'nonsense' for which the Tutor of Mathematics would have needed no decanal guidance is the treacle-well of the Dormouse's story, which must owe its origin to his discovery, after arrival at Oxford – for they are not known to be in abundance all over the British Isles – of St Fridewide's treacle-well at Binsey. It may well have been the effigies of the Saxon patron saint of Oxford in Christ Church Cathedral that pointed him in its direction. With his interest in words ('Language is worth a thousand pounds a word', as Alice is told by her fellow passengers in the railway-carriage), he is sure to have known that *treacle* derives from θηριακή 'an antidote for poison'. The foregoing is common enough knowledge such that any schoolboy might be thought, by an optimistic schoolmaster, to know. There are, however, at least two other Hellenogenic items in the *Alice* books which smack strongly of the abstruse erudition of the classical scholar, in particular, perhaps, one engaged in the compilation of an explanatory Greek dictionary. It may be that the Tutor of Mathematics, his mind readily receptive of all things intrinsically nonsensical, came across them in his own reading. On the other hand, they are the stuff of lexicography and High Table conversation, in which words and strange phenomena flutter as thickly as bats in a belfry, so that it is very possible that they were the channels by which they entered Dodgson's creative imagination; his own degree in Classics, commendable for a mathematician as it may have been, was not such that he might not benefit from the Dean's superiority (he got a First, of course) in such matters.

The game of croquet as practised by the Queen of Hearts comprised rules that spelt considerable discomfort, not to say cruelty, for the flamingoes and, perhaps to a somewhat lesser extent, since they could curl themselves up into a prickle-protected ball, for the hedgehogs. One can understand that the animals were none too co-operative:

> The chief difficulty Alice found at first was in managing her flamingo: she succeeded in getting its body tucked up, comfortably enough, under her arm, with its legs hanging down, but generally, just as she had got its neck nicely sraightened out, and was going to give the hedgehog a blow with its head, it *would* twist round and look up in her face, with such a puzzled expression that she could not help bursting out laughing: and when she had got its head down, and was going to begin again, it was very provoking to find that the hedgehog had unrolled itself, and was in the act of crawling away.

[13] *Odyssey*, ix, 366, 408.

In our day this sport could get one into serious trouble, and rightly so, with the Society for the Protection of Birds. The Greeks were not so squeamish. A particularly unpleasant fate was reserved for the ἴυγξ or 'wryneck'. According to Liddell and Scott[14] this bird derived its name ultimately from its cry (ἰῦ 'exclamation of surprise'), which, like the 'puzzled expression' of Alice's flamingo, makes great sense considering the use, or rather misuse, to which it was put. 'The ancient wizards and witches used to bind it to a wheel, which they turned round, believing that they drew men's hearts along with it and charmed them to obedience; hence it was much used to recover unfaithful lovers. This operation was called ἕλκειν ἴυγγα ἐπί τινίν to set the magic wheel or bird a-going against some one; the English expression 'to put a jinx on someone' derives from this. In Pindar,[15] ἴυγγα τετράκναμον is probably the wheel with the wings and legs of the wryneck spread out so as to form four spokes, spread-eagle fashion'. Pindar[16] relates that the device was invented by Aphrodite to help Jason win the love of Medea, though it seems not to have left its mark on Caucasian culture. Whether the legend that Iuno was a nymph who used magic spells to seduce the Father of the Gods and was jealously turned into the wryneck by Hera is responsible for Pindar's legend, or vice versa, is not a question over which to lose any sleep. In Theocritus' *Idylls*[17] Simaetha, in an unsuccessful attempt to win back her lover, is said to use the device, which Gow[18] interprets to be a simple wheel with two holes through which a string is passed which, when twisted, causes the wheel to spin. It seems at least plausible that Dodgson had this, another ancient gadget that could never have worked, at least for its declared purpose (I call upon Sextus Empiricus for moral support), in mind when inventing the Queen of Hearts' variant of the game of croquet. But his source was unlikely to have been directly Pindar himself who was, though introduced to the highest form at Winchester in the 1820s, 'too difficult', 'not a poet to whom schoolboys could be introduced at an early age'.[19] Liddell however quotes him extensively in the *Lexicon*, and may have quoted him at High Table. Dodgson's source could have been lexicographical or oral.

Another passage which may have had its origin in an item of classicial knowledge that cannot be expected to have been known to every schoolboy (unlike Porson's law of the final cretic)[20] appears in Alice's conversation with that dogmatic master of language, Humpty Dumpty:

> 'There's glory for you!' (says Humpty Dumpty).
> 'I don't know what you mean by "glory",' Alice said.
> Humpty Dumpty smiled contemptuously.

---

[14] *Op. cit.* (n. 5), p. 716.
[15] *Pythian Odes*, v, 205.
[16] *Ibid.*, iv, 214.
[17] No. 2 *Pharmakeutria*, 'L'empoisonneuse'.
[18] A.S.F. Gow, 'ἴυγξ ῥόμβος, Rhombus, Turbo', *The Journal of Hellenic Studies*, liv, 1934, pp.1–13.
[19] Ogilvie *Op. cit.* (n. 8), p. 86.
[20] *Ibid.*, p. 87.

'Of course you don't – till I tell you. I meant, "There's a nice knock-down argument for you!"'
'But glory doesn't mean "a nice knock-down argument",' Alice objected.
'When *I* use a word' Humpty Dumpty said in rather a scornful tone, 'it means just what I choose it to mean – neither more nor less.'
'The question is' said Alice, 'whether you *can* make words mean different things.'
'The question is' said Humpty Dumpty, 'which is to be master – that's all.'

One of the more fantastic etymologies in Plato's discourse on the nature of language, entitled *Cratylus*, in which Socrates does not show himself to be the wisest of men for all time, concerns the word δόξα, one principal meaning of which, developing from 'expectation, opinion, the opinion others have of one, repute', is 'glory', particularly in the Septuagint and the New Testament, though Demosthenes also used it in this sense. In *Cratylus*[21] Socrates is represented as deriving this from δίωξις 'pursuit', or alternatively from 'a shot from a bow' (τῇ ἀπὸ τοῦ τόξου βολῇ).

'"Glory" a shot from a bow, Dean?', one can imagine the Tutor in Mathematics saying as Dr Liddell related this at High Table, with an indulgent post-Boppian chortle, as one of the etymologies that was *not* going into his Lexicon. 'Now that's what I call a good knock-down argument for you!'

Humpty Dumpty was right. A lot depended on who was Master.

Or Dean.

[21] 420B

The *Perpetuum Mobile* of Cornelis Drebbel

Jennifer Drake-Brockman

Tria sunt arcana majora, quibus homo nil majus hac in mortali vita desiderare potest. videlicet, motus perpetuus, ars transmutatoria, medicina universalis.[1]

Cornelius the Dutchman the most admired man of Christendome for singuler Invention and Arte.[2]

Cornelis Jacobzoon Drebbel was born in Alkmaar in 1572, the year made famous by the appearance of the 'new star' in the constellation of Cassiopeia. In the eyes of at least one contemporary the two events were not unconnected: Gerard Schagen, a fellow-townsman of Drebbel's, in a letter introducing the book in which Drebbel published his invention of the *perpetuum mobile*,[3] remarks that, as Jesus Christ had a star to herald his birth, so the heavens again blazed forth to announce the nativity of the great inventor. These are exceptional credentials, even for a seventeenth-century philosopher; Schagen's letter captures the millennial optimism that was a significant aspect of the intellectual climate in which Drebbel lived.

Drebbel's apprenticeship to the multi-talented artist and alchemist Hendrik Goltzius at nearby Haarlem must have been at least as much a formative intellectual experience as a technical education. With the painters Karel van Mander and Cornelis Corneliszoon,

[1] P. Mormius, *Arcana totius naturae secretissima, nec hactenus unquam detecta, a Collegio Rosiano in lucem produntur*, Leiden, 1630, p. 13.

[2] Bodleian Library MS Tanner 89, f. 52, reproduced in *Notes & Queries*, ser. ii, x, 1914, pp. 481–482. The context is a prospectus, dated 1620, for the building of an amphitheatre in London for the staging of all kinds of shows.

[3] Schagen's letter appears on sig. A3^{r-v} of *Wonder-vondt van eeuwighe beweghingh...*, Alkmaar, 1607, followed by Drebbel's dedication of the *perpetuum mobile* to James I (sigs A4^{r}–B2^{r}). The *Wonder-vondt* is a curious compilation containing, in addition to Drebbel's and Schagen's letters, matter relating to Archimedes' supposed ether-powered *perpetuum mobile* (Claudian's epigram, references from Du Bartas, Cicero and Lactantius) and, making up the bulk of the volume, texts ascribed to Hermes Trismegistus.

From *Learning, Language and Invention: Essays presented to Francis Maddison*, ed. W.D. Hackmann and A.J. Turner. Published by Variorum, Ashgate Publishing Ltd, Gower House, Croft Road, Aldershot, Hampshire, GU11 3HR, Great Britain.

Goltzius had founded an academy of art in Haarlem in 1583. The presiding artistic ethos of this academy was Italian mannerism, and its members had links with artists across Europe. Drebbel was thus introduced at an early age into an international intellectual milieu, but one that was not necessarily 'learned' in the sense of being 'Latin-learned'.

Evidence for Drebbel's early activities in Holland shows a strong aptitude for mechanical invention and an ingenious turn of mind. In June 1598 he obtained a patent from the States General for an improved water pump and a self-winding, self-regulating clockwork; in February 1602 his invention of a chimney with a strong draught was likewise patented. In the meantime he had adorned the town of Middelburg with a fountain and associated waterworks.

The motive for Drebbel's removal to England is unknown, as is its exact date. An important piece of evidence is a letter from John Speed, written from Ely, to Henry Briggs, professor of geometry at Gresham College, London:

> Mr Briggs not withstanding my much busines and troblesom observation, I have sent you the rud draught of the most strangest worke that ever my eyes beheld. A gentle man being Duche born and dwelling at Ipswich hath made a continuwall motion of this biggnes and sise as ner as I will gesse,[4] the worke is this a ball or round globe ever standing without moving, and upon the north and south sydes a diall within lik unto a clock or sunne diall both which moving and shewing the courses of the heavens. Round about the east and west parts doth a ring or hollow trunke of christall stand and that without moving and the same filled to his halfe with fayre water, which without any instrument that can be perceived doth ebb and flow with the Seas in every part of the world, my self stayed so longe that I sawe it ascend up the trunk a good height and left the lower compasse of the ringe empty, the man is very religious and [has] an exceding good report of the inhabitants, and himself to me affyrmed on his fayth that it shold so ever remove without any more help of man for hundreds of yeres: yf it wer not broken, and that both coniunctions and esclips for many hundreds or thousands of yers yf the world shold contynew shold be and is therin now sene.[5]

The Dutchman is not named, but the sketch [Figure 1] that Speed included with his letter, when compared with that in Thomas Tymme's account (see below), is indisputably of a Drebbelian *perpetuum mobile.* Also authentic Drebbel is the good impression that the inventor made on Speed – and the citizens of Ipswich – and the patter about the durability of the machine.

Speed's letter is not dated beyond a scrawled 'Sunday 3 June'; however, there are only two years in the first decade of the seventeenth century when 3 June fell on a Sunday: 1604 and 1610. By 1610 Drebbel was attached to the household of Prince Henry, whose investiture as Prince of Wales took place on 4 June that year and his *perpetuum mobile*, then installed at the royal palace at Eltham, was a tourist attraction. External evidence is thus strongly against 1610 as the year of Speed's letter, as are the

[4] The diameter of the whole device (excluding protective stand) in Speed's sketch is 152mm.; of the sphere alone 124 mm.

[5] Bodleian Library MS Ashmole 1813, f. 374.

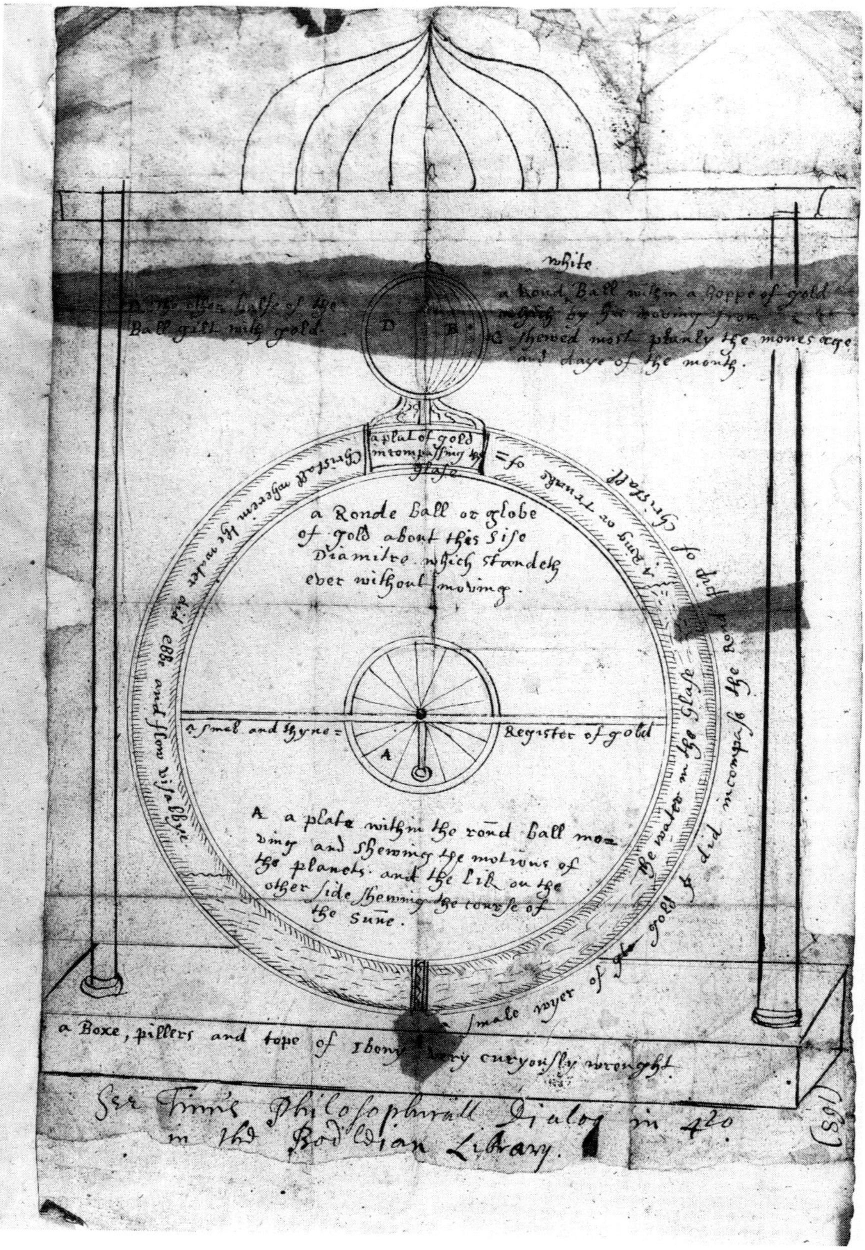

Figure 1. John Speed's sketch of Drebbel's perpetuum mobile *from Bodleian MS Ashmole 1813 (Copyright Bodleian Library, Oxford)*

Figure 2. Hiesserle von Chodaw's sketch from Raiss Buch und Leben *(Copyright Prague National Museum)*

terms in which the letter itself is cast. If the letter does indeed date to 1604, it is the first description of Drebbel's *perpetuum mobile.*

Drebbel's presentation of his *perpetuum mobile* to King James I took place in 1607. An eye-witness account of the occasion survives in the travel diary of a Bohemian gentleman, Heinrich Hiesserle von Chodaw, who spent four months at the English court in 1607. Again, Drebbel's name is not given, but the accompanying sketch shows a Drebbelian *perpetuum mobile.*[6]

> A Dutchman of not at all a poor appearance came to your Majesty and at the table where one was eating, knelt down and offered his service to your Majesty in this form: he had discovered perpetual motion, which all philosophers have pursued and to all of whom (except himself) God had revealed nothing.
>
> If it cost him his life, he wished to give this to your Majesty to see and to understand, with the result that he should be believed. Whereupon the king, as a learned man, was much astonished and began to laugh, but gave this reply: the speech surprised him not, but that this great secret which had remained hidden to all learned men from the beginning of the world, should be revealed to him alone. However, as he offered to do this, he should carry on and should he do justice to his claim, it would then be appreciated. Whereupon he thanked [the King] and straightway took out from under his servant's coat a globe, in which was the perpetual motion, which he then offered to your Majesty in the presence of many courtiers, among whom I also stood. As the globe was perfectly formed, I think it worthwhile in this report to show drawings of it with letters to clarify [Figure 2].
>
> AA is a round gilded globe, which indicates by two different hands the month and the day and in which signs of the zodiac the sun and moon are. B is a round globe representing the moon, and how it waxes and wanes, how old it is, and the connection between high and low water. CD is a glass ring. C is above and D is below water, which moves all the time, up and down, twice every twenty-four hours, like the sea. All these movements proceed of their own accord, and without one doing anything, which is to be seen as the most wonderful thing on earth.
>
> Hereupon the King asked him what sort of water was in the glass. He answered that it was common water, but very salty so that it should not degenerate. The king then further asked what imparted the motion to the water and heavenly bodies. He then answered that it was this that was the perpetual motion, which lay inside, as manifest art, hidden, yet attainable, and which made everything move; and that your Majesty should have all philosophers come, and that they should study this and speculate on it, and say whence the motive power comes, and what the perpetual motion is. He however would not wish to reveal the secret so quickly. Your Majesty alone should seal it up and lock it away as long as you like, so that no man shall come near it, then will you see that it has a continual movement. Then the King asked him: For how long? To which he replied: As long as the world exists, or as long as it was not broken. The King then asked again from what the perpetual motion derived its power, to which he answered shortly, saying that it was the Air, the principal element, which made all things move. Hereupon the King asked whether the motion could be used for any other purpose. To which he answered it could

[6] Heinrich Hiesserle von Chodaw, 'Raiss Buch und Leben', Prague Nat. Mus. MS vi A 12, ff. 48^{v}–50^{r}; this extract translated from the German by Phil Rogers.

be put to any purpose and he would make it into a wonderful instrument for the young prince, which would last as long as the world, and which would be made musical by the moon by night and by the sun by day, and that it would then chime as long as one wished. Likewise he would put a plough in the fields that would not depart from them, and also a mill that would turn without ceasing, as long as the world existed...

At this the King was very astonished and in consequence had the *perpetuum mobile* brought over to his palace in London and ordered the young prince to seal and lock it up in his room with his own hands, which he did. Two months later, when the King came to see whether the *perpetuum mobile* had stopped or not, and he found that it was still going, with the water in the glass flowing like the tide in London, ebbing and rising, and likewise the position of the heavenly bodies on the globe corresponding to their position in the sky. The King was very astonished at this, and would not stop asking the man whence it came. For which reason he sent for him, and had sufficient explanation from him, which the King then believed, and announced that he had never looked for this in such a man. The whole court then rang with the news that this *perpetuum mobile* had been demonstrated to the King in fact. In my opinion the secret was revealed to the King under the condition that he should reveal it no further. That then, at the present, is all that I have seen and heard of the *perpetuum mobile.*

Although much more detailed, Hiesserle von Chodaw's account corroborates Speed's in important respects, not only concerning the design of the machine, but also the impression made by Drebbel himself.

The other contemporary document relating to the 1607 presentation is Drebbel's own letter to James.[7] The starting point is the familiar topos of God's revelation of himself in two kinds: through the holy scriptures and through the natural world. There follows a flowery, but perhaps basically truthful, report of his researches over a number of years into the *primum mobile.* (Drebbel uses this phrase in both Dutch and Latin versions of the letter, but clearly means by it the *primum mobile* as manifested in nature rather than Aristotle's remote sphere.) In the 'various excellent little fountains', which he says were the product of his investigations into water there may be a reference to the Middelburg fountain. Realizing that water was not the key to his quest, Drebbel eventually hit upon it by 'observing the wonderful nature of fire':

at last I accomplished that which till then had involved me in error: that which is the cause of the primum mobile, by which the sky, the stars, the planets, water and earth are moved, and by which means the earth is borne up in the midst of the air, the water encompasses the mass of earth in a circle, and all things seek their centre, except fire.

Although he had now discovered the principle of the *primum mobile,* Drebbel continues, he was not content until he was able to demonstrate his discovery by 'working models' ('levendige instrumenten') as well as by natural reason. This leads him to a description of the machine itself and his claims for it:

This is the offshoot of the tree of perpetual motion (den eeuwigh-bewegenden Boom) grafted upon a true knowledge of the elements, serving as an eternal reminder to inves-

[7] The Dutch version was published in *Wonder-vondt van de eeuwighe bewegingh* (n. 3).

> tigators into the nature of things that they should grasp the wonder of nature and what we can achieve by her gift. Therefore I am prepared to make other demonstrations, hoping thereby that I shall offer to many that pleasing sweetness which is perceived from the hidden causes of things. For experience teaches me that there is no pleasure to be set against the understanding of real nature, since this shows to us the perfect goodness, wisdom and power of God.

All this, Drebbel maintains, he achieved through his own observations, with 'neither the help of ancient writers nor of anyone else'. (He does, however, allude to the celebrated sphere of Archimedes 'which moved perpetually according to the course of the ether'.)

Passing references to the *perpetuum mobile* and its inventor over the next three years show that it became an established part of the London scene and also that it fuelled intense excitement and speculation in alchemical circles. Despite its fame and Hiesserle von Chodaw's evidence of the initial acclaim accorded to the *perpetuum mobile*, it is doubtful that Drebbel's bid for royal patronage met with practical reward from James, who was chronically short of money.

Even at the 1607 presentation to James, Drebbel was, by Hiesserle von Chodaw's account, apparently hedging his bets when he offered to make the *perpetuum mobile* 'into a wonderful instrument for the young prince'. The circle surrounding the heir to the throne was a far livelier and more innovative milieu than the court proper; Drebbel may also have found its uncompromisingly Protestant ethos congenial. In late 1609, when Henry began to form his own independent household, Drebbel seems to have sought and found employment in it – though in what capacity is uncertain. Records of Prince Henry's privy purse expenses include two payments of £20 to 'Cornelius the Dutchman' on 18 December 1609 and 29 March 1610. It is tempting to surmise that Drebbel was a technician in the team that produced the chivalric show of the *Barriers*, held on Twelfth Night 1610 to mark the fifteen-year-old prince's formal debut in public life, and the masque *Tethys' Festival* staged in June the same year to celebrate the prince's investiture; if so, the payments may have been in connection with these shows.[8]

A begging letter from Drebbel to James, undated but presumably written sometime in 1610, complains that about three years had elapsed since James promised him remuneration for his labours in developing and making the *perpetuum mobile*, but that nothing had come of these promises. The letter throws light on what happened to the device following its presentation: after having been set up in Whitehall for four or five months, it was taken from there to the royal palace at Eltham; 'then the Queen

[8] For Prince Henry's 1610 shows and his connection with Drebbel, see R. Strong, *Henry Prince of Wales and England's Lost Renaissance*, London, 1986, pp. 141–158, 216–217. Drebbel's own estimate of his proficiency in the staging of optical illusions exists in his letter to Ysebrandt van Rietwijck, written from Westminster probably in 1608 or 1609 and preserved among the Huyghens MSS (bundle XLVII) in the Koninklijke Nederlandse Akademie van Wetenschappen, Amsterdam; the text is printed as appendix VIa in F. M. Jaeger, *Cornelis Drebbel en zijne tijdgenooten*, Groningen, 1922. In the abortive amphitheatre proposal of 1620 (n. 2), Drebbel was to have had responsibility for 'Sea Fights, Prospectiues [*sic*], Nocturnalls, Driades, Naides, Fire and Water-workes'; all the other participants named by the project's three entrepreneurs were musicians.

touched it with her curious hand, with the result that all actions ceased'.[9] The letter also shows Drebbel no longer relying only upon the *perpetuum mobile* as his claim to notice, but also advertising other devices, including a telescope that could read letters at a distance.

Allusions to the *perpetuum mobile* in the years following 1607 hint at its contemporary fame. In Ben Jonson's comedy *Epicoene* (first acted in 1609), the beleaguered Morose exclaims: 'My very house turnes round with the tumult!...The Perpetuall Motion is here and not at Eltham!' (V. iii). It appears even more briefly in Jonson's epigram 'On the New Motion' as 'the Eltham-thing'. On 1 May 1610 Drebbel himself was at Eltham showing off his perpetual motion to the German prince, Lewis Frederick of Württemberg, one of whose entourage recorded the visit in his diary:

> His Excellency went to Eltham Park to see the perpetual motion: the inventor's name was Cornelius Trebel, a native of Alkmaar, a very fair and handsome man, and of very gentle manners, altogether different from such-like characters; we also saw there virginals which played of themselves.[10]

Henry Peacham, who dedicated his emblem-book *Minerva Britanna* to Prince Henry in 1612 and had close links with the prince's circle, alludes to 'that Heavenly Motion of Eltham' in his 'Sights and Exhibitions in England', prefixed to Coryate's *Crudities* (1611).

Another visitor who saw the *perpetuum mobile* in these early years was Johann Ernst Burggrav, who was to publish a Latin translation of Drebbel's *Van de Natuyre der Elementen* at Frankfurt in 1628. Burggrav wrote an account of the machine to his correspondent Marcellus Vranckheim, then at Padua. Vranckheim's reply, dated 20 November 1609, was ecstatically enthusiastic: 'You write that you have seen in England that stupendous and prodigious globe of a certain Dutchman (Cornelius Drebbel by name) about which I had heard in Belgium.'[11] Vranckheim unhesitatingly identifies the motive power as Paracelsus' 'Hylech of Nature', pursued like the Snark by mystical alchemists under a variety of fanciful names. The terms in which Vranckheim couches his discussion of the *perpetuum mobile* and the authorities he invokes (Democritus, Pythagoras, Plato, Apollonius (of Tyana), the Brahmans and Gymnosophists) are strongly suggestive of early Rosicrucian texts, with which his letter is nearly contemporary.[12] Although this letter says nothing practical about the

[9] 'postquam ei curiosam manum admovisset Regina, ita vt omnes actiones cessarent'. The letter, preserved among the Balfour MSS, is printed in Abbotsford Club *Miscellany*, vol. 1, 1837, pp. 111–113.

[10] The journal of Hans Jacob Wurmsser von Vendenheym, quoted from W. B. Rye, *England as Seen by Foreigners*, London, 1865, p. 61.

[11] 'Stupendam illam & prodigiosam, de qua in Belgio inaudiveram, Batavi istius (Cornelio Drebbelio ei nomen,) te in Anglia Sphaeram vidisse scribis', in J. E. Burggravius, *Biolychnium seu Lucerna*, Frankfurt, 1611, sig. D2^r. Vranckheim's letter occupies sigs D1^r–*4^v. Burggrav apparently had also reported seeing the self-playing virginals (sigs D4^v–5^r).

[12] Among the dignitaries worthy to be patrons of Drebbel's machine Vranckheim mentions Christian of Anhalt, adviser to the Elector Palatine and identified by Frances Yates in *The Rosicrucian Enlightenment*, London and Boston, 1972, as the chief architect of the anti-Hapsburg policies out of which she believed that the Rosicrucian movement grew.

perpetuum mobile itself, it is eloquent of the expectation and speculation that it raised in alchemical quarters.

One final piece of evidence for the *perpetuum mobile*'s early years is contained in a book published in 1612 by Thomas Tymme, rector of Hasketon in Suffolk, who had a reputation as a translator of hermetic texts. Its title gives the flavour of Tymme's preconceptions and intentions: *A Dialogue Philosophicall. Wherein Natures Secret Closet is Opened, and the Cause of All Motion in Nature Shewed, out of Matter and Forme, tending to mount mans minde from Nature to Supernaturall and Celestiall promotion: And how all things exist in the number of three. Together with the wittie inuention of an Artificiall perpetuall motion, presented to the Kings most excellent Maiestie.*[13]

One of the things that Tymme sets out to demonstrate is that Copernicus got it wrong about the motion of the earth around the sun, and that the existence of a working model of the Ptolemaic universe was 'proof' to this effect. The machine is illustrated, but is also described in rather more detail than is shown by the picture. Tymme's description adds little to the Speed and Hiesserle von Chodaw accounts, although he does mention 'diuers wheeles of brasse, carried about with mouing, two pointers on each side of the Globe'. But his explanation of the motive power expands interestingly on what was probably, as the following extract suggests, Drebbel's own line:

> You haue heard before that fire is the most actiue and powerfull Element, and the cause of all motion in nature. This was well knowne to *Cornelius*, by his practise in the untwining of the Elements, and therefore to the effecting of this great worke, he extracted a fierie spirit, out of the minerall matter, ioyning the same with his proper Aire, which encluded in the Axeltree, being hollow, carrieth the wheeles, making a continuall rotation or reuolution, except issue or vent be giuen to the Axeltree, whereby that imprisoned spirit may get forth. I am bolde thus to coniecture, because I did at sundry times pry into the practise of this Gentleman, with whom I was very familiar. Moreouer, when as the King our Soueraigne, could hardly beleeue that this motion should be perpetuall, except the misterie were reuealed vnto him: this cunning *Bezaleel*,[14] in secret manner disclosed to his Maiestie the secret, whereupon he applauded the rare inuention. The fame hereof caused the Emperour to entreate his most Excellent Maiestie to licence *Cornelius Bezaleel* to come to his Court, there to effect the like Instrument for him, sending vnto Cornelius a rich chaine of gold.

Hasketon is only a few miles from Ipswich, so it is possible that Tymme, as rector there, became 'very familiar' with Drebbel during the latter's Ipswich days.

This possibility is given some backing by the apparent existence of at least part of the *Dialogue Philosophicall* in another form, either as an unrecorded publication or a now vanished manuscript. Elias Ashmole's alchemical notes[15] include a page headed

---

[13] For the account of Drebbel's *perpetuum mobile*, see pp. 60–63.

[14] The craftsman appointed by God to make the Israelites' tabernacle and ark of the covenant; see Exodus 31:1–4.

[15] Bodleian Library MS Rawl D 864, f. 224. An almost identical note of the title (with the substitution of 'the Soule' for 'mans Soule') appears in Bodleian Library MS Ashmole 1417 f. 3, again linking Drebbel's name with the work; the other works mentioned on this page are one by Miles Blomefield and 'Chymische Hochzeit Christiani Rosencreutz' (*The Chemical Wedding of Christian Rosencreutz*), an interesting but perhaps chance collocation of Drebbel with an important early Rosicrucian text.

'Tho: Tymme. Cor. Drebil of Alcmaer in Holland', which begins with quotations in longhand from (or closely following) Drebbel's 1607 dedication of the *perpetuum mobile* to King James:

> All things except ffire tend to their proper Center / I giue to many that sweete & plesant relish wch is in the hidden cause of thinges... I explaine & examine the examples by the truth...as by the lydian Touchstone.

There follows in Ashmole's cypher a list of 'Theoremes' that set out various relationships between the four elements ('Fire hath 3 qualities sharpness thinness and moving / air borroweth 2 qualities of fire namely thinness and moving' and so forth). At the foot of the page is a title:

> Demonstracon, Theologicall Philosophicall & Mathematicall of a Triple Ternary supernaturall & Naturall Wherein is set forth the perpetuall Mocon of the Vniversall [*sic*] & the cause thereof, together with a Declaracon of the immortality of mans Soule Published by Tho. Tymme minister.

Next to this title is a tiny sketch of a Drebbelian *perpetuum mobile* [Figure 3], which seems from the figures that support it[16] to be the same machine as that of which there is an engraving in Tymme's *Dialogue Philosophicall*, but with additional details (the arbor to the moon globe, the rings or dials on the central sphere) not shown in the printed illustration [Figure 4].

Unfortunately, in the absence of any identifiable copy of this 'Demonstracon', either in print or manuscript, its whole relationship to Drebbel can only be surmised. The phrase 'Published by Tho. Tymme minister' does however suggest that the 'Demonstracon' was not Tymme's own original composition. Could this page of Ashmole's notes be evidence for a translation by Tymme of one of Drebbel's own works? As far as Ashmole's knowledge of the appearance of the *perpetuum mobile* is concerned, he may either have been copying a detailed drawing in the 'Demonstracon' or it is possible that he had seen an actual example of the *perpetuum mobile* and so was drawing from life or at least personal observation.

Tymme's *Dialogue* is near-contemporary evidence for Emperor Rudolf II's having made the flattering first approach to Drebbel, rather than vice versa.[17] On the evidence

[16] The sphinxes are shown in Hiesserle von Chodaw's sketch, but instead of the spear-holding Atlas, the bottom of the globe rests on top of a metal ring set vertically on the base, which encircles but does not touch the glass tube. This form is repeated in the Breughel and Staben paintings (see nn. 49 and 50 below). Speed shows neither sphinxes nor Atlas and a much smaller ring apparently clasping the glass tube and attaching it to the box base; this may be because he was sketching in haste and including only what appeared to him to be the essentials or because the *perpetuum mobile* he saw did not have supporters – which would again indicate an early or basic version of the device.

[17] Later evidence, derived from the suspect Kuffler source, is in Carpentras Bibl. Inguimbertine MS 1776, f. 409^r: 'L'Empereur Rodolfe ayant sceu que Drebbel auoit tant de belles inuentions et qu'il scauoit faire un globe de verre dans lequel il enfermit les quatre elemens, et que cela auoit moouuement, enuoya prier par homme exprez le Roy d'Angleterre de le luy enuoyer pour faire un des ses globes, ce que le Roy d'Angleterre luy permit.'

Figure 3. Ashmole's sketch of the perpetuum mobile, *MS Rawl D 864 (Copyright Bodleian Library, Oxford)*

Figure 4. Engraving of Drebbel's perpetuum mobile *in Thomas Tymme's* Dialogue Philosophicall, *1612 (Copyright Bodleian Library, Oxford)*

of the begging letter quoted above, a mood of mutual disillusionment may well have existed between Drebbel and the parsimonious James by autumn 1610; Rudolf's intervention may have been a relief to both parties. On 3 October 1610 Rudolf's vice-chancellor issued a passport for Drebbel and his family to travel to Prague.[18] Later the same month the Tuscan ambassador to Prague, in correspondence with Galileo, informed him that there had arrived in the city a Fleming who was credited with being able to construct a *perpetuum mobile.*[19]

The most detailed account of Drebbel's adventures in Prague is that given by Abraham and Aegidius Kuffler to the French philosopher Peiresc and preserved, along with other information supplied by the Kufflers in the early 1620s, among Peiresc's papers at Carpentras.[20] As the events described occurred a decade before the Kufflers came into Drebbel's life, their authority for them must have been Drebbel himself. According to the Carpentras MS, Drebbel surpassed himself in the masterpiece that he made for Rudolf.[21] It incorporated in a single device the already familiar Drebbelian inventions: glass globe, fountain and clockwork powered by perpetual motion. An additional feature in this machine was an 'artificial sun'.

Unluckily for Drebbel, by the time he came to Prague, the ageing Rudolf was finally losing his grip on power, and Bohemia was sliding into religious and political turmoil. Rudolf's younger brother, the Archduke Matthias, took control, leaving Rudolf helpless to prevent his coronation as King of Bohemia in May 1611. Drebbel himself narrowly escaped being executed, and, doubtless considering that a meagre living in London under the tight-fisted James was preferable to an executioner's block in Prague, he headed back to London.

If Drebbel's next begging letter to King James is to be believed, it was actually while he was en route for England that he heard of the death of Prince Henry, who had died of typhoid on 6 November 1612. This letter differs from the Carpentras manuscript in the account that it gives of the motives behind Drebbel's departure from Prague: Drebbel tells the king that it was Prince Henry 'of blessed memory' who insisted upon his return in the face of Matthias' reluctance to let him go. Henry of course was

[18] It may be no more than coincidence that it was at the end of September 1610 that Prince Henry was making confidential overtures to the Grand Duke of Tuscany to see if he could be persuaded to part with the services of the brothers Tommaso and Alessandro Francini, who were renowned for '*si belli artifizi*' as their stage machinery for the French 'ballet de cour'. The search resulted in the appointment of the unsatisfactory Florentine Constantino de' Servi (*c.* 1554–1622); see Strong *Op. cit.* (n. 8), pp. 88–105.

[19] G. Tierrie, *Cornelis Drebbel (1572–1633)*, Amsterdam, 1932, p. 5. Galileo, who had only very recently (September 1610) resigned the chair of mathematics at Padua, may already have known of Drebbel through Vranckheim, whose letter to Burggrav (n. 11) alludes familiarly to Galileo in the context of the invention of the telescope and the claims of Drebbel's fellow-countryman Metius to be its inventor.

[20] Carpentras MS (n. 17), ff. 409^{r}–410^{r}.

[21] Carpentras MS (n. 17), f. 409^{v}: 'Estant auec l'Empereur, il luy fit ce globe de verre, et entreprint de luy faire en une place une fontaine qu'il fairoit monter mil pieds s'il vouloit, dont le structure seroit fort haulte, et mettroit au dessus son mouuement perpetuel, qui fairoit aller un horloge, Et au milieu de la machine il faisoit un soleil artificiel qui auroit tousiours esclairé iour et nuit, qui estoient trois grandes et bien rares inuentions.'

no longer around to confirm or deny this version of events; if Drebbel hoped to stir a sense of obligation in Henry's father, there is no record of his having been successful.

Drebbel's movements now become obscure for some years, but his *perpetuum mobile* continued to excite interest. In February 1612, while Drebbel himself was still in Prague, D. Antonini wrote two letters to Galileo from Brussels with a description of the principles of a Drebbelian machine.[22] In his first letter Antonini repeats the common report about the machine presented to King James, namely that the fluid ('certa acqua') in the glass tube ('canale de vetro') ebbs and flows like the sea. Antonini, being undisposed to seek or see an 'explanation' in any magical correspondence between the tides and the *perpetuum mobile*, realized that the common report was a blind:

> Thinking about it, it occurred to me that it was nothing to do with the ebb and flow, but that men said it was so to conceal the true cause, and that the truth was that this motion came about through the alteration of the air, being caused by heat and cold.[23]

Antonini then set about building a device out of metal and glass that would reproduce the effect described. In his letter he accompanied his account of the construction with a sketch to show the relationship between the glass and the metal tubes and the tiny hole communicating with the air outside. Apparently as an embellishment of his own, since no such feature is mentioned in any earlier account of a Drebbelian *perpetuum mobile*, Antonini calibrated his device. 'Then I placed behind the [glass] tube a tablet on which thick lines, placed equidistantly, were drawn, together with numbers, so that the movement could be noted.'[24] This was a quantum leap in thinking about Drebbel's machine; from uncritical wonder at the mere fact of the observed movement, or speculation on magical or quasi-magical sympathies and correspondences, Antonini moved the debate into an area that is recognizably scientific in the modern sense through his insight that the fluctuations in the machine are linked to external phenomena in a way that demanded to be quantified.

News of Antonini's device reached the Archduke Albert, the youngest brother of Rudolf II and governor of the Netherlands. Antonini not only demonstrated it before him; he also gave it to him as a gift. The process of copying – and thus demystifying – Drebbel's machine was under way. The Archduke encouraged Antonini to try to find a practical application for his machine. Antonini, explaining that the difference between the movement of his device differed from that of a watermill only in that its cause was invisible, whereas that for the mill was visible, saw that the problem in both cases was the same: 'to convert this unregulated motion to a regular by making it run a clockwork'.[25] The first letter ends with a promise that if his present work on this

[22] Letters dated 4 and 11 February 1612/13 in Jaeger *Op. cit.* (n. 8), appendix X, pp. 115–118.

[23] *Ibid.*, p. 115: 'Sopra il che considerando io, caddi in pensiero che questo non fusse altrimenti flusso et reflusso, ma cosi si dicesse per coprir la vera causa; et la verità fusse che questo moto fusse dalla mutatione del'aria, cioé di caldo et fredo fosse causato.'

[24] *Ibid.*, p. 116: 'Dietro poi al canale ho posta una tavoletta, attraversata con spesse linee uqualimente distanti, con i loro numeri segnate, acció si possa notare il movimento.'

[25] *Ibid.*: 'd'applicar questo moto irregolare ad un regolare, per far caminar un horologio'.

ingenious device ('machina assai artificiosa') is a success he will not omit to send Galileo a drawing of it.

A week later Antonini wrote again to Galileo, saying that he had forgotten to include the dimensions of his device in his previous letter and that the extent of the glass tube was about two armslengths. In this letter he includes a sketch of the workings of the *perpetuum mobile* belonging to the King of England, showing the hollow metal sphere surrounded by a glass ring and the way in which they were connected. Antonini is confident that it will be plain to Galileo that the cause of the motion is 'the rarefaction and condensation of the air enclosed in the metal sphere'.[26] The letter concludes with remarks on the stir caused by the *perpetuum mobile* among the Brussels intelligentsia.

Antonini's account is of particular interest in that it shows an analytic and critical approach to the *perpetuum mobile* very different from the more mystical enthusiasm of the machine's earlier admirers. From then on, those who commented upon the *perpetuum mobile* took increasingly divergent viewpoints. On the one hand were those philosophers, including Drebbel himself, who explained the machine in mystical or alchemical terms, and whose mind-set might be described as the Rosicrucian tendency; their thinking was ultimately grounded in the Aristotelian universe, the building blocks of which were the four elements. On the other hand were the exponents of the scientific tendency, whose efforts were directed towards an understanding of observed physical phenomena and to whom the Aristotelian worldview was increasingly an irrelevance; indeed the perception of the inadequacy of Aristotle's account of motion (among other things) was a factor liberating the nascent scientific tendency from the straitjacket of ancient dogma.

The next definite sightings of Drebbel himself are in 1619. Willem Boreel, sent to England in that year on the business of the Dutch East India Company, informed Pierre Borel that he saw Drebbel in London with a new invention: a microscope.[27] Further evidence is the 'Album Amicorum' of Joachim Morsius, which Drebbel signed at London in November 1619.[28] Morsius published a Latin version of Drebbel's tracts on the nature of the elements and the quintessence, together with Drebbel's letter to King James about the *perpetuum mobile*, at Hamburg in 1621.

However, the main sources for Drebbel's activities in the early 1620s are the *Journal* of the Middelburg savant Isaac Beeckman[29] and Peiresc's papers mentioned above; in both cases their major informants were the Kuffler brothers. In December 1622 Peiresc was also soliciting information about Drebbel from his English correspondents, John Selden and William Camden.[30] Another of Peiresc's friends and correspondents,

---

[26] *Ibid.*: 'la rarefation et condensatione del'aria chiusa nella sfera metalica'.

[27] P. Borel [Borellus], *De vero telescopii inventore*, The Hague, 1655, p. 34.

[28] See Jaeger *Op. cit.* (n. 8), p. 45. For Morsius, see H. Schneider, *Joachim Morsius und sein Kreis*, Lübeck, 1929; the people who appear in his 'Album Amicorum' are listed on pp. 79–109 (Drebbel's death date is incorrectly given as 1643). Among the other signatories are Ysebrandt van Rietwijck (n. 8) and Peter Lauremberg who translated Drebbel's tracts into Latin.

[29] Isaac Beeckman, *Journal*, ed. de Waard, 4 vols, The Hague, 1939.

[30] William Camden, *Epistolae*, London, 1691, pp. 332–333, letter to Camden from Paris, 21 Dec. 1622; *ibid.*, pp. 385–387, letter to Selden from Paris, same date.

Peter Paul Rubens, was apparently also canvassed for information and was instrumental in getting a design of the *perpetuum mobile* sent to Peiresc. In a letter from Antwerp dated 3 August 1623 Rubens mentions that he would like to have copies made of two of Drebbel's inventions – the 'little mirror' and the perpetual motion – and alludes to a design of the latter which had been sent to Peiresc, 'accurately done and with the sincere intention of communicating the true secret to you'.[31]

Whatever the information contained in this design, it was soon supplemented by a personal communication from Abraham and Aegidius Kuffler in Paris in early September 1624.[32] The two Kufflers had been dispatched to the Continent to publicize Drebbel's inventions. The account of the *perpetuum mobile* in the Carpentras MS[33] apparently combines Peiresc's own notes and sketches of an actual *perpetuum mobile* and its parts, based on information from the Kufflers. What is plain is that the clockwork device described here is a complicated water- or mercury-powered Heronic automaton that needed to be placed against a sunny wall, rather than the original simple, portable instrument described by Speed, Hiesserle von Chodaw and Tymme:

> He said that at the age of eight he [Drebbel] first discovered perpetual motion, which he has since brought to perfection, and this was while he was playing, having made a little fountain out of a knuckle-bone and straw, and that that was the foundation of his whole invention, which consists of nothing more than finding the means to raise water or any other substance that one might wish to use for these movements an inch or two, because having once made it rise it raises itself to greater effect by its fall. When it falls back again still greater force is created, and gives further movement to the instrument which he wishes to drive, for which no great force is necessary as it is nothing more than a time-keeper. This has its movement from two wheels, of which the larger is divided into 32 teeth, and the little one I do not remember, and above the two wheels there is a large balance...
>
> It had also some other little thing that I do not remember; the little wheel made the hand of the movement turn. I saw a small piece of wood on which there was a small instrument of brass ... which was supported at the centre. The end of it was made to strike, as if on an anvil, on a piece of brass set in the wood, and the other end fell by a counterweight of brass which was attached to it and entered a hole worked expressly for it in the wood, and tipped from one to the other.
>
> I believe I also saw a small piece of wood in which there was a small tube of glass...
>
> When he wanted to make a model of this perpetual motion here to show to the king he wanted to have made a glass bottle...It is two feet long overall and four fingers wide, with a neck four fingers long. He ordered glass tubes of one aune long but that they were very narrow and he said it was with this that he made his instrument. And the substance that he used for the movement was quicksilver [mercury], because it did not consume itself like water and any other sort of liquid which can dry up...

---

[31] Peter Paul Rubens, *Letters*, tr. and ed. Ruth S. Magurn, Harvard, 1971, no. 52, p. 91.

[32] Heading of Carpentras MS (n. 17), f. 407.

[33] Carpentras MS (n. 17), ff. 407–408^{v}. There is a partial transcription of this section of the MS in Anthony Turner and Nadine Gomez, *Pierre Gassendi: explorateur des sciences* (catalogue of the exhibition to commemorate the 400th anniversary of Gassendi's birth, Musée de Digne, 19 May – 18 October 1992), Digne-les-Bains, 1992, p. 133.

> The painted model that he had made to demonstrate to the king was nothing more than a timekeeper mounted on a large base which also supported two small figures on either side. Of these, one was a satyr who held a horn, from which emerged a small fountain which fell back into a large shell at his feet, and the other was a young child who saw emerging at his feet a fountain that spouted on high and fell back into a similar shell and re-entered the base. These two fountains ought to be of quicksilver and should not move anything larger than the iron of a needle. I do not know if this is not to take wind or air to make the first elevation and by this means to give force to the movement…
>
> Kuffler told me that the said Drebbel had made one of these everlasting movements for the Prince of Wales.[34]
>
> And there was another excellence: It was necessary to position this instrument against a wall that had to be pierced, and that this wall was on the side giving onto the sun and that there should be placed in the hole a small wooden instrument like that used in the baking of bread,[35] and to set it in such a way that the sun could touch it at least once a month. There was a large balance attached to the end of this instrument and he said that when the sun [illegible: ?played upon] this instrument that adjusted the hand to its true noon so that when it positions itself on three o'clock the hand springs back of itself to noon. This instrument of wood is rather like what follows and is [illegible: ?supported] by the middle.
>
> This is what I understood from Kuffler with regard to the perpetual motion.

Later that same year, Rubens' correspondence shows him making good his promise to have a copy of the *perpetuum mobile* made for Peiresc. (Possibly he also took the opportunity then of having a copy made for himself, as the picture of his studio by Staben, mentioned below, suggests.) Peiresc having by then left Paris, Rubens wrote on 12 December to Peiresc's brother Valavez, who was acting as intermediary, to say that he had dispatched the *perpetuum mobile*, carefully packed, but he would be grateful if Valavez would check on the condition of the glass tube, which was the part of the device most vulnerable to damage in transit. 'There is also a little glass half-filled with green water, and with the same water I have filled the tube as much as is necessary for its operation'.[36] On 10 January 1625 Rubens acknowledged Valavez's confirmation that the machine has arrived safely. If it successfully completed its onward journey to Aix, Peiresc would have possessed a *perpetuum mobile* much closer to the original type of Drebbelian machine than to the device apparently shown to him by the Kufflers in Paris. It seems that in later versions of the *perpetuum mobile* the demon of amplification and elaboration had taken over, obscuring the real, if limited, efficacy of the original device.

---

[34] It is unclear whether this was Drebbel's former patron Henry or the then Prince of Wales, Charles. Later on, however, the MS refers to an invention made by Drebbel 'when the Prince of Wales went to Spain' (f. 412^{v}), which can only refer to Charles's unfortunate excursion in the spring and summer of 1623.

[35] Peiresc's sketch of this implement shows that he was thinking of the long-handed paddle used by bakers to convey loaves of bread into an oven for baking.

[36] Rubens, *Letters* (n. 31), no. 58, pp. 97–98.

Rubens himself eventually became completely disillusioned with Drebbel's inventions, though his letter to Peiresc from London, written on 9 August 1629, shows that Drebbel's personal charisma remained undimmed:

> The famous philosopher Drebbel I have seen only on the street, where he exchanged a few words with me in passing. He is staying in the country, some distance from London. This man is like those things of which Machiavelli speaks, which, in the popular opinion, appear greater at a distance than at close range. Here they tell me that in all these years he has invented nothing except that optical instrument with the perpendicular tube which greatly magnifies objects placed under it. As for the perpetual motion apparatus in the glass ring, that is only nonsense ... But I do not want to rely on public gossip, to the detriment of so illustrious a man. I shall visit him at home, and talk with him intimately, if possible. I do not recall ever having seen a physiognomy more extraordinary than his, something remarkable, I do not know what, emanates from the ragged man, and not even his thick cloak makes him ridiculous, as it would a lesser person.[37]

Like Peiresc and Rubens, Isaac Beeckman made it his business to discover what he could about Drebbel's inventions. His *Journal* contains numerous allusions to various Drebbelian devices, and he certainly knew of the *perpetuum mobile* by the summer of 1622. However, his main accounts of the principle of the device date from 1626.[38] It is clear from his comments that for practical men of science the *perpetuum mobile* was no longer a mystery and also that its so-called motion was now generally seen to be erratic, relying as it did on the expansion and contraction of air under the influence of temperature. No one had been able to set up Drebbel's device so that it would deliver perpetual motion ('een eeuwich roersel'). Beeckman's sketch of the principle of the 'tidal' ring of the *perpetuum mobile*, for instance, is accompanied by the comment that the movement over twelve hours is attributable to Drebbel's regulation ('regieren') of the warmth in the globe.

Reaction to the revelation of the principle underlying Drebbel's basic device of tube and globe varied. To some the exposure of the secret obviously rendered the discovery trivial. Writing at the end of the decade, the Italian Jesuit Nicolo Cabeo reflects on his difficulties in ascertaining the truth about Drebbel's machine and charts his progress from Drebbel's own pseudo-explanation to an understanding of the genuine scientific value of the device.[39] Cabeo's conclusion ('the thing became cheap because it was easy, however it was none the less estimable') suggests a sophisticated understanding at work on the implications of Drebbel's *perpetuum mobile*; although

---

[37] *Ibid.*, no. 196, pp. 322–323.

[38] Beeckman, *Journal* (n. 29), vol. 2, pp. 344, 372.

[39] Nicholas Cabeus, *Philosophia magnetica*, Ferrara, 1629, book I, ch. x, p. 36:

> nouissimis enim Rodulphi Imperatoris temporibus quidam Anglus aquam arte quadam sublimasse dicitur, quae vitro inclusa perpetuo fluctuabat, si verum est, quod ego ex oculato teste viro fide dignissimo audiui; quamuis cum hoc ex alio oculatiore teste quaesierim, inveni in aqua illa nullum latere propemodum artificium, sed solum ex alteratione aeris inconstanti, ascendere, et descendere per fistulam vitream aquam, ea ratione, qua etiam hic in Italia paucis ab hinc annis artificium vulgatum est, quo singulorum cubiculorum in aestate frigoris gradus metiri poterat quisque: quae res viluit, quia facilis erat, non tamen minoris erat aestimanda.

Drebbel's own 'explanation' was debunked, the operation of the device afforded scope for speculation of an entirely new kind. No longer seen as a conjuring trick to be exposed, the *perpetuum mobile* was gaining credibility among people whose interests were increasingly removed from the magical agenda of its inventor and his uncritical admirers like Tymme.

Despite the increasing scepticism surrounding Drebbel and his device *qua perpetuum mobile* in the 1620s, the idea of perpetual motion retained its fascination. The link between the Drebbelian machine and the sphere of Archimedes, made by Drebbel himself in his 1607 letter to King James, endowed the *perpetuum mobile* with a venerable pedigree, and the two often appear together in seventeenth-century texts. The *locus classicus* for the sphere of Archimedes is Claudian's epigram on the subject, of which there were at least three seventeenth-century English translations; the earliest of these to be published was Nathanael Carpenter's in his *Geography* (1625), with a comment explicitly linking the device with Drebbel's.[40]

Michael Maier, physician to Rudolf II and later to the Landgrave of Hesse, mentioned the Archimedean globe and linked Drebbel with it in an admiring reference in a Rosicrucian tract published in 1624.[41] Only slightly later, and arising in a similar intellectual milieu (the writer's dedication to members of the Palatinate nobility is signed from Leyden in November 1632), is John Jonston's *History of the Constancy of Nature*, written to show that the world was not going from bad to worse as many people held. In support of this ameliorist position he cites a number of ancient and modern instances in nature and art, including '*Cornelius Drebbils* wonderfull Sphere ... wherein he did by vertue of a perpetuall Motion, represent the constant and most apparent Motions and Law of the Heavens and the Stars, and the Predestinations of Times, and Motions in them'.[42] Jonston was a friend of Jan Amos Comenius and Samuel Hartlib, and, like them, shows an openness to modern trends in practical science geared to the benefit of society.

Comenius too started from a comparison between Archimedes' globe and Drebbel's, but adds some interesting insights into the later history of Drebbel's quest for perpetual motion. The title of Comenius' account of perpetual motion shows that he was interested in the Drebbelian machine over a seven-year period in the 1630s, also that he was writing with an eye firmly on the utilitarian potential of the discovery.[43] Although

---

[40] Nathanael Carpenter, *Geography*, Oxford, 1625, book I, ch. vii, p. 162. Other versions are by Thomas Randolph and Henry Vaughan. The latter also translated Grotius' epigram on Drebbel's machine, 'In organum motus perpetui quod est penes Maximum Britannicum Regem Iacobum' (in Hugo Grotius, *Poemata*, 1639, p. 371) for Thomas Powell's *Humane Industry*, 1661.

[41] M. Maier, *Silentium post clamores*, Frankfurt, 1624, ch. viii, p. 49: 'Eiusmodi machinamenta aut iis similia mirae artis quidam Teutonicus Com. [*sic*] Dreppels iam pridem effecit et maiora machinatur.'

[42] John Jonston, *An History of the Constancy of Nature*, London, 1657, pp. 110–111. The book was originally published in Latin at Amsterdam in 1634.

[43] The Hague, Koninklijke Bibliotheek, Huyghens MS K.A. XLVII, ff. 134–144: 'De arte spontanei motus quem perpetuum vocant septennio vestigati iamque tandem, quod felix faustumque sit generi humano, Deo autem sapientiae lumen nobis multiplicanti gloriosum, feliciter inventi historica ad amicum interim relatio. Anno 1639 decurrente.' The manuscript was identified as being by Comenius by Julie Nováková, and is printed in Comenius' *Works*, Hildesheim, 1975, vol. 3, pp. 218ff.

he himself had not seen a Drebbelian machine at this stage, relying instead on the numerous witnesses who had, his conclusion was that it was more 'subtle than useful' ('subtilius quam utilius') and of 'no use beyond the pleasure of observing it' ('praeter spectandi voluptatem usus nihil').[44] Moreover the machine was not a true *perpetuum mobile*, nor did it even work of its own accord, because in the globe 'there lurked a fraud, a burning lamp that, resolving oil into air, infused the globe and drove it slowly round'.[45] This burning lamp, which is probably the 'artificial sun' of the Rudolfian version, has Rosicrucian resonances in the artificial sun or ever-burning lamp that was said in the *Fama* of the order to have been discovered in the tomb of the founder.

It is clear that in Comenius' opinion the failure of Drebbel's *perpetuum mobile* did not invalidate the inventor's claim to be taken seriously, for, having destroyed the unsatisfactory device 'he attempted to achieve perpetual motion by other means' thereby gaining a reputation as a man greatly skilled in natural and mechanical matters.[46] The preference here is for honest scientific endeavour, even if ultimately unsuccessful, over the heigh-presto of the conjurer. Comenius was informed by a countryman of his that even as late as 1632 Drebbel was 'still sweating along on the trail of perpetual motion, but in vain', still vowing that he would be the one to discover it.[47]

The prolific Jesuit Athanasius Kircher also cited the Archimedes–Drebbel connection, quoting Claudian's epigram to support his contention that the central globe of the device was made of glass. The principle of the Drebbelian device is then shown incorporated into an elaborate mechanism that owes much to the ingenuity of Hero of Alexandria and his followers like Salomon de Caus: an engraving of the machine shows that the expansion of air in the globe as it was warmed forced water up a tube to fill a small receptacle, from which it fell down to drive a wheel.[48] This device, if ever made, would very soon have been brought to a standstill by evaporation and friction, however carefully it was adjusted. Kircher specifies that it must be set up against a south-facing wall ('meridionalem ... plagam') to catch the warmth of the sun, as its action on the air in the globe was the means by which all the other operations in the machine were set in motion. This is reminiscent of the Carpentras MS account of the *perpetuum mobile* demonstrated by the Kufflers to Peiresc.

Evidence from paintings of the period bears out the role of the sun's warmth operating on the simple globe-and-ring device of Drebbel's original invention. Henri Michel reproduces a detail from a picture by the Antwerp artist Frans Francken II (1581–1642) showing the visit of the Archdukes Albert and Isabella to the cabinet of

---

44 Comenius, *Works* (n. 43), vol. 3, p. 222.

45 *Ibid.*, pp. 222–223: 'in qua latuit fraus, Lampas ardens quae oleum in aërem resolvens, globo inspirabat, eumque lente circumagebat'.

46 *Ibid.*, p. 223: 'destruxisse dicitur machinam suam, alliisque viis motum perenne venari aggressus: Quo sane nomine sagacissimi hominis ingeniose multa in naturalibus et mechanicis excogitandis industria laudanda est'.

47 *Ibid.*, p. 225: 'Ante septennium hoc Anno 1632 reversus ex Anglia vir generosus, gentilis meus, inter alia retulit, Drebbelium eatenus perpetui mobilis vestigationi insudare, sed frustra, quanquam praesumat et jactet: ex ore eius auditam sibi referre vocam: se repeturum, aut neminem.'

48 Athanasius Kircher, *Magnes*, Rome, 1641, book II, part iv, ii, pp. 341–342, 352 and illustration opposite.

an antiquary.[49] On a table by the window stands a *perpetuum mobile* very close in appearance to the one drawn by Hiesserle von Chodaw. A similar but not identical machine is seen in a similar position on a table by a sunny window in a later picture by Hendrik Staben showing the visit of the rather older Archdukes to Rubens' studio.[50]

These pictures bear witness to the continuing appeal of the original Drebbelian device and are all the more useful as by the 1630s Drebbel's own experiments and the propaganda of the Kufflers make it impossible in many instances to know for certain what version of the machine is being alluded to. For instance, Beeckman records in August 1633 that Johann Sibertus Kuffler had been trying to interest a client in a *perpetuum mobile* containing mercury, but this is hearsay rather than a eye-witness description.[51] What however did become increasingly clear is that effort expended on trying to achieve perpetual motion by harnessing the Drebbelian device to drive a Heronic automaton (as in the Kircher machine) was so much wasted ingenuity.

Possibly the final glimpse of the original device presented to King James is in John Wilkins' *Mathematicall Magick*, published in 1648. In it he numbers 'the famous motion invented by *Cornelius Dreble*' among the attempts to achieve perpetual motion by chemical means. After quoting Tymme's account of the machine (the anti-Copernican conclusion of which he ridicules), Wilkins observes:

> What strange things may be done by such extractions, I know not, and therefore dare not condemn this relation as impossible; but me thinks it sounds rather a chymicall dream, then a Philosophicall truth. It seems this imprisoned spirit is now set at liberty, or else is grown weary, for the instrument (as I have heard) hath stood still for many years.[52]

The 'chymicall dream' continued to find favour in some quarters,[53] but the future of Drebbel's invention lay with those who saw that the crux was the temperature-related alterations in the air – the 'rarefaction and condensation of the air', to use Antonini's phrase. In particular it lay with those who appreciated that the device was not a machine in the sense that it generated or delivered energy, but that it could be the basis of an instrument for taking measurements related to the degree of heat or the pressure of the air, as foreshadowed by Antonini. If Henri Michel is correct, a version of the Drebbelian *perpetuum mobile* was already being used as a barometer in 1621 by Ghysbrecht Dronckere, engineer to the Archdukes, to predict changes in the weather.[54]

---

[49] Henri Michel, 'Le Mouvement perpetuel de Drebbel', *Physis*, xiii, 1971, pp. 289–294. The picture, in the Walters Art Gallery, Baltimore, is also attributed to Jan Breughel the Elder (1568–1625).

[50] 'Albert and Isabella in Rubens' Studio', in the Museés Royaux des Beaux-Arts, Brussels.

[51] Beeckman, *Journal* (n. 29), vol. 3, p. 302.

[52] John Wilkins, *Mathematicall Magick*, London, 1648, book II, ch. 9, pp. 229–230. Wilkins's ironic reference to 'this imprisoned spirit' echoes Tymme's words, which derive ultimately from the 'inclusus ... spiritus' of Claudian.

[53] For instance, another Jesuit, Gaspar Schott (*Mechanica hydraulicopneumatica*, Würzburg, 1657, part II, ch. ii, pp. 377–379) claimed that Drebbel 'had enclosed in the glass ring two mutually contrary and extremely antipathetic fluids, which waged perpetual war against each other' ('annulo vitreo inclussisse dicitur duos sibimet contrarios et summe anthipaticos [*sic*] liquores, qui perpetua inter sese decertabant pugna').

[54] Michel *Op. cit.* (n. 49), pp. 293–294.

In England Drebbel's reputation was kept alive not only by his sons and sons-in-law, but also by the philosophers belonging to the various groups out of which eventually came the Royal Society. How far Drebbel's tireless experimentation was in harmony with the ethos of the nascent Society is suggested by a passage in a letter from William Petty to Henry More early in 1649:

> To bee sure I accompt him the best philosopher I know most of, and is best acquainted with nature's Workes, and knoweth how to rule them in order to provarment of good help to Mankind, wherefore God made them. Now whether or noe or how much des Cartes his philosophy conduseth here unto, wee may know...by Enquiring how much the Telescopes are mended, since the writing of his diopticks; whether any man can foretell the weather by his discourse of Meteors. ...Or who, that hath followed his principles, hath drawne any new usefull or pleasant art or reall conclusion from them. Till Monsieur des Cartes hath approved himself a philosopher in this Sense, I shall preferre Cornelius Drebbel before him, though he understood no Latin as one that hath done more though said less.[55]

Petty's choice of grounds for comparison between Drebbel and Descartes is interesting: telescopes and foretelling the weather. Drebbel's much-trumpeted claims in the matter of telescopes have been mentioned, but a method of, or instrument for, weather forecasting was not apparently in his repertoire of inventions. Petty appears to be crediting Drebbel with something to which the inventor himself did not lay claim. Within Drebbel's own lifetime Dronckere had grasped the utility of the *perpetuum mobile* in its humble role of weather forecaster; this letter suggests that posterity was likewise willing to credit Drebbel with the invention of a weather-glass – which was none other than the failed *perpetuum mobile.*

An unsigned and undated letter to Hartlib, probably from John Beale, deals with 'those rare instruments they call thermometers, or weather-glasses' and discusses the need to optimize their design and construction. The writer goes on to list Drebbel's discoveries relating to the device:

> Yet it is certaine that Drebble that greate singular Learned Mechanicke did by the helpe of this Instrument make a dyall continually to move of itselfe Regularly showing both the tyme of the day and other Motions of the Heavens. Did also make an Automatous Instrument of the Musick and found out a furnace which he could governe to any degree of heate.[56]

The phrase 'thermometers or weather-glasses'[57] occurs again in the context of another person who might have possessed a copy of the Drebbel *perpetuum mobile* or a device

[55] University of Sheffield, Hartlib Papers VII, 123, printed in C. Webster, 'Henry More and Descartes: some new sources', *British Journal for the History of Science*, iv, 1969, pp. 367–368.

[56] University of Sheffield, Hartlib Papers bundle 26/56/1. This letter was printed in Robert Boyle's posthumously published *General History of the Air*, London, 1692, title xiii.

[57] The two words were both introduced into English in the 1620s. They were often linked as synonyms in the scientific literature of the seventeenth century, indicating that the instrument's potential for measuring the two different, though related, phenomena of temperature and atmospheric pressure was not properly appreciated until Boyle's researches. See *O.E.D.* s.v. Barometer, Thermometer, Weather-glass.

based upon it: the English occult philosopher and alchemist William Backhouse of Swallowfield near Reading, who became Elias Ashmole's 'father' in alchemy in 1651. Backhouse's father, Sir Samuel Backhouse, had at some time had direct contact with Drebbel over an alchemical recipe written in Dutch,[58] so there was even a tenuous personal link. In 1656 Ashmole informed Hartlib that Backhouse possessed 'a rare Termometer or Weather glasse the heate alw. descending being sealed in both sides close and as long almost as a Cane which stands to the Wall in his roome'. Four years later Hartlib again refers to this instrument:

> one of the best Thermometers or Weather-glasses which also K. Charles saw and was mightily pleased wth it. Hee [i.e. Backhouse] hath also a long Gallery wherin are all manner of Inventions and Rarities.

This type of thermometer, called by Robert Boyle 'New or Hermetical', apparently made use of a more sophisticated design than that of the instrument attributed to Galileo.[59]

Because the extensive documentation relating to Drebbel's *perpetuum mobile* spans approximately the first half of the seventeenth century, it offers an exceptional opportunity to study the process of science at the interface of the old and the new philosophies. The *perpetuum mobile* was the product of a thoroughgoing Aristotelian cosmology; it was validated by its links with Archimedes' globe and the Ptolemaic arrangement of the universe. Coupled with this was a new intellectual fashion which I have called the Rosicrucian tendency, the adherents of which were characterized by an intense millenarian enthusiasm and a compulsion towards utilitarianism. As Drebbel himself boasted for his device:

> whatever may be performed by the falling of waters, by wind, by the strength of men or beasts, by weights or by revolving steel, the same can be done by this invention in perpetuity, or until the instruments are worn out.[60]

Such claims could not of course be realized; there was no way that the genuine phenomenon that Drebbel observed and incorporated into his *perpetuum mobile* could ever be magnified, geared up or duplicated to perform the feats promised. Hence the inventor's lifelong unavailing pursuit of the chimaera of a practical form of perpetual motion; hence the elaboration of his device when the original simple version

---

[58] C.H. Josten, *Elias Ashmole*, 5 vols, Oxford, 1966, vol. 1, plate 1, shows Ashmole's cypher note concerning an alchemical recipe for making gold; Sir Samuel Backhouse deciphered it and, 'finding it to be written in Dutch, Cornelius Dreble was gotten to translate it'.

[59] *Ibid.*, vol. 2, p. 684 and n. 13; vol. 1, p. 76 and n. 10. Boyle's phrase is of interest in that it is the earliest use given in *O.E.D.* of the word 'Hermetical' in the sense of 'air-tight', as opposed to the earlier sense of 'pertaining to occult philosophy or alchemy'; see *O.E.D.* s.v. Hermetic, Hermetical. Both senses may have been present to Boyle's mind; it seems plausible that it was the device under discussion that mediated the displacement of the 'occult' sense of 'hermetic' by the 'air-tight' sense in scientific discourse.

[60] Abbotsford Club *Miscellany*, vol. 1, p. 112: 'et quidquid descensu aquarum, vento, hominum bruterumque robore, pondere, circumvoluto chalybe praestatur, idem hac inventione fit in perpetuum, aut eousque dum instrumenta deteruntur'.

failed to deliver the required effects; but hence too the honing of his mechanical skills – and his inventions along the way of some of his numerous other devices – of which his contemporaries wrote with such respect.

There are many ironies in all this. It is ironic that Drebbel got the answer right in 1607 when James I asked what was the cause of the movement in the *perpetuum mobile*; however, the form in which the reply was cast shows how fatally Drebbel was committed to the old explanations, the soon-to-be-superseded cosmology – 'the Air, the principal element, which made all things move'. This, coupled with the confusion as to whether it was the air or the more observable water that was the motive force, vitiated all Drebbel's attempts to go beyond his original insight. The doomed Rosicrucian obsession with grandiose practical schemes ultimately brought this particular line of approach to the device to a dead end. Drebbel's failure to see the humble but serviceable weather-glass in his derided *perpetuum mobile*, a genuine discovery among the multiplicity of bogus claims, is the crowning irony of all.

It took someone from the circle of Galileo's correspondents to realize that the whole value and interest of the so-called *perpetuum mobile* lay in itself, not in any putative utilitarian spin-offs. Once it was accepted that the phenomenon observable in the device was linked with external phenomena in a way that demanded explanation, and that such an explanation could not be furnished by the old philosophy, an entirely new set of questions could be asked. It is this change in the agenda that warrants the rescue of Drebbel's *perpetuum mobile* from obscurity.

Jan van der Straet (Stradanus) and the Origins of the Mariner's Compass

Willem Hackmann

W'are not to *Ceres* so much bound for bread,
Neither to *Bacchus* for his clusters red,
As Signior *Flavio* to thy witty triall,
For first inventing of the Sea-mans dyall,
Th'use of the needle turning in the same,
Divine device, O admirable frame!
Whereby thorow th'Ocean in the darkest night
Our hugest Carrackes are conducted right,
Whereby w'are stor'd with trouch-man, guide & Lampe,
To search all corners of the watery Campe.
Whereby a ship that stormy heav'ns have whorld
Neere in one night into another world,
Knowes where she is, and in the *Card* discries
What degrees thence the *Aequinoctiall* lies.[1]

[1] Quoted from Du Bartas in G. Hakewill, *An Apologie or Declaration of the Power and Providence of God In the Government of the World*, 3rd edn, London, 1635, p. 326. For the original, see Guillaume de Saluste du Bartas, *His Divine Weekes and Workes,* London, 1621, pp. 67–68, and T.H. Haight, *The Divine Weeks of Josuah Sylvester. Mainly translated from the French of William de Saluste, Lord of the Bartas,* Waukeska, 1908, p. 63, though the standard modern edition is *The Divine Weeks and Works of Guillaume de Saluste Sieur de Bartas translated by Joshua Sylvester*, edited with Introduction and Commentary by Susan Snyder, 2 vols, Oxford, 1979, vol. 1, lines 973–985, p. 199. There are many minor variations in the text of the various editions. Du Bartas' emblematic title-page of the English translation by Sylvester is reproduced (and discussed) in M. Corbett and R. Lightbown, *The Comely Frontispiece. The Emblematic Title-Pages in England 1550–1660,* London: Routledge, 1979, pp. 99–104. I hope that Francis Maddison will find much in this paper to remind him of our many hours of stimulating conversation on these topics.

From *Learning, Language and Invention: Essays presented to Francis Maddison*, ed. W.D. Hackmann and A.J. Turner. Published by Variorum, Ashgate Publishing Ltd, Gower House, Croft Road, Aldershot, Hampshire, GU11 3HR, Great Britain.

Stradanus and a Crucial Episode in the History of Navigation

I have long been fascinated by one of the plates in a series executed by Jan van der Straet in the 1570s under the rubric *Nova reperta* ('New Discoveries').[2] It is a typical Renaissance scene, repeated in dozens of paintings and prints, of the lone scholar [Figure 1] sitting at his table, in his bachelor room, intent on his studies and unaware of the activity taking place in the sunshine outside.[3] Behind him is his four-poster bed, beside him a sleeping dog,[4] and above the empty grate at the opposite end of the room is an intriguing allegorical device, or perhaps a coat of arms, consisting of an upright mongoose (or perhaps a dog) fighting with a snake.[5] From his arched window we see bystanders watching a rowing boat being manoeuvred towards the shore, and a small sailing vessel being prepared for a journey, its unfurled sail just about to catch the wind. A third vessel is riding at anchor. The model of the ocean-going galleon with full sail, suspended from the ceiling, links the activity in the study to what is happening in the world outside. The maritime theme is reinforced by the magnetic and navigational devices scattered around the room on every available surface. More or less identifiable items are an armillary sphere, navigator's quadrant, folding or set square, rule, magnetic compass, and what are probably intended to be a rolled-up chart and navigational books, resting on the table near the empty grate, while on the scholar's desk we can identify a celestial globe, pocket sundial, sand-glass, and another set square. The scholar appears to be checking (or perhaps marking out by means of the dividers)

---

[2] Reproduced in Burndy Library, '*New Discoveries' The Sciences, Inventions and Discoveries of the Middle Ages and the Renaissance as represented in 24 engravings issued in the early 1580s by Stradanus*, Norwalk: Burndy Library Publication no. 8, 1953, pl. 2. The Introduction is by Bern Dibner, and the Latin captions were translated by Edward Rosen. On dating, see below.

[3] Examples popular at this time were of the Humanist scholar Jerome portrayed in his study, for instance, in a copper engraving by Dürer (1514), or in paintings by Quentin Massys and his pupils. A fine example in Düsseldorf, dated 1520, by one of Massys' pupils, shows Jerome in reflective mood with the index finger of his left hand on a skull, and on the wall behind him the inscription 'Cogita mori' (inv. no. D. 1/1953/14). Through the window can be seen daily life, with a resting camel train and a fully laden donkey, but in a setting of a German town. Nobert Schneider, *The Art of the Still Life*, Cologne: Benedikt Taschen, 1990, pp. 77, 81 (illus.). See also Christiane Wiebel, *Askese und Endlichkeitsdemut in der italiensichen Renaissance. Okonologische Studien zum Bild des Heiligen Hieronymus*, Weinheim, 1988.

[4] The dog is often seen in such a setting. An intriguing example in the Museum's collection, is the plate in Leonhard Zubler, *Novum instrumentum scioterucum. Das ist, kurtzer und grundlicher Bericht, wie nicht allein aller hand Sonnenuhren...*, Zurich, 1609. The maker is setting out a sundial in his workshop. This plate is obviously copied from B[urckhard] L[eemann], *Instrumentum instrumentorum: horologiorum scioticorum. Erstlich werden gelehrt uffryssen die vier hauptsonnen uhren ohne einiche verenderung dess Circkels*, Zurich, 1604, which has a similar arrangement but with a different instrument, and without the sleeping dog in the left-hand corner. The dog in such cases may well be the symbol of fidelity. The animal is also associated with melancholy, one of the 'Four Temperaments', most notably in Dürer's famous alchemical engraving 'Melancholy' (1514).

[5] It would have been interesting to have been able to interpret this scene. It is not associated with Stradanus' ducal patron in Florence, nor with the arms of Amalfi. Such scenes were, however, often depicted in natural history books and paintings, for instance in Jan van Kessel's allegorical paintings entitled 'Asia' (1664–6) and 'America' (1666) in the Alte Pinakothek in Munich, and which were based primarily on accounts by

Figure 1. Stradanus' Nova reperta *Pl. 2: 'Flavio' in his study*

Figure 2. *Stradanus'* Nova reperta *title-plate*

the wind-rose on a marine compass by comparing this with the text in the open book to which he is pointing with his left hand in order to prevent him from losing his place. Dominating the scene in the left foreground is a natural magnet or lodestone, floating on a raft of wood in a large basin of water. Lastly, there is an intriguing instrument placed near the scholar's left hand. This appears to be a pivoted compass needle, but it has been suggested that it may well be a copy of William Gilbert's *versorium*, about which more later.

It could well be that the intention of the artist in this allegorical scene is to demonstrate the practical fruits produced by an unworldly scholar, and this is reinforced by the Latin caption which translates as 'THE LODESTONE, THE MAGNET. That stone reveals to Flavio its secret love for the pole, but he revealed it to the sailor'.[6] The intriguing question is not that the artist considered the mariner's compass to be one of the great inventions that fuelled the Renaissance, but that he singled out the Amalfi historian Flavius Blondus as its inventor. His reasons were based on traditional knowledge concerning the diffusion of the marine compass, and on the time and geographical location when he made this plate. That he definitely considered Flavius to be the inventor is also shown by the title-plate to this series, published by Philips [Philippe] Galle,[7] and which features the main navigational and technological discoveries [Figure 2] according to Stradanus. At the top left (I) America is shown with Christopher Columbus of Genoa as its discoverer and Americus Vespuccius of

travellers. One of the scenes in the former is of snakes and other reptiles advancing on a herd of large white mammals (bears?), while in the latter two snakes are threatening a group of elephants. They are reproduced in Schneider (*Op. cit.* (n. 3), pp. 164–165). Mongooses were observed to be bold predators, several species killing poisonous snakes by darting upon the snake's head and cracking the skull with a single powerful bite. Thus, this allegorical scene was probably intended to illustrate the possibility of travel to exotic countries made possible by the compass. The snake *motif* in the engraved title-page to the *Nova reperta* series [Figure 2] is better understood. Both the woman entering the picture at the left and the man leaving the picture at the right are holding a snake biting its tail. This is based on a religious image from ancient Egypt, signifying *uroburos*, the symbol of the eternal return (rebirth) of life. This image became especially potent to Linnaeus who used it in the frontispieces to his *Hortus Cliffortianus*, Amsterdam, 1737, and more potently in *Fauna svecica*, 1746, in which a 'Nordic Diana' is holding the circular snake in her right hand and the moon (or egg) in her left hand. The frontispiece to Philip Miller, *Introduction to the Sexual System of Linnaeus*, 1779, has the portrait of Linnaeus framed by *uroburos*. On this potent symbol in alchemy, see H.J. Sheppard, 'The *ouroboros* and the Unity of Matter in Alchemy: a Story in Origins', *Ambix*, x, 1962, pp. 83–93. The story of Kekulé's discovery of the ring-structure of benzene after a day-dream of a snake biting its own tail is well-known. It is perhaps not too fanciful to suggest that he may well have known about the *uroburos* image – at least in his subconscious. For a fuller discussion of the snake symbolism, see James Hall, *Dictionary of Subjects & Symbols in Art*. Introduced by Kenneth Clark, London: Murray, 1992 (reprint), pp. 285–286.

[6] 'LAPIS POLARIS MAGNES. Lapis reclusit iste Flavio abditum Poli suum amorem, at ipse navitae.'

[7] Philips Galle (1537–1612) moved from Haarlem where he was the printer and publisher of Maarten van Heemskerck to Antwerp in 1570, the centre of print production. He joined the famous Saint Lucas guild and collaborated with the outstanding engravers of his day such as the Collaert family (Adriaen Collaert was his son-in-law). Galle was joined by his sons Theodor (1571–1633) in 1596 and Cornelius I. (1576–1650) in 1605/8. See A. von Bartsch, *The Illustrated Bartsch vol. 56: Netherlandish Artists: Philips Galle* by Arno Dolders, New York: Abaris, 1987; and Ilja M. Veldman, 'Philips Galle, een inventieve prentonderwerper', *Oud Holland*, cv, no. 4, 1991, pp. 262–290.

Florence as the source of the name,[8] and at the top right (II) is illustrated the magnetic compass's wind-rose with 'Flavius of Amalfi' as its inventor. The other key inventions depicted here are (III) gunpowder,[9] (IV) the printing press,[10] (V) the mechanical clock,[11] (VI) guaiacum or lignum vitae (from America) as a specific against venereal disease,[12] (VII) distillation,[13] (VIII) silk,[14] and (IX) the stirrup.[15] The Southern Cross is shown between the words Nova and Reperta.[16]

Stradanus and the *Nova reperta* Series

Of great interest would be a discussion of the sources, literary, iconographical, and others, which led Stradanus to his choice of inventions and the manner of their depiction. For instance, plate 18 which has the caption: 'Astrolabe – Americus Vespuccius found the cross with the four stars in the silent night', was inspired by the famous words which Vespuccius allegedly quoted from Dante's *Divine Comedy*.[17] However, the aim of this paper is much more humble and intends to trace only the

[8] This refers to pls 1 and 18 of the set. In pl. 1 Americus Vespuccius of Florence, who landed on the South American coast in 1497, is shown disembarked, holding a banner with the Southen Cross and an astronomical ring. In the hammock before him is a native girl rising from her sleep ('awakened America'), and around him are representations of the flora and fauna of America, including an ant-eater, a tapir and a sloth. More sinister, what appears to be an idyllic scene are natives roasting a human leg on a spit. In pl. 18 Stradanus celebrates Vespuccius' discovery of the Southern Cross. Stradanus' original drawing of this plate is reproduced in Gunther Thiem, 'Studien zu Jan van der Straet, Genaennt Stradanus', *Mitteilungen des Kunsthistorischen Institutes in Florenz*, vol. 8, parts 1–6, October 1957–May 1959, pp. 91–92, fig. 2. Of interest is that Vespuccius is not using an astrolabe but an armillary sphere which is generally considered to be a teaching device, although a small Italian Ptolemaic armillary sphere in the Museum of the History of Science in Oxford, dated *c.* 1500 (inv. 55–64), and one of the oldest surviving armillary spheres, can be used as an observational instrument. Its sphere can be placed in the horizon circle of the stand to represent the heavens from any latitude; its base has engraved an eccentric zodiac/calendar scale with rotatable index, and an inset simple magnetic compass showing the four cardinal points; a central groove cut into the zodiac band may have held an index or 'sight'; and there is a means of suspending a plumb-line and bob. See F.R. Maddison, *Medieval Scientific Instruments and the Development of Navigational Instruments in the XVth and XVIth Centuries*, Agrupamento de Estados de Cartografia Antiga, vol. 30, Coimbra, 1969, pp. 10–11, fig. 2. Large armillary spheres were used as observational (astronomical) instruments, for instance, by Tycho Brahe. Despite Stradanus' plate to the contrary, it is very unlikely that such an instrument was ever used for navigating at sea.

[9] Pl. 3. See also Jacopo Coppi, 'The Invention of Gun Powder', (n. 13), executed at the same time.

[10] Pl. 4. The original design by Stradanus is in the Windsor collection, see Leo van Puyvelde (n. 24), item 158, p. 24.

[11] Pl. 5, see nn. 16 and 24.

[12] Pl. 6, see nn. 16 and 24.

[13] Pl. 7 on distillation is most intriguing as it has no alchemical overtones. It is probably of the Grand Duke of Florence, Francesco I, in his laboratory, surrounded by his assistants, and is loosely based on a painting by Stradanus painted for the Grand Duke in 1570, for the latter's 'Studiolo' in the Palazzo Vecchio. This is one of four paintings in this small room lined with pictures with a scientific/technological theme, representing activities concerned with the use of fire: Stradanus' 'The Alchemist's Laboratory' (essentially of distillation), Jacobo Coppi's 'The Invention of Gun Powder' (the same topic as Stradanus's pl. 3), Giovanni Maria Butteri's 'The Glass Factory' and Alessandro Fei's 'The Goldsmith's Laboratory'. See A.A.A.M. Brinkman, *De Alchemist in de prentkunst*, Amsterdam: Rodopi, 1982, pp. 37–40, a reprint of his paper

sources that may have inspired Stradanus to attribute the origin of the magnetic compass to Flavius Blondus. In the course of this certain problems concerning the early transmission of the magnetic compass, which are still unresolved after a great deal of scholarly attention, will be highlighted.[18]

Jan van den Straet, who latinized his name to Johannes Stradanus, but who was better known in Italy as Giovanni della Strada, was born in Bruges in 1523. He was taught by his father with whom he worked until the age of twelve, became a master at twenty-two and then began his travels as was the custom of artists and students. He first went to Lyon, then to Venice, but finally settled in Florence in *c.* 1553. He became a member of the Academy of Florence and among his patrons were the Grand Dukes Cosimo I and Francesco I. Initially, he was the assistant of Giorgio Vasari with whom he worked on the design of the famous tapestries in the Palazzo Vecchio.[19] He also worked for the Vatican in Rome on paintings and frescoes. He died in Florence in 1605.[20]

'Chemie en kunst XV Stradanus' "Nova Reperta"', *Chemie en Techniek*, xxx, 1975, pp. 1–4; and A.A.A.M. Brinkman, *Chemie in de Kunst*, Chemie en Techniek Cahier, Nr 1, Amsterdam: Rodopi, 1975, pp. 13–15, reprinted from 'Noord Italië', *Chemie en Techniek*, xxv, 1970, pp. 460–462. Stradanus' painting is also described in Jacques van Lennep, *Alchemie*, Brussels: Gemeentekrediet, 1984, p. 361, fig. 87, who erroneously states that this painting was for the Grand Duke Cosimo, a confirmed alchemist. Its drawn design is in the Windsor collection, see Leo van Puyvelde (n. 24), no. 168, p. 26.

[14] Pl. 8 is a variation of a drawing on this subject which forms part of the *Vermis sericus* series of six plates. Indeed several of the subjects of the *Nova reperta* series reoccur in other series. See the two pen and ink drawings on silk in the Windsor collection, items 162 and 163, p. 25 in Leo van Puyvelde (n. 2). See also R.A. Scorza, 'A Modello by Stradanus for the "Scala di Penelope" in the Palazzo Vecchio', *Burlington Magazine*, cxxvi, no. 976 (July 1984), p. 434.

[15] Pl. 9, see nn. 16 and 24.

[16] Not alluded to in this title-page are plates 10 (watermill), 11 (winged mill), 12 (olive oil), 13 (sugar), 15 (spectacles), 16 (longitude, but see n. 55), 17 (polished armour), 18 (astrolabe) and 19 (engraving on copper). Yet another extremely interesting plate not referred to in the title-page is 14 (oil colours). It purports to be of Jan van Eyck's (*c.* 1390–1441) studio. Van Eyck is credited with the invention of oil paints, and he is shown here painting a 'Saint George and the Dragon' while his brother and assistants are making the paints. Two paintings have been associated with Van Eyck's lost St George: the one in the National Gallery in Washington is ascribed by some to Rogier van der Weyden, the other in the Museo Arquelogico in Palma, Majorca, is by Piedro Nisart. A third, unsigned, painting in the Alte Pinakothek in Munich has similarities to the lost painting as described in a letter written by Pietro Summonte in 1524. Stradanus may well have known Van Eyck's painting as his is the closest to Summonte's description. See E.J.K. Reznick, 'De overlevering van een verloren gegaand schilderij van Jan van Eyck', in A. Horodish, ed., *De arte et libris Festschrift Erasmus 1934–1984*, Amsterdam: Erasmus Antiquariaat en Boekhandel, 1984, pp. 387–393.

[17] For the original drawing, see n. 8. Stradanus quoted under his drawing 'Purgatory', Canto I, 22–27 which has been reproduced in an ornate box at the side of the plate (obscuring part of the original drawing) in both Italian and its Latin translation. It reads:

> Then to my right I turned to contemplate
> the other pole and there saw those four stars
> the first man saw, and no man after him
>
> The heavens seemed to revel in their flames.
> O widowed Northern Hemisphere, deprived
> forever of the vision of their light!

Stradanus' drawings are a remarkable record of the scientific, technical and colonial expansion that occurred during the Renaissance. He had his drawings mass produced by means of copperplate engravings, taking full advantage of the commercial possibilities of this technology. He sent his drawings by courier from Florence to Antwerp, at the time the greatest centre of print production in Europe.[21] Initially he formed a partnership with Hieronymus Cock who prepared and issued the engravings, but shortly after the latter's death he joined forces with the Galle family. He drew the *Nova reperta* plates after the success of the one hundred and four prints making up his *Venationes* series.[22]

The first issue of the *Nova reperta* consisted of eight plates. Later twelve more were added, and this was followed by the four of the *Americae retectio* set, representing

Dante, *The Divine Comedy Vol. II: Purgatory*, translated by Mark Musa, Penguin Classics, 1984, p. 2. This reference is possibly to the Southern Cross which Dante insists had only been seen previously by Adam and Eve. Allegorically, the four stars represent the four cardinal virtues: Prudence, Temperance, Justice, and Fortitude. See also Dante, *The Divine Comedy*, World's Classics, OUP, 1993, p. 574, note 23. Note that pl. 3 of his *Americae retectio* series depicts Vespuccius on board a galleon using a quadrant.

18 The literature is vast, but the following are useful starters: Heinz Balmer, *Beiträge zur Geschichte der Erkenntnis des Erdmagnetismus*, Veröffentlichungen des Schweizerischen Gesellschafft für Geschichte der Medizin und der Narurwissenschaften, vol. 20, Aarau, 1956; A. Crichton Mitchell, 'Chapters in the History of Terrestrial Magnetism', chapter I, *Terrestrial Magnetism and Atmospheric Electricity*, xxxvii, 1932, pp. 105–146, chapter II, *ibid.*, vol. xlii, 1937, pp. 241–280, chapter III, *ibid.*, xliv, 1938, pp. 77–80, chapter IV, *ibid.*, li, 1946, pp. 323–351; S. Chapman, 'Archaeologica geomagnetica', *Terrestrial Magnetism and Atmospheric Electricity*, xlviii, 1943, pp. 1–2, 77–78, introducing H.D. Harradon, 'Some Early Contributions to the History of Geomagnetism – I'; *ibid.*, pp. 3–17, II and III, pp. 79–91, IV, pp. 127–130, V, pp. 197–199, VI, p. 202; J. Needham, *Science and Civilisation in China*, vol. 4:1, Cambridge, 1962, pp. 229–334; D.W. Waters, *The Art of Navigation in England in Elizabethan and Early Stuart Times*, London, 1958, pp. 22–32; and Francis Maddison (n. 8), and his surveys (a) 'A Consequence of Discovery: Astronomical Navigation in Fifteenth-Century Portugal' in T.F. Earle and Stephen Parkinson, eds, *Studies in the Portugese Discoveries I*, Proceedings of the first Colloquium of the Centre for the Study of the Portugese Discoveries, Warminster, 1992, pp. 71–110 (magnetic compass pp. 75–76), and (b) 'Marine Compass' in A. Bedini, ed., *The Christopher Columbus Encyclopedia*, New York, 1991, vol. 1, pp. 205–207. For a brief survey of this literature, see also W.D. Hackmann, 'Magnetism in Navigation', in K. Thrower and P. Leggatt, eds, *Papers Presented at the Twentieth IEE Weekend on the History of Electrical Engineering*, London: IEE, 1993, pp. 9/1–9/22.

19 Scorza *Op. cit.* (n. 14), pp. 432–435, 437–438.

20 J.A.F. Orbaan, *Stradanus te Florence 1553–1605*, Rotterdam, 1903; and Thiem *Op. cit.* (n. 8), pp. 88–111, 155–166.

21 E. Cockx-Indestege, *Christoffel Plantijn en de exacte wetenschappen in zijn tijd*, Brussels: Gemeentekrediet, 1989. See also the introduction (Dutch and English) in J.P. Filedt Kok, W. Halsema-Kubes and W. Th. Kloek, *Kunst voor de beeldenstorm. Noornederlandse kunst 1525–1580*, Gravenhage: Staatsuitgeverij, 1986, which has an extensive bibliography.

22 Approximately four hundred engravings are known of Stradanus' drawings devoted to different broad themes, roughly one-third to sport, another third to religious subjects, and the final third to a miscellany including still-life, domestic scenes, allegory (such as on human nature, the senses, the seven liberal arts, the eight practical arts, the seven planets, the four elements, the eight wonders of the world), and historical scenes. See Dibner *Op. cit.* (n. 2). The Windsor plates are bound in a plain book issued by Galle in *c.* 1600. See Leo van Puyvelde *Op. cit.* (n. 24).

Columbus, Vespuccius and Magellan on their ships of discovery in allegorical settings.[23] The plates were sold individually and also bound in book form. The first edition probably dates from *c.* 1600.[24] The main title-page of the second edition, also issued by Ioannes Galle, is entitled *Speculum diversarum imaginum speculativarum*, and dated Antwerp 1638. It is a reverse copy of a drawing by Maarten van Heemskerck.[25]

The Lodestone and the Origins of the Mariner's Compass

Before discussing the sources that may have inspired Stradanus it may be of interest to give a brief survey of the present history of the magnetic compass, concentrating on its cultural transmission.[26] There has been an enormous antiquarian interest in the history of this instrument since the seventeenth century, but until recently histories have been distorted by the legendary Chinese south-pointing chariot, which was not a magnetic compass at all and, perhaps more critically, by Viking legends of the supposed

[23] The title-page (pl. 1) is too full of allegory to be able to do it justice in a note. The globe showing America, Africa and Europe is carried by a rather weary looking 'Oceanus', flanked by 'Flora' and 'Janus'. Suspended from the top, flanking the title *Americæ retectio*, are cartouches of Americus Vespuccius Florentinus and Christophorus Columbus Genuensis, and three instruments: a navigational quadrant, pair of dividers and set square. The coast shown is from Livorno to Savona, with Florence placed prominently inland in the left-hand corner. The engraving reminds me of Du Bartas' lines:

> Yea, though *Vespucio* (famous *Florentine*)
> *Marke Pole*, and *Columb*, brave Italian Trine:
> *Our (Spaynes-Dread) Drake, Candish, and Cumberland*
> *Most valiant Earle, most worthy High Command*:
> And thousand gallant moderne *Typhies* else,
> Had never brought the North-Poles Paralels
> Under the *South*; and sayling still about,
> So many *New-worlds* under us found out.
> Nay, never could they th'*Artike* Pole have lost,
> Nor found th'*Antartike*; if in every Coast
> Sea liquid Glasse round-bow'd not every-where,
> With sister Earth to make a perfect Sphere.

Snyder *Op. cit.* (n. 1), vol. 1, lines 415–426, p. 184. The same quadrant is depicted in pl. 3 entitled in translation, 'Americus Vespuccius of Florence in a marvelous expedition to the west and to the south opened up two parts of the earth, greater than the shores which we inhabit and known to us in no previous age, one of which by common consent of all human beings is called by his name, America. An. Sal. MIIID (1497).' The fourth plate in this series shows Magellan sailing through the straits named after him, making calculations by means of an armillary sphere. Unlike the small armillary sphere held by Vespuccius in pl. 18 (n. 8), this is a table model.

[24] The Burndy set consists of the 20 plates of *Nova reperta* (including the introductory plate), of which, according to Dibner (n. 2), 17 were engraved by Ioan. Galle (introductory plate and Pls 2 – 14, 16, 19), three by Ioan. Collaert (pls 15, 17, 18) and one (pl. 1) by Theodor Galle, father of Ioannes, and the four plates entitled *Americae retectio* were engraved by Adrianus Collaert. Very few complete sets have survived. There is some confusion concerning the actual engravers of these plates as of only pls 1, 15, 17 and 18 is the engraver given, while pls 15 and 17 name both Ioan Collaert ('sculp.') and Ioan. Galle ('excudit'), and the others only give Ioan. Galle 'excudit'. The Ashmolean Museum in Oxford has fifteen prints of the *Nova reperta* set. These are without the signature of the engraver. Some are of the first state issued by Philips

use of magnetic compasses on long sea journeys by the Norsemen. As it turns out, the written versions of these legends post-date the first European references to the magnetic compass.[27]

The words used for magnetism, the lodestone, and the mariner's compass in the different cultures give a fascinating insight into the possible transmission of this knowledge. They highlight several intriguing puzzles. In ancient Greece the lodestone was known as *lithos Herakleia* ('Heraklean stone'), which was either derived from a geographical location – the town of Heraclea near Mount Latmus in Caria on the borders of Lydia – or from the mythological figure Heracles. Another name was *lithos magnetis*, after one of two towns named Magnesia in Lydia, one of which is near Heraclea.[28] The Romans called it *adamas* derived from the Greek, meaning 'unconquerable'; a word applied to the hardest objects known such as iron and diamond. It survived in English in the fifteenth century. According to a contemporary inventory, two English ships had, respectively, '1 sailing piece [*sc.* lodestone]', and '12 stones, called adamants, called sailstones…'.[29] It also led to the French derivation of *aimant*, 'loving stone', because of confusion with the Latin word 'adamo'. Intriguingly, the most common Chinese word for lodestone was *tzhu shih* ('loving stone'). Perhaps this

Galle, and others of the second issued by his nephew Johan Galle. According to F.W.H. Holstein, *Dutch and Flemish Etchings, Engravings and Woodcuts ca. 1450–1700*, vol. 4, Amsterdam: Menno Hertzberger, 1951, p. 213, the publisher of the twenty plates (entries 129–148) is Philips Galle. Ten plates were engraved by Philips Galle and the other ten by Hans (Jan Baptist Ioan.) Collaert. The set of twenty plates were also issued by the Deutsches Museum (1972), ten full-size and ten reduced. According to Leo van Puyvelde, *The Flemish Drawings In the Collection of His Majesty the King at Windsor Castle*, London: Allen and Unwin and Phaidon, 1942, items 157 and 158, p. 24, these plates were bound in a small volume, undated, and issued by Philips Galle in *c.* 1600. Incidentally, item 157 identified with pl. 19 of *Nova reperta*, is, in fact, a quite different composition of the same theme. There is also a second plate dealing with magnetic navigation, and that is pl. 16 depicting a 'Stevin-type' azimuth compass, dealt with in the text.

25 This title-page is a reverse copy by Johannes Galle of a print by Heemskerck entitled 'Democritus and Heraclitus, or Allegory of the Vanity of Human Passions', issued by Hieronymous Cock, and with the texts from two prints by Cornelus Cort depicting the busts of Democritus and Heraclitus, but retaining the title and text from the central cartouche of the Heemskerck print. Another reversed copy but without the frame is in T. de Bry's *Emblemata saecularia* (1596), reproduced with Introduction by F. Warnecke, Berlin and Paris, 1894, and with a modified motto in De Bry's 1627 edition. See I.M. Veldman, *Maarten van Heemskerck and Dutch Humanism in the Sixteenth Century*, Amsterdam: Maarsen, 1977, pp. 76–79, especially nn. 90, 91 and 105.

26 This survey of the early history of the magnetic compass is primarily based on the references listed in n. 18, including my paper presented at the Twentieth IEE Weekend on the History of Electrical Engineering.

27 E.G.R. Taylor, 'The South-Pointing Needle', *Imago Mundi*, viii, 1951, pp. 1–7; Jean Randier, *Marine Navigation Instruments*, London, 1980, p. 15, resurrects this hoary myth; C.E.N. Bromstead, 'Ships' Lodestones', Mineralogical Magazine and Journal of the Mineralogical Society, xxviii, 1947–9, p. 429. H. Winter, 'Who Invented the Compass?', *Mariner's Mirror*, xxiii, 1, 1937, pp. 95–102. See also n. 48.

28 D.H.D. Roller, *De magnete of William Gilbert*, Amsterdam: Menno Hertzberger, 1959, pp. 13–32. On the classical texts dealing with the lodestone, see Albert Radl, *Der magnetestein in der antike. Quellen und zusammenhänge*, Stuttgart, 1988; Crichton Mitchell *Op. cit.* (n. 18), IV, pp. 323–351.

29 Waters *Op. cit.* (n. 18), pp. 22–23. Chaucer, too, refers to 'adamants' in *Romaunt of the Rose* and *Assembly of Foules*, quoted in Fleury Mottelay (n. 47), p. 62.

is the origin of the tale in Marbodeus Gallus' *Liber lapidum* (*c.* 1120) in which an unchaste wife would fall out of bed when a lodestone was applied to her head:

> This Stone doth reconcile the man and wife,
> And her recal that from her husband goes:
> Under her head, when that she sleeps, it shows:
> For she that's chast, will presently embrace
> Her husband whilst she sleepeth; but a whore
> Falls out o'th' bed, as thrown out with disgrace,
> With stink o'th'Stone, which shows this, and much more.[30]

As is well known, lodestones were used on board ship to re-magnetize the compass needles.

Turning to European sources for 'compass', an early (mid-thirteenth century) use of the word *compasso* means 'sailing directions', comprising the *portolano* (strictly speaking, the *written* guide) and the nautical chart of the Portuguese navigators.[31] Later, this word came to be applied solely to dividers and related instruments for making plots on nautical charts, such as *compasso di riduzione*, or 'proportional compass' and *compasso di proporzione* or 'sector', while the Italian *bussola* (or French *boussole*) originally referred to the magnetic needle supported by the floating straw type of arrangement.[32]

The first component of the word 'lodestone', or 'loadstone', is derived from an Old English word for 'way' or 'journey'. Similarly, the German *leiderstein* and Swedish *seglestein* originates from its directive (or compass) property. The Dutch term *zeilsteen* (or 'sailstone') denotes its use as a compass with sailing ships. The Italian *calamita* (magnet) suggests its origin from the Greek *kalamos*, which has two suggested meanings, both are early forms of magnetic compasses: (1) a small piece of reed which allowed the magnetic needle to float, or (2) a frog or tadpole float incorporating a small piece of lodestone. This latter term, which occurs in Pliny, is especially intriguing as the same is found in several early Chinese texts.[33]

---

[30] Balmer *Op. cit.* (n. 18), p. 242, quotes (in German) Marbode's hexameter taken from the Latin 1644 edition of Porta's *Magia naturalis* (see n. 77). This quote is taken from the English edition of 1658 (n. 78). The same idea is expressed evocatively in lines 935–950, p. 198 of Du Bartas in the Snyder edition (n. 1). See also Roller *Op. cit.* (n. 28), p. 27, who also gives a brief summary of the ideas of the medical powers of the lodestone; S.P. Thompson, *Notes on the De magnete of Dr. William Gilbert*, London: privately printed, 1901, p. 18.

[31] R. Motzo, *Il Compasso da navigare*, Cagliari, 1947. See also Maddison *Op. cit.* (n. 8), pp. 15–17; and (n. 18, a), pp. 75–76; (n. 18, b) pp. 435–436.

[32] There are many illustrations in physics textbooks from the 1600s onwards of floating lodestones, but there would be great difficulties in using such an arrangement as a compass. Henri Michel, 'Notes sur l'histoire de boussole', *Communications de l'Académie de la Marine de Belgique*, v, 1950, pages not numbered. Baylak al-Qipjāqi, an Egyptian scholar, observed during a voyage from Tripoli to Alexandria in 1242/3, that on starless nights the seamen used a floating straw compass: see Eilhard Wiedemann, *Aufsätze zur arabischen Wissenschaftsgeschichte*, ed. Wolfdietrich Fischer, Hildesheim & New York, 1970, pp. 36–37. Peregrinus described a similar device in his *Epistola de magnete* (1269).

[33] Needham *Op. cit.* (n. 18), p. 273: '*Hsia-ma tzu*, the tadpole is also called the 'mysterious needle', (*hsüan chen*), or the 'mysterious fish' (*hsüan yü*), and another name for it is the 'spoon-shaped beastie' (*kho tou*). Its shape is round and it has a long tail'.

Is the similarity between Pliny and these early Chinese texts a coincidence, or does it point to a diffusion of knowledge of the directive properties of magnetism between these cultures? Chinese compasses were used for geomantic (divinational) purposes. They were South-pointing, although the Chinese realized that they could be North-pointing as well depending on the orientation of the magnet. They had probably introduced the magnetized iron needle (obtained by stroking it with a lodestone) by the seventh or eighth century. This gave greater directional accuracy, but these needles were still only used in geomantic compasses, which remained in this form for many centuries[34] and became prized by European colletors.[35]

Both China and Classical Europe knew of the basic magnetic properties of the lodestone in about the first century BC, but the Chinese went ahead of Europe when they developed their geomantic compasses based on their discovery of magnetic orientation.[36] References to this magnetic property only appear in Western Europe in the twelfth century AD, which led to the development of the European mariner's compass. The route by which this knowledge reached the West from China is at present not known. There are no clues in contemporary Arabic, Persian, or Indian sources. One term common to all is 'fish-shaped' to describe the compass. It could be significant that the Arabs singled out the compass's ability to point South, and that the Persians and Turks called the compass 'South-pointer'. It has been suggested that the water-bowl compass preceded the dry-pivot version, but evidence is certainly not conclusive. In 1280 Bailak al-Qibjāqī described such a compass seen during a voyage from Tripoli to Alexandria in 1242/3, but the earliest surviving Islamic instrument has a dry-pivoted compass. This is an astronomical compendium (a simple astrolabe and sundial for finding Muslim prayer-times, and the direction of Mecca from various places) made by Ali b. ash-Shātir. As it dates from 1366/7[37], it does not help us with establishing

34 *Ibid.*, pp. 271–273.

35 European collectors became very interested in these compasses in the seventeenth century judging by extant specimens at the Museum of the History of Science in Oxford and the Istituto e Museo di Storia della Scienza in Florence (inv. no. 2483), described by Isaia Iannaccone, 'Il "Feng Shu" e la bussola geomantica dell'Istituto e Museo di Storia della Scienza di Firenze', *Nuncius*, v, 1990, pp. 205–217, especially pp. 213–216. The Oxford specimen, John Selden's *pixis sortilega sinesium*, came into his possession in about 1600. For a detailed late-eighteenth-century description of such compasses, see Sir George Staunton, *An Authentic Account of an Embassy from the King of Great Britain to the Emperor of China; including Cursory Observations made, and Information obtained.... Together with a Relation of the Voyage...*, London, 1797, vol. 1, pp. 441–447. For further details, see the entries on Chinese compasses in W.D. Hackmann, *Catalogue of the Pneumatic, Magnetic, Electrostatic, and Electromagnetic Instruments in the Museo di Storia della Scienza*, Florence: Giunti, 1994.

36 Needham *Op. cit.* (n. 18), pp. 239–245. Related to (magnetic) directivity is the concept (and observation) of (magnetic) 'polarity'. For a discussion how this important concept was derived from cosmology in classical antiquity, was used by Peter Peregrinus in his study of magnetism in the thirteenth century, and was then developed as a concept in science generally in the seventeenth century onwards, see J. Rosmorduc, 'Les tribulations d'un terme scientifique: "pôle" de la cosmologie de Platon à l'optique du XIXe siècle', *Revue d'histoire des sciences et leurs applications*, xxix, 1976, fasc. 1, pp. 51–62.

37 AH 767. He was an important astronomer and *muwaqquit* of the Umayyad Mosque in Damascus. This instrument is now in the Awqāf Library, Aleppo. Its importance was recognized by S. Reich and G. Wiet, 'Un astrolabe syrien du XIVe siècle', *Bulletin de l'Institut français d'Arcéologie Orientāle du Caire*, xxxviii,

any transmission route; nor does the Arabic word *al-kombas*, which must have gained currency after the compass had been established in the West. There are no other surviving magnetic compasses which can be dated before those incorporated towards the end of the fifteenth century, in sundials and a globe, and none of these instruments was intended for nautical use. A suggestion that a pivoted fish decoration at the centre of a Sutton Hoo Saxon hanging bowl was a primitive compass can almost certainly be discounted.[38]

Various routes have been suggested by which the Chinese dry-pivoted geomantic compass came to Europe. Transmission through Islam is suggested by a mid-ninth century Arabic poem referring to the nautical use of the lodestone,[39] but other historians have favoured an overland westward transmission across Central Asia. An important route at that time was the 'Silk Route', a 4 000-mile caravan track which started at Sian, followed the Great Wall of China to the North-West, by-passed the Takla Makan Desert, climbed the Pamir Mountains, crossed Afghanistan, and so on to the Levant, from where merchandise was shipped to the Mediterranean. This route became increasingly unsafe after the gradual decline of the Roman Empire and rise of Islam. In the thirteenth century the route was revived by the Mongols, and was followed by Marco Polo. He claimed to have reached the Khan's court in 1275 and returned to Venice some twenty years later.[40]

1939, pp. 195–202, reprinted in E.S. Kennedy & Imad Ghanem, eds, *The Life and Work of Ibn al-Shâtir*, Aleppo, 1976, pp. 36–44. This instrument is also described in detail in Maddison & Turner (Bibliography no. 32), no. 83, pp. 159–161, in which it was pointed out for the first time that this instrument contains the pivot for a compass needle. See also Louis Janin and David King, 'Ibn al-Shātir's Sandūq al-Yawāqit: An Astronomical Compendium', *Journal for the History of Arabic Science*, 1, 1977, pp. 187–256, pls 1–6, reprinted in David A. King, *Islamic Astronomical Instruments*, London, 1987, chapter XII. The magnetic compass is rarely mentioned in the known astronomical texts prior to Ibn al-Shātir; the only example known to David King is a later thirteenth century treatise on the astrolabe and the sundial by the Yemeni Sultān al-Ashraf: see his *Mathematical Astronomy in Medieval Yemen. A Bibliographical Survey*, Malibu, 1983, p. 97, pl. 7, and 'Astronomie im mittelalterlichen Jemen', in Werner Daum, ed., *Jemen*, Innsbruck, 2nd edn, 1988, pp. 278–279. For a further discussion on the significance of *al-kombas*, see Balmer *Op. cit.* (n. 18), pp. 53–54. Some scholars have suggested that the compass was transmitted from Christian Europe to Islam rather than the reverse.

[38] M.S. Tite, 'Report on a Feasibility Study of the Use of the Fish in the Large Sutton Hoo Hanging-Bowl as a Magnetic Compass', Appendix A, in Bruce-Mitford, ed., *Sutton Hoo*, 1986, pp. 296–8, and Bruce-Mitford's description of this bowl and archaeological parallels in the same volume, pp. 221–227, 239–244. It is unlikely to have been a rotating magnet on the Chinese pattern, and note its similarity to the traditional Islamic *phiala*, of which an eighteenth-century Turkish specimen in the Ashmolean Museum has at its centre rotating fishes and birds, and a central movable fish that appears to 'swim' on the water when the bowl is filled: see J.W. Allen, *Metalwork of the Islamic World. The Aaron Collection*, Sotheby's, 1986, pp. 42–46, especially fig. 34.

[39] Eilhard Wiedemann, *Aufsätze zur arabischen Wissenschaftgeschichte*, vol. 1, ed. by Wolfdietrich Fischer, Hidesheim and New York, 1970, pp. 36–37; Maddison *Op. cit.* (n. 18, b).

[40] L. Boulnois, *The Silk Road*, London, 1966; Jean-Pierre Drège, *Marco Polo et la Route de la Soie*, Gallimard, 1989, with many interesting pictures; H.E. Kiewe, *Civilisation on Loan*, Oxford, 1973, pp. 95–101, a rather rambling account of the influence of China on western design, but nothing worthwhile on the magnetic compass. According to the most recent scholarship Marco Polo may never have reached the Khan's court and note that some of his account could have been based on contemporary travellers' tales.

It has been argued that the voyages of Chinese sailors to the Persian Gulf in the eighth century could not have been accomplished without some form of compass. The first recognizable description of a Chinese navigational, or marine compass dates from between 1111–1117, describing events that took place in 1086.[41] This pre-dates the first known European description by a hundred years. The Chinese common form, which lasted for many centuries, consisted of a needle floating on water in a brass case. In a surviving specimen, South is at half past 10 o'clock, and the line at the bottom of the case seems to make allowance for declination as well as polarity – pre-dating European knowledge by centuries.[42] The dry-pivoted marine compass certainly did not become common on Chinese vessels until it was introduced from the West in the sixteenth century.[43]

There is literary evidence that the marine compass was in use among North and Mediterranean seamen by the twelfth century; the needle either floating or dry-pivoted. A port that becomes prominent in this context in the literature is Amalfi on the Mediterranean.[44] The first known mention of the use at sea of a compass is by Alexander Neckham who lived in the last quarter of the twelfth century. In his *De utensilibus* he writes about mariners finding their course when Sun or stars were invisible, by means of a North-pointing needle, and in his *De naturis rerum*, he

[41] Needham *Op. cit.* (n. 18), pp. 279–280, but see also p. 230. William Gilbert thought erroneously that the ship's compass was transmitted to the west by Marco Polo, or one of his contemporaries. See n. 72.

[42] *Ibid.*, pp. 286–287, fig. 333. However, one of the puzzles are two ceramic bowls found in Liria (Valencia), Spain which are said to be inscribed in the Iberic language written in Chalcidian script, with 36 or 48 wind directions and instructions to float a needle or a lodestone on mercury in the bowl to find direction at night or in an overcast sky. One has been dated possibly from 300 to100 BC, and the other around the 1st century AD. They await further investigation, as does their possible relationship in the transmission of the compass. See A.J. Turner, *Mathematical Instruments in Antiquity and the Middle Ages. An Introduction*, London: Vade Mecum, 1994, pp. 119–120, for further references.

[43] *Ibid.*, pp. 288–292, fig. 337. Three such compasses have been examined by me: one in the Istituto e Museo di Storia della Scienza in Florence (inv. no. 125), probably acquired in the eighteenth century. It consists of a wooden circular case, made from a single piece of lacquered wood, with steel needle on brass pivot, glass cover, and thin brass rod indicating south-north. The inner ring has the four cardinal points with south (coloured red), and the outer one the 24 compass-points, which is the normal form, and dates from the early Qing period. Two similar specimens at the Museum of the History of Science in Oxford lack the inner scale of four cardinal points, as does the example illustrated in A. Schück, *Der Kompass*, 1911–1918, vol. 1, pl. 1, fig. 5. The smaller one signed on its lid 'Hong Shun Zao', probably meaning 'Made by Hong Shun', but with no surname, and with the batch no. '70' (in Chinese), was acquired by Captain Benjamin Langlois Lefroy, RN, at the siege of Peking in 1857; the other one (inv. no. O.17) is signed 'Zheng Hong Yuan', and was probably also acquired by a collector in the nineteenth century. 'Zheng' is probably the surname and *yuan* means 'source' or 'origin'. The lid of the Florence specimen is signed *lao shi zi*, most likely meaning 'the follower of the master' (i.e. 'apprentice'), or the name of the philosopher Laozi.

[44] Waters *Op. cit.* (n. 18), p. 21; Lynn White, (a) *Medieval Religion and Technology*, University of California Press, 1978, pp. 54–56, 80–81; and (b) *Medieval Technology and Social Change*, Oxford, 1962, pp. 131–133; Crichton Mitchell *Op. cit.* (n. 18), I, pp. 105–146, especially pp. 127–128. The frequently repeated assertion in the past that evidence for the Amalfian origin of the compass is demonstrated by the representation of the compass in the arms of the town is incorrect.

describes the needle as being placed on a pivot.[45] A few other references date from this time. In about 1204 Guyot de Provins referred in his poem, *La Bible*, to a floating needle pointing to a 'star' (*sc.* the Pole Star) after it had been touched by a lodestone,[46] and shortly before 1218, Jacques de Vitry, Bishop of Acre, wrote that the compass was 'very necessary' to those navigating on the sea.[47] By 1225, the navigational use of the compass was commonplace in Iceland.[48]

The most detailed early account is Peter Peregrinus de Maricourt's *Epistola de magnete* of 1269. He describes both the floating needle and a rather complicated dry-pivoted compass consisting of two pivots, above and below the vertical arbour carrying the needle.[49] It might be assumed that the use of a pivot for the compass-needle was common by the end of the fourteenth century, but there are no descriptions, illustrations or artifacts of marine compasses until the appearance of a compass-box and, possibly, a compass-rose drawn in the margin of a manuscript, written in the first half of the fifteenth century, of *La Sfera*, a cosmological poem, by either Gregorio or Leonardo Dati. It is not absolutely clear that the 'star' in the picture represents a compass-rose on a fly, rather than just the *Stella Maris* on the lid of the box. The *Stella Maris* became the

---

45 Maddison *Op. cit.* (n. 8), p. 39 for further references. Crichton Mitchell *Op. cit.* (n. 18), II, p. 241, states it very succinctly: 'In Chapter I it was shown that the compass had been brought into use in north-western Europe for the purpose of navigation by the year 1187 (i.e. the date he takes for Neckham's publications), that its first mention in the literature of that year does not refer to it being of recent introduction, that the discovery of the directive property of a magnet in the Earth's field must have preceded this application by some considerable time, but that of this time it was not possible to frame any reliable estimate.' Anthony Turner *Op. cit.* (n. 42), p. 122, has suggested that in northern seas the magnetic compass remained relatively crude as it was used as a standby instrument, while the more sophisticated dry compass enclosed in a box with a wind-rose was developed in the Mediterranean, and used together with the Portulan chart. This is a useful hypothesis to work with.

46 *Ibid.*, p. 15 quotes the relevant part of Guyot's poem; translated in German in Balmer *Op. cit.* (n. 18), p. 443.

47 Jacques de Vitry, *Bibliotheca geographica Palestinae*, ed. Rohrich, Berlin, 1890, pp. 48–50. He saw a compass in use in 1204; Crichton Mitchell *Op. cit.* (n. 18), I, p. 126; P. Fleury Mottelay, *Bibiographical History of Electricity & Magnetism Chronologically Arranged*, London: Charles Griffin, 1922, pp. 30–33.

48 White *Op. cit.* (n. 44, b), p. 132, but see also Waters *Op. cit.* (n. 18), p. 25, note 2, quoted by Maddison *Op. cit.* (n. 8), p. 16. See the detailed analysis of the sources suggesting the early use of the mariner's compass in northern Europe in Crichton Mitchell *Op. cit.* (n. 18), pp. 122, 124. See also G.J. Marcus, *The Conquest of the North Atlantic*, Woodbridge, 1980, pp. 17–20; and W.E. May, *A History of Marine Navigation*, Henley-on-Thames, 1973.

49 P. Radelet de Grave and D. Speiser, 'Le *De magnete* de Pierre de Maricourt. Traduction et commentaire', *Revue d'histoire des sciences et leurs applications*, xxvii, 1975, fasc. 3, pp. 193–234; Edward Grant, ed., *A Source Book in Medieval Science*, Cambridge, Mass., 1974, pp. 368–376, no. 59, 'The First Systematic Description in Europe of the Properties of the Lodestone – Peter Peregrinus (fl. 1269)' [translated by Brother Arnold, 1904; annotated by Grant); H.D. Harradon, 'Some Early Contributions to the History of Geomagnetism – I', *Terrestrial Magnetism and Atmospheric Electricity*, xlvii, 1942, pp. 3–17. The first widely circulated edition of Peter Peregrinus was A.P. Gasser, ed., *Petri Peregrini Maricortensis De Magnete seu rota perpetui motus*, Augsburg, 1558. In this context should also be noted the rare translation by Silvanus P. Thompson, [London], 1902, limited to 240 copies. Thompson compared the various editions with an MS copy dated 1391 which he owned.

traditional decoration of the compass-rose, when it was customary to mark North by the fleur-de-lys and East by a cross (indicating the direction of Jerusalem).[50] The mariner's compass decorating fifteenth-century maps show a round box and rose (but no needle) and similar examples are found in a few near-contemporary portraits.[51] There is no sign of any gimbals mounting to compensate for the roll of the ship, although the idea of such a mounting was known in medieval technology, and its use with the compass was suggested by Leonardo da Vinci in *c.* 1500.[52]

By the fifteenth century sailors realized that the compass did not point true North, but to the East of North. Various techniques were pioneered by the Portuguese navigators to determine declination at sea.[53] This led to the development of the azimuth (or altazimuth) compass,[54] which was in common use when Stradanus drew his plates for the *Nova reperta*. Stradanus' two compasses in his 'Flavio' plate are relatively crudely drawn and cannot be identified, but in appearance are like the simple version found as decoration on contemporary nautical maps. His portrayal of an early azimuth

---

[50] Maddison *Op. cit.* (n. 8), pp. 15–17, and (n. 18, b).

[51] See, for instance, the portrait of Edward Fiennes, Lord Clinton and Say, the Lord High Admiral, dated 1562, in the Ashmolean Museum, Oxford, illustrated in Maddison *Op. cit.* (n. 8), fig. 31, and Waters *Op. cit.* (n. 18), fig. IV.

[52] An interesting drawing of a marine compass and rose (but not showing gimbals) is da Vinci's red chalk on brown/grey paper, entitled, 'Allegory with a Wolf and Eagle', *c.* 1515–16 in the Royal Library, Windsor Castle, illustrated as item 82, pp. 155–156, in Martin Kemp and Jane Edwards, *Leonardo da Vinci*, New Haven and London: Yale University Press in Association with the South Bank Centre, 1989. On the origin of gimbals, see Waters *Op. cit.* (n. 18), pp. 28–29; Maddison *Op. cit.* (n. 8), p. 39; Needham *Op. cit.* (n. 18), vol. 4:2, 1965, pp. 228–236, for a survey of Chinese and European references to the gimbal mounting; and Schück *Op. cit.* (n. 43), I, p. 12, pl. 24, fig 8a and b, depicts an early (1571) gimbal-mounted compass by Hans Göbe. An eccentric attempt to insulate the observer from the motion of the vessel was the gimbal-mounted marine chair in Jacques Besson, *Livre des instruments mathématiques et mechaniques*, Paris, 1571–2, reproduced as fig. 34, p. 94, in E.G.R. Taylor and M.W. Richey, *The Geometrical Seaman. A Book of Early Nautical Instruments*, London, 1962, and in Randier *Op. cit.* (n. 27), pp. 78–79, fig. 87. It should be noted that virtually all western references to gimbals are pre-dated by surviving examples of portable Persian incense burners in which the incense is burned in a gimbaled dish enclosed by open-work metal spheres. There are two fourteenth-century examples in the Barghello Museum in Florence (if the date is to be believed), and a fifteenth-century version in the Ashmolean Museum in Oxford (private communication, Anthony Turner).

[53] A. Bedini, 'Declination of the Compass', *Columbus Encyclopedia* (n. 18, b), vol. 1, pp. 207–208; Crichton Mitchell, 'Chapters in the History of Terrestrial Magnetism. Chapter II – The Discovery of the Magnetic Declination' (n. 18), pp. 241–280; and *idem*, xliv, 1938, pp. 77–80; Muammer Dizer, 'The Dā'irat al-Muᶜaddal in the Kandilli Observatory, and Some Remarks on the Earliest Recorded Islamic Values of the Magnetic Declination', *Journal for the History of Arabic Science*, i, 1977, pp. 257–260. For Columbus' observations with the magnetic compass, see Francis Maddison, 'Tradition and Innovation; Columbus' First Voyage and Portugese Navigation in the Fifteenth Century', in Jay A. Levenson, ed., *Circa 1492. Art in the Age of Exploration*, (Washington: National Gallery of Art, New Haven and London: Yale University Press, 1991, pp. 89–94; S. Chapman, 'Edmond Halley and Geomagnetism', *Terrestrial Magnetism and Atmospheric Electricity*, xlviii, 1943, pp. 131–144.

[54] Maddison *Op. cit.* (n. 8), pp. 35–41. For a translation of the description of Pedro Nunes' 'shadow instrument' taken from his *Estromento de sombras*, Lisbon, 1537, see Harradon *Op. cit.* (n. 48), V, pp. 197–199.

compass in a later plate in this series appears to be much more realistic [Figures 3 and 4]. Because of Stradanus' reference to Plancius, he probably made the original drawing in 1599 or shortly thereafter.[55]

Stradanus' Possible Sources

So what would have been Stradanus' sources in the 1570s which caused him to attribute the invention of the marine compass to Flavio Blondus? He would have found no clue in Dante's *Divine Comedy*, but he may well have known the Canto in which the compass is used allegorically:

> then, from the heart of one of those new lights
> there came a voice that drew me to itself
> (I was a needle pointing to the star);[56]

It may well have been in the port of Amalfi where Peter Peregrinus first observed the mariner's compass,[57] but it was the historian Flavius Blondus who, in about 1450 claimed that this device was invented and first used by navigators from his town of

[55] Stradanus' pl. 16 in the *Nova reperta* series is more technical than his 'Flavio' plate as it actually depicts a navigator using an early azimuth compass placed on a table on the poop deck. It is interesting enough for it to be the subject of another paper. The detail of this plate is reproduced in Maddison *Op. cit.* (n. 18, a), fig. 6, and described on p. 76 and n. 26 (p. 107). Stradanus' caption reads when translated: 'LONGITUDES OF THE EARTH FOUND FROM THE DEVIATION OF THE MAGNET FROM THE POLE – By means of the magnet, which always deviates a little to either side, Plancius makes it possible to find port everywhere.' As it turned out, the determination of longitude by observation of the variation of magnetic declination proved not to be successful. What is intriguing is why Stradanus should have attributed this technique to the Flemish politician and astronomer, Petrus Plancius (1552–1622), who developed a Mercator-type projection in the early 1590s (see Fleury Mottelay *Op. cit.* (n. 47), pp. 533–534; Balmer *Op. cit.* (n. 18), pp. 341–361), and became well known for his table of magnetic declination reproduced in Simon Stevin, *Portuum investigandorum ratio* or *Havenvindingh*, published by Christoffel Plantijn at Leiden (see n. 21) in 1599, and which appeared simultaneously in Dutch, Latin, English and French. The azimuth compass in Stradanus' plate looks rather similar to fig. 25 (Dutch text) of such an instrument described in Stevin's book. It is the combination of two instruments [Figure 4]: Reyner Pieterszoon's azimuthal quadrant placed north-south on top of the compass. The quadrant was hinged at its base and had a very heavy counterweight, so that it would remain perpendicular on the swaying ship. Stevin acknowledges Plancius as the authority for these declination-values (see Harradon (n. 18), VIII, p. 63). This work brought him to the notice of Stradanus. The discoverer of magnetic declination was Georg Hartmann in 1544, but this work remained hidden until 1841. The discovery of this phenomenon is generally attributed to Robert Norman (1576), but see Harradon (n. 18), vii, p. 185 on the determination of magnetic declination by Joâo de Castro in 1538 using Pedro Nunes' method (previous note). See the references in n. 53 and also Crichton Mitchell *Op. cit.* (n. 18), III, pp. 77–80; Harradon *Op. cit.* (n. 18), IV, pp. 127–130 which reproduces Hartmann's letter to Duke Albrecht of Prussia, dated 4 March 1544, announcing his discovery. Note that Plantijn (Plantin) was not only involved with books on scientific matters but also with the marketing of instruments made in the Arsenius workshop: see Maddison *Op. cit.* (n. 8), p. 22, note 29, and Colin Clair, *Christopher Plantin*, London, 1960, pp. 198–199.

[56] Musa translation (n. 17), 'Paradise', Canto XII, 28–30. According to E.H. Plumtree, *The Commedia and Canzoniere of Dante Alighieri*, 2 vols, London, 1886, vol. 2, p. 66, n. 30, Dante's source could be either his master Guido Guinicelli, who alludes to the magnetic needle in almost the same way, or Roger

Amalfi, without specifying when this occurred, or the name of the inventor.[58] Nor did Flavius substantiate his statement in any way. Perhaps he was inspired by Peregrinus. However, his claim was repeated by a number of Italian authors during the next hundred years, the tale unfortunately – like gossip – growing in the telling. Antonio Beccadelli, otherwise known as Antony of Bologna or 'the Panormitan', wrote a popular poem in which he repeated Flavius' claim that the Amalfians were the first to use the marine compass, but without giving them credit for the invention.[59] His poem achieved wide circulation at the time Stradanus arrived in Florence.[60] Among those who attributed the compass to Amalfi were Baptista Pio (1511),[61] Pandolfo Collenuccio (1539)[62] and the famous humanist scholar Lulius Gregorius Giraldus, who also credited the Amalfians with using a circular compass card attached to the needle.[63]

The story was taken one significant stage further when Giraldus in 1540, misquoting Baptista Pio, ascribed the invention of the marine compass to a person named Flavio,

---

Bacon who dwells on it as a 'miraculum in parte notum'. Plumtree's translation is more evocative than Musa's:

> Then, from the heart of one of those new lights,
> There came a voice which made me turn to see
> E'en as the star the needle's course incites.

Dante and Beatrice are standing in the centre of the circles, and on hearing the voice from one of the spirits, Dante turns towards it in the same way as a compass needle is attracted to the North Star; see Musa, 30, p. 150. According to Crichton Mitchell *Op. cit.* (n. 18), I, pp. 126, 145, Dante's tutor Brunetto Latini anticipated Petrus Peregrinus when referring to the polar property of the magnet, but Paget Toynbee (revised by Charles S. Singleton), *A Dictionary of the Proper Names and Notable Matters in the Works of Dante*, Oxford: Clarendon Press, 1968, points out that the assumption that Latini was Dante's teacher is based on a misunderstanding of 'Inferno', Canto XV, 82–85. See also Mottelay (n. 47), pp. 43–44.

[57] See note 49.

[58] *Blondi Flavii forlivensis de Roma triumphante libri decum... Romae instauratae libri iii, Italia illustrata. Historium ab inclinita. Rom. imperio decades iii*, Basle, 1531, p. 420: 'Sed fama est qua Amalphitanos audivimus gloriam, magnetis usum, cujus adminiculo navigantes ad arctem diriguntur, Amalphi fuisse inventus quicquid vero habeat in ea re veritas, certum est id noctu navigandi auxilium priscis omnino fuisse incognitum.' Also quoted in Crichton Mitchell *Op. cit.* (n. 18), p. 127. See also the Italian translation, p. 238 in *Flavio Blondo, Roma ristaurata, et Italia illustrata di Biondo da Forti. Tradotte in buona lingua vulgare per Lucio Fauno*, Venice, 1543.

[59] 'Prima dedit nautis usum magnetis Amalphis', quoted in Balmer *Op. cit.* (n. 18), p. 592.

[60] According to Crichton Mitchell *Op. cit.* (n. 18), p. 127, with the publications of Torello da Fano's *Digestorum seu Pandectorum ex Florentinis pandectis repraesentatis* Florence, 1553, and Scipione Mazzella's *Descrittione del regno di Napoli*, Naples, 1586, but on the latter, see n. 68.

[61] 'Amalphi in Campania veteri magnetis usum inventus a Flavio traditur', in Baptista Pio, *In Caium Lucretium poetam commentarii*, Bonon, 1511, p. ccvii; quoted in Crichton Mitchell *Op. cit.* (n. 18), I, p. 127.

[62] Pandolfo Collenuccio, *Compendio dell'historia del regno di Napoli*, 2 vols, Venice, 1557, vol. I, lib. I, pp. 9A–9B.

[63] Crichton Mitchell *Op. cit.* (n. 18), p. 127, referring to Giraldus' *De re nautica Caelii Calcagnini commentatio* [n.d.]. For a brief survey of the early wind-rose, see Balmer *Op. cit.* (n. 18), but with no reference to Giraldus.

Figure 3. *Stradanus'* Nova reperta *pl. 16: Determining longitude at sea*

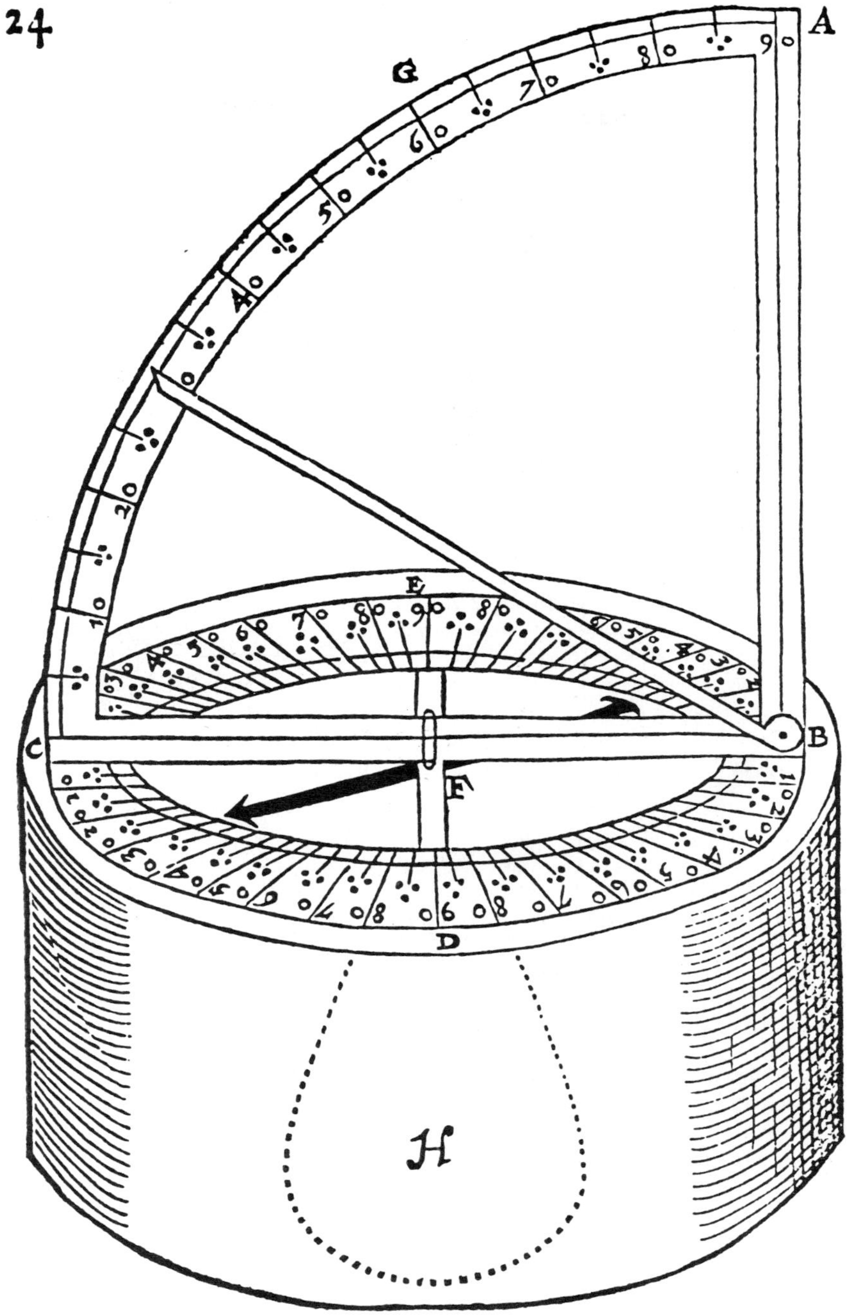

Figure 4. Azimuth compass published by Stevin in 1599

whereas, as we know, Flavio Blondus had merely been the originator of the statement that this device had been invented by the Amalfians.[64] The new attribution was widely circulated at the time of Stradanus, for instance, by Pierre Belon,[65] and in particular by such authorities as Hieronymous Cardanus,[66] and Achilles Pirminius Gasser who referred to it in the Preface to his edition of Peregrinus published in 1558.[67] It was this edition that brought Peregrinus's ideas on magnetism to the fore and influenced the work of William Gilbert.

The tale took one further twist. Giraldus had only referred to a person named 'Flavio', which was generally assumed to be Flavio Blondus. Stradanus probably took it as such, but in 1586 Scipione Mazzella named the supposed inventor Flavio Gioja, giving no hint of sources for such an attribution.[68] He may well have had in mind an actual person, now forgotten, but the similarity with Guyot de Provins of *La Bible* fame has been noted; Mazzella may have come across this somewhere and given the name the Italian form of 'Gioja'.[69] Further confusion occurred when in the early 1600s Abraham Ortelius gave the name of the inventor as Giovanni Gioja or Giri, and shifted the scene of the invention from Amalphi to Melphi.[70] All these versions persisted in the

---

[64] 'Sed & non multis retro seculis Amalphis in Ca*m*pania oppido, antiquis naviga*n*di usus incognitus per magnetem & chalybem, quorum indicio naute ad polos diriguntur à Flavio quodam excogitatus traditur...', L.G. Giraldus, *Ferrariensis, de re nautica libellus, admiranda quadam & recondita refertus, nunc primum & natus & aeditus*, Basel, 1540, pp. 5–6; quoted (almost verbatim) by Crichton Mitchell *Op. cit.* (n. 18), I, p. 127, but dating the book 1580, thereby destroying his argument.

[65] 'Comperimus eum qui primus huius lapidis usum invenerit, Flavū' fuisse nuncupatum.' in P. Belon, *Plurimarum singularium & memorabilium rerum Graecia...*, Antwerp, 1589, lib. II, cap. XVI, p. 213. First edition Paris, 1553.

[66] Hieronymus Cardanus, *De subtitilitate Libri XXI*, Paris, 1551, lib. VII, pp. 157A-158B; Leiden, 1559, lib. VII, p. 277.

[67] 'Presertim quando Flavius Blondus diserte in illustrata Campania, Amalphitanos haud perperam gloriari prodat, Magnetis usum, cuius adminiculo navigantes ad Arcton diriguntur, apud sese primo inventum este', Achilles P. Gasseri, 'Praefatio in Epistolam Petri Peregrini de Magnete', in *Epistola de magnete*, Augsburg, 1558, p. 8. See Harradon *Op. cit.* (n. 18), I, p. 4.

[68] 'In Amalfi nell'anno 1300 fu (gloria) degli Amalfitani) ritrovata da Flavio di Gioia la Bussola tanto necessaria a naviganti; per il che Giovanni Pontano chiamò detta città Magnetida: & Antonio Panormita in lode d'esta fece questo verso. Prima dedit nautus usum magnetis Amalphitis.', in Scipione Mazzella, *Descrittione del regno di Napoli*, Naples, 1586, p. 41. For a synopsis of the variations of this name in the literature, see Balmer *Op. cit.* (n. 18), pp. 730–731. See also T. Bertelli, 'Discussione della legenda di Flavio Gioia, inventore della bussola', *Rivista di Fisica, Matematica e Scienza naturale*, ii, 1901, pp. 529–541.

[69] Crichton Mitchell *Op. cit.* (n. 18), I, p. 129; also Fleury Mottelay *Op. cit.* (n. 47), pp. 56–58. This book is a mine of information, but even more frustrating than Balmer *Op. cit.* (n. 18), and has to be used with caution.

[70] A. Ortelio, *Theatro del mondo*, Antwerp, 1612, cited by Crichton Mitchell *Op. cit.* (n. 18), I, pp. 128 and 146, but I have been unable to find this edition. A typical quote is from William Whiston, *The Longitude and Latitude Found by the Inclinatory or Dipping Needle; Wherin the Laws of Magnetism are also disclosed*, London, 1721, p. ii: after discussing the improbability that the compass originated in China or Arabia, he writes:

> Others therefore, with much greater Probability, ascribe this noble Invention, which is one of the greatest for the Use of Mankind that ever was yet made, to a Neopolitan, John Goia by Name, AD 1300, who is, however, owned to be the First that apply'd it to the Guidance of Ships in the Mediterranean.

Whiston makes no references to Stevin or Plancius.

literature until the detailed study by Timoteo Bertelli in the late nineteenth century began a serious reassessment of the compass's history.[71] Although if you asked any Italian child today who was the inventor of the compass, he would almost certainly respond 'Flavio Gioja'.

William Gilbert is among the first to compound the various errors in his *De magnete* (1600):

> The people of Melfi, in the kingdom of Naples, first, 'tis said, constructed a mariner's compass; and, as Flavius Blondus says, the townsmen do not without reason boast, they were taught by one Joannes Goia, a fellow-citizen, in the year 1300.[72]

Gilbert remained one of the chief sources used by others to perpetuate this myth in histories and physics textbooks until the late nineteenth century. He is cited by George Hakewill in his commentary on that portion of Du Bartas' verse dealing with the compass,[73] by the historian John Stow in his Annales, or a *Generalle Chronicle of England* in that portion devoted to 'the life and raigne of Queene Elizabeth',[74] and by textbook writers, of which Henry Noad's *Lectures on Electricity* from the 1840s is a typical example:

> A Neapolitan named Flavio Gioia, who lived in the thirteenth century, has been regarded by many as the inventor of the compass. Dr. Gilbert affirms that Paulus Venetus brought the compass from China to Italy in 1260; and Ludi Vestomannus asserts, that about 1500, he saw a pilot in the East Indies direct the course by a magnetic needle like those now in use.[75]

It has not been possible to identify the sources that may have inspired Stradanus. He stays silent on the matter. The most convenient would have been an encyclopaedic work that dealt with all the topics of the *Nova reperta* series, but this is unlikely. Stradanus certainly had access to fine libraries in Florence (either the Grand Duke's or that of the Academy of Florence of which he was a member), so literary sources would have been no problem for him. Identifiable sources certainly would have helped with the dating of these plates. They were probably executed from about 1570, the date of

---

[71] Apart from Bertelli's paper referred to in n. 68, see the detailed listing in Crichton Mitchell *Op. cit.* (n. 18), I, n. 103, pp. 138–139. There was, however, contemporary scepticism as well, as voiced by William Barlowe, *Navigator's Supply*, London, 1597, p. A3^{r}: '...the lame tale of one *Flavius* at *Amelphis* in the Kingdome of Naples, for to have devised it [the compass], is of very slender probabilitie'.

[72] F. Fleury Mottelay, *William Gilbert of Colchester, Physician of London on the Loadstone and Magnetic Bodies, and on the Great Magnet the Earth*, London: Bernard Quaritch, 1893, p. 6. This is effectively the 5th edition of *De magnete* (see Roller *Op. cit.* (n. 28), pp. 176–177). For the full title in Latin, see n. 82. I have used the 2nd edn of London, 1628.

[73] Hakewill agrees with Gilbert that the knowledge was brought from China by Paulus Venetus (Marco Polo), but that it may well be that 'Flavius the Melvitan was the first inventor of guiding the ship by the turning of the needle to the North...', see (n. 1), p. 326. But concerning Marco Polo's journey to China, see also n. 40.

[74] John Stow, *Annales, or a Generalle Chronicle of England*, London, 1631, pp. 809–810. See Mottelay *Op. cit.* (n. 47), p. 211.

[75] H.M. Noad, *Lectures on Electricity, comprising Galvanism, Magnetism, Electro-Magnetism, Magneto- and Thermo-Electricity*, new enlarged edn, London, 1844, p. 281.

Stradanus' oil painting dealing with the same subject as his 'distillation' plate, and the 1590s.[76]

An encyclopaedic work that achieved great popularity at this time was Giambattista della Porta's *Magiae naturalis*, first published in Naples in 1558.[77] Eighteen separate editions appeared in different languages prior to 1600. According to della Porta:

> *Flavius* saith, an Italian found it out first, whose name was *Amalphus*, born in our Campania. But he knew not the Mariners Card, but stuck the needle in a reed, or a piece of wood, cross over: and he put the needles into a vessel full of water, that they might flote freely: then carrying about the Loadstone, the needles would follow it...Now the Mariners Compass is made, and a needle touched with the Loadstone, is so fitted to it, that by discovering the pole by it, all other parts of the heavens are known.[78]

Stradanus had plenty of versions of the Flavio myth to choose from in the 1570s and early 1580s. If only he had also put the surname on his drawing, as his 'Flavio' can as easily refer to Gioja as to Blondus. If he had specifically mentioned Gioja, then his engraving would have to be dated post 1586.[79] Whoever he intended, Stradanus depicts the scholar in the act of examining (or perhaps having just completed) the wind-rose of a compass. It is after all the dry-pivoted needle and compass card that distinguishes in Stradanus' mind the mariner's compass from the primitive floating-needle compass which would have been difficult to use at sea. It is the dry-pivoted compass which Stradanus would have regarded as Flavius' contribution.

The *versorium*

Undoubtedly, the most intriguing device depicted by Stradanus is the small pivoted needle on the corner of the desk near the Flavio's left hand. At first sight it might have helped with dating this engraving. Large, horizontally-pivoted magnetized needles are commonly depicted in engravings of this period, for instance in the plate accompanying Porta's section on magnetism. Here, too, the various magnetic devices are in a room, including on a table, a rather crudely-drawn large magnetic needle, in appearance like an arrow.[80] Such needles (either magnetized or not) were used in experiments with lodestones. A similar device 'ex quoius metallo' three or four fingers long, was described by William Gilbert in his *De magnete* for the detection of weak electrostatic attraction produced by rubbed amber [Figure 5]. As such it was the first

[76] See n. 13 and 55.

[77] The earliest editions: Naples, 1558; Antwerp, 1560; Cologne, 1562; Antwerp, 1564, only go up to 'libri iiii', but are not simply the first four books of the later editions, and have no material on lodestones or the magnetic compass. The first edition with all twenty books is probably the Naples edition of 1589, see next note, and may well post-date Stradanus' 'Flavio' engraving.

[78] J.B. della Porta, *Natural Magic in XX Bookes*, London, 1658, book 7, ch. 32, p. 206. 'Curius inventio Itali suit Amalphi oriundi nostrae Caspaniae, ut à Flavio traditur.', in *Magicae naturalis libri xx*, Naples, 1589, lib. 7, ch. 32, p. 141.

[79] See n. 68.

[80] N. 78, bound between pp. 190–191. The Naples edition of 1589 only has the small woodcuts and not the plate of the English 1658 edition. See also note 77 concerning the earliest editions.

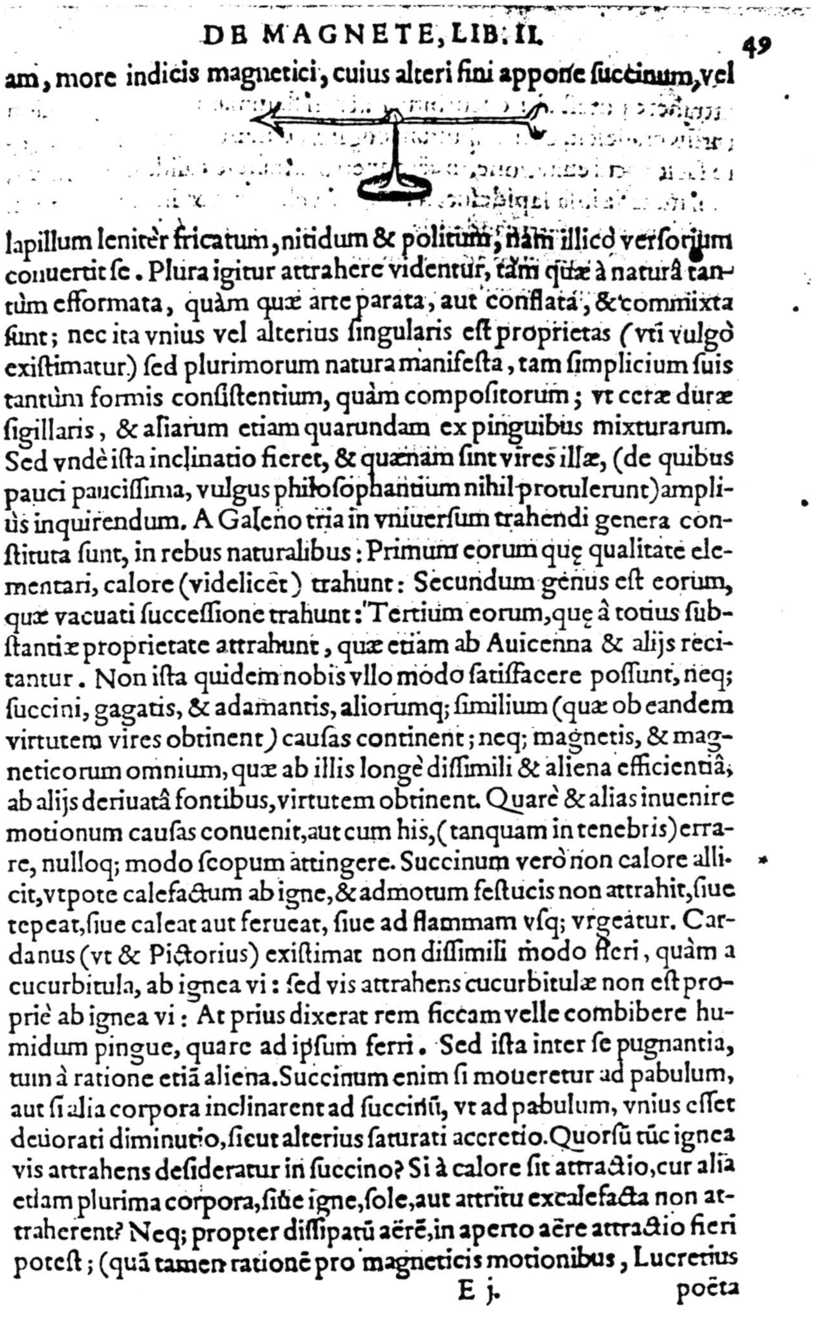

DE MAGNETE, LIB. II. 49

am, more indicis magnetici, cuius alteri fini appone ſuccinum, vel lapillum leniter fricatum, nitidum & politum, nam illicò verſorium conuertit ſe. Plura igitur attrahere videntur, tam quæ à naturâ tantùm efformata, quàm quæ arte parata, aut conflata, & commixta ſunt; nec ita vnius vel alterius ſingularis eſt proprietas (vti vulgò exiſtimatur) ſed plurimorum natura manifeſta, tam ſimplicium ſuis tantùm formis conſiſtentium, quàm compoſitorum; vt ceræ duræ ſigillaris, & aliarum etiam quarundam ex pinguibus mixturarum. Sed vndè iſta inclinatio fieret, & quænam ſint vires illæ, (de quibus pauci pauciſſima, vulgus philoſophantium nihil protulerunt) ampliùs inquirendum. A Galeno tria in vniuerſum trahendi genera conſtituta ſunt, in rebus naturalibus: Primum eorum quę qualitate elementari, calore (videlicet) trahunt: Secundum genus eſt eorum, quæ vacuati ſucceſſione trahunt: Tertium eorum, quę à totius ſubſtantiæ proprietate attrahunt, quæ etiam ab Auicenna & alijs recitantur. Non iſta quidem nobis vllo modo ſatiſfacere poſſunt, neq; ſuccini, gagatis, & adamantis, aliorumq; ſimilium (quæ ob eandem virtutem vires obtinent) cauſas continent; neq; magnetis, & magneticorum omnium, quæ ab illis longè diſſimili & aliena efficientiâ, ab alijs deriuatâ fontibus, virtutem obtinent. Quarè & alias inuenire motionum cauſas conuenit, aut cum his, (tanquam in tenebris) errare, nulloq; modo ſcopum attingere. Succinum verò non calore allicit, vtpote calefactum ab igne, & admotum feſtucis non attrahit, ſiue tepeat, ſiue caleat aut ferueat, ſiue ad flammam vſq; vrgeatur. Cardanus (vt & Pictorius) exiſtimat non diſſimili modo fieri, quàm a cucurbitula, ab ignea vi: ſed vis attrahens cucurbitulæ non eſt propriè ab ignea vi: At prius dixerat rem ſiccam velle combibere humidum pingue, quare ad ipſum ferri. Sed iſta inter ſe pugnantia, tum à ratione etiã aliena. Succinum enim ſi moueretur ad pabulum, aut ſi alia corpora inclinarent ad ſuccinũ, vt ad pabulum, vnius eſſet deuorati diminutio, ſicut alterius ſaturati accretio. Quorſũ tũc ignea vis attrahens deſideratur in ſuccino? Si à calore ſit attractio, cur aliã etiam plurima corpora, ſiñe igne, ſole, aut attritu excalefacta non attraherent? Neq; propter diſſipatũ aërẽ, in aperto aëre attractio fieri poteſt; (quã tamen rationẽ pro magneticis motionibus, Lucretius

E j. poẽta

Figure 5. Gilbert's versorium *in* De magnete *(1600)*

instrument invented for the study of electrical phenomena, but Gilbert's *versorium* was obviously inspired by these magnetic needles.[81] It is not known whether Stradanus has drawn his needle to scale, but it is surprisingly similar to the device illustrated in Gilbert's book.

Stradanus' engraving pre-dates Gilbert's book by more than twenty years, so the question is whether the artist could have been inspired by Gilbert's instrument. Nothing is known of Gilbert's movements after leaving Cambridge in the early 1570s and settling in London some years later to practise medicine. He obtained a grant of arms in 1577, and sometime before 1581 became a member of the Royal College of Physicians. He probably studied magnetic phenomena shortly after leaving Cambridge and before he became too busy with his medical practice. Indeed, there is evidence that Gilbert completed the work for his *De magnete* some seventeen years before his book was published in 1600, that is, in about 1583.[82] It has been suggested that Gilbert may well have gone abroad before he settled in London, that is sometime between 1570 and 1573. If he had gone on the Grand Tour, he would almost certainly have visited Florence, where he may have shown his *versorium* to the cognoscenti, including Stradanus' who was obviously interested in matters scientific, and who could have depicted it in his engraving. Alas, there is no reliable evidence for such a scenario.[83] According to the foremost authority on Gilbert's electrical studies, 'Gilbert was quite certainly using the *versorium* before the Stradanus engraving was published', and by implication it is his *versorium* that is illustrated in that engraving.[84] This would be a very neat solution, but not a likely one. Stradanus' engraving was probably executed sometime after 1570 and possibly before 1586 (when surely he would have alluded to Gioja). The needle in Stradanus' engraving can only be Gilbert's *versorium* if

---

[81] On Gilbert's *versorium*, see W.D. Hackmann, 'Eighteenth-Century Electrostatic Measuring Devices', *Annali dell'Istituto e Museo di Storia della Scienza di Firenze*, iii, 1978, p. 12, and on the problems of such a device, Hackmann, *Electricity from Glass: The History of the Frictional Electrical Machine 1600–1850*, Alphen aan den Rijn: Sijthoff & Noordhoff, 1978, p. 18–19, and in particular his 'The Relationship Between Concept and Instrument Design in Eighteenth-Century Experimental Science', *Annals of Science*, xxxvi, 1979, p. 7. There has been some discussion in the literature concerning the origin of the word '*versorium*'. According to Crichton Mitchell *Op. cit.* (n. 18), I, p. 113, it was invented by Gilbert in his *De magnete* (see bk 2, ch. 2, p. 48; or in Fleury Mottelay translation (n. 72), p. 79), and introduced into the literature by Nicolaus Cabeus in 1629. Hakewill *Op. cit.* (n. 1), p. 326 was among those to suggest that when Plautus, in *Mercator*, V, p. 2, writes 'Hic ventus nunc secundus est, cape mode versoriam', he did not mean a mariner's compass but probably a rope or rudder for steering. For full details, see Sylvanus Thompson's notes in his translation of Gilbert's *De magnete*, Chiswick Press, 1900 [but appeared in 1901].

[82] Roller *Op. cit.* (n. 28), p. 71, based on the statement 'Optimis igitur auspicijs (doctissime D.D. Gilberte) in lucem prodeat Magnetica philosophia tua, non in nonum tantum annum (quod Horatius preacipit) sed in alterum iam fere novennium pressa…', in the preface to the *De magnete* by Edward Wright, *Gulielmi Gilberti Colcestrensis, Medici Londinensis, de magnete magnetisque corporibus, et de magno magnete tellure; physiologia nova, plurimis & argumentis, & experimentis demonstrata*, London, 1600, p. v^v. For the translation, see Fleury Mottelay (n. 72), p. xliv.

[83] For a detailed discussion of the lack of evidence concerning Gilbert's foreign tour, see Roller (previous note), pp. 66–71.

[84] *Ibid.*, p. 100. A detail of Stradanus' engraving is reproduced as fig. 14.

Stradanus had been shown this instrument during Gilbert's putative visit to Florence in the early 1570s. The illustration in Gilbert's *De magnete* came too late for Stradanus' engraving. It has to be concluded that Stradanus was most probably inspired by the type of magnetic needle illustrated in Porta's work on the lodestone, rather than by Gilbert's instrument, despite their extraordinary similaritities. This brings us to experiments with the lodestone which captured the imagination during the final years of Stradanus' life in Florence.

Lodestones in Florence

The claims that Gilbert may have met Galileo or that he worked on magnetism in Italy can also not be substantiated.[85] For one thing Galileo was too young; for another he did not have to go to Italy to practice science as there existed a flourishing experimental tradition in England. As is well known, Galileo was impressed by *De magnete.* In the 'Galileo Room' of the Istituto e Museo di Storia della Scienza in Florence are six small armed lodestones which, according to tradition, were used by Galileo in his experiments on magnetism inspired by Gilbert's book.[86] They were afterwards used by members of the Accademia del Cimento.

Galileo, like Stradanus, must have been touched by the sheer magic of the lodestone and the magnetic compass, as another of his natural magnets, probably dating from shortly after Stradanus' death, consists of a lodestone suspended from a wooden gibbet. Its keeper is secured by screws to the lid of a thick iron casket with crescent feet, with at each side a small brass lion's head with ring. According to Benedetto Castelli, Galileo presented this lodestone to the Grand Duke Ferdinand II.[87] It was inspired by the fable of Moḥammed's tomb suspended in mid air between the ceiling and the floor of magnetic rock [Figure 6]. The inspiration for instruments, too, could come from legends.[88]

---

[85] *Ibid.*, p. 69, especially n. 57.

[86] M. Loria, 'William Gilbert e Galileo Galilei. La terrella e le calamite del Granduca', in Carlo Maccigni, ed, *Saggi su Galileo Galilei,* Florence: G. Barbèra, 1972, vol.3, pp. 208–246. All the lodestones discussed in this section of the paper are described in detail in Hackmann (n. 35). See also P. Galluzzi, 'Gli strumenti di Galileo', in M. Miniati, ed., *L'età di Galileo. Il secolo d'oro della Scienza in Toscana,* Catalogue: Istituto e Museo di Storia della Scienza, 1988, item IV.13, p. 61.

[87] Galluzzi (previous note), item IV.12, p. 60.

'Ho visto un pezzo di calamita di sei once solamente armata di ferro con esquisita diligenza dal Sig.r. Galileo e donata al Ser.mo Gran Duca Ferdinando, la quale tiene sospese quindici libbre di ferro lavorato in forma di sepolcro', *Discorso sopra la calamita,* Ms. Galil. 111, c 303^v, Biblioteca Nazionale Centrale, Florence.

[88] Gilbert *Op. cit.* (n. 72), pp. 3–4, reports on this and other tales. Notable are Pliny's account of the arched roof of the temple of Arsinoë at Alexandria which was made of lodestone by the architect Chinocrates so that the goddess' effigy would be suspended in mid air. In the meantime the architect died, as did Ptolemy who had the work done in honour of his sister. In the Temple of Serapis there was an iron chariot suspended in a similar way, and which was dashed to pieces when the lodestones were removed. In Thomas Browne, *Pseudodoxia epidemia,* London, 1658, bk, 2, p. 79, is also recounted the tale of Bellerophon's horse, which, 'framed in iron, was placed between two Loadstones, with wings expansed pendulous in the ayr'.

There are two other large lodestones in the present Museum which originate from the ducal collection. One is a large suspended lodestone in a plain brass box attached to a gilt block of wood,[89] while the other is an armed rectangular lodestone in a wooden box painted blue, its soft iron caps held in place by large wooden screws projecting from the sides of the box. The lodestone can be seen through a lozenge-shaped window at the front. The large wrought-iron keeper attracted to the pole-pieces is suspended from a pulley with at the other end a heavy triangular alabaster counter-weight.[90] This may well be the same lodestone Galileo advised Cosimo II to purchase in 1609, and which is illustrated in its uncapped state by the Dutch traveller Cornelius Meyer [Figure 7] in his book, *Nuovi ritrovamenti* (1696).[91] In the background is the large suspended lodestone, seen here in this engraving in its original state.

Guilio d'Alfonso Parigi's Version of the Stradanus' 'Flavio' Engraving

Some of the finest instruments in the ducal collection were celebrated in art in the room of the Uffizi Gallery which originally displayed these devices. The walls were painted with maps, and the ceiling, in the grotesque style, was decorated by Guilio

---

These stories were very widespread, for instance, in William Barlowe, *Magnetical Advertisements*, London, 1616, p. 45; A. Kircher, *Magnes; sive de arte magnetica*, Cologne, 1643, lib. 2, pars 4, prob. 6, pp. 312–314; Albert Magnus, *De mineralibus & rebus metallicis libri quinque*, Venice, 1541, lib. 2, cap. 6, p. 227, and Porta (n. 78). Particularly widespread were stories about mysterious magnet mountains fatally attracting ships in far away seas. One such story is 'The Tale of the Third Kalandar' in *The Book of the Thousand Nights and One Night*, vol. 1, pp. 89–103 of the translation of J.C. Mardus by Powys Mathers, Routledge, 1993. The Kalandar's ship, after losing its way in a tempest, approached a magnetic mountain, which drew out all its nails so that the ship fell apart and only the story-teller was not drowned. He managed to neutralize the mountain's magic by shooting a brass statue from its summit into the sea. Balmer (n. 18), pp. 527–528 points out that the Arabian Nights stories appeared in the tenth century at the time when Ptolemy was translated into Arabic. For a detailed discussion on the origins of these stories, see the delightful book by Robert Irwin, *The Arabian Nights: A Companion*, London: Allen Lane, the Penguin Press, 1994, in particular pp. 37–40 (his opinion on the Mardus 'improved' translation); pp. 14–17 (on the first European translations), and 48 (the stories' origins). Although the first European translation was by Antoine Galland in the early 1700s, some stories had circulated since the Middle Ages. Stradanus or Galileo may not have been directly influenced by the Arabian Nights, but thrilling tales of magnetic mountains, etc. were common at that time and undoubtedly fuelled their imagination.

[89] Its present-day suspension and wrought-iron keeper, with its curved ends from which is suspended a heavy iron chain, are not original. *Galluzzi Op. cit.* (n. 86), item IV.16, p. 61.

[90] This specimen was probably armed and placed in the wooden box in 1775 by Felice Fontana, the then Director of the Royal Museum. See Hackmann *Op. cit.* (n. 35). M. Miniati, 'Origini della specola fiorentina', *Giornale di Astronomia*, 1984, pp. 209–220; F. Fontana (attrib. to), *Saggio del Reale Gabinetto di Fisica*, Rome, 1775.

[91] C. Meyer, *Nuova ritrovamenti*, Rome, 1696, pp. 2–3. Meyer writes on p. 2: 'Giovanni Goja, ô come altri vogliono Gira da Amalfi ritrovò, e fece mediante la Calamita la Bussola da navigare, onde disse il Panormitano, *Prima dedit nobis usum Magnetis Amalphis.*' He describes the large lodestone as follows: 'Il Serenissimo Gran Duca di Toscana conserva nella sua inestimabile Galeria in Fiorenza una gran calamita che prima tirava libre 25 di ferro, e la quale poi per non essere posta in luogo opportuno haveva perso part del suo vigore, in modo che non tirava che libre 20 in circa: onde mi fu commesso da quell'Altezza Serenissima da rilegare essa Calamita con un modo da me insinuato per dare à vedere quanta forza potevo farla riacquistare, in maniera che dopo rotrovati da me li veri punto pilari, e fatta ogn'altra operatione necessaria

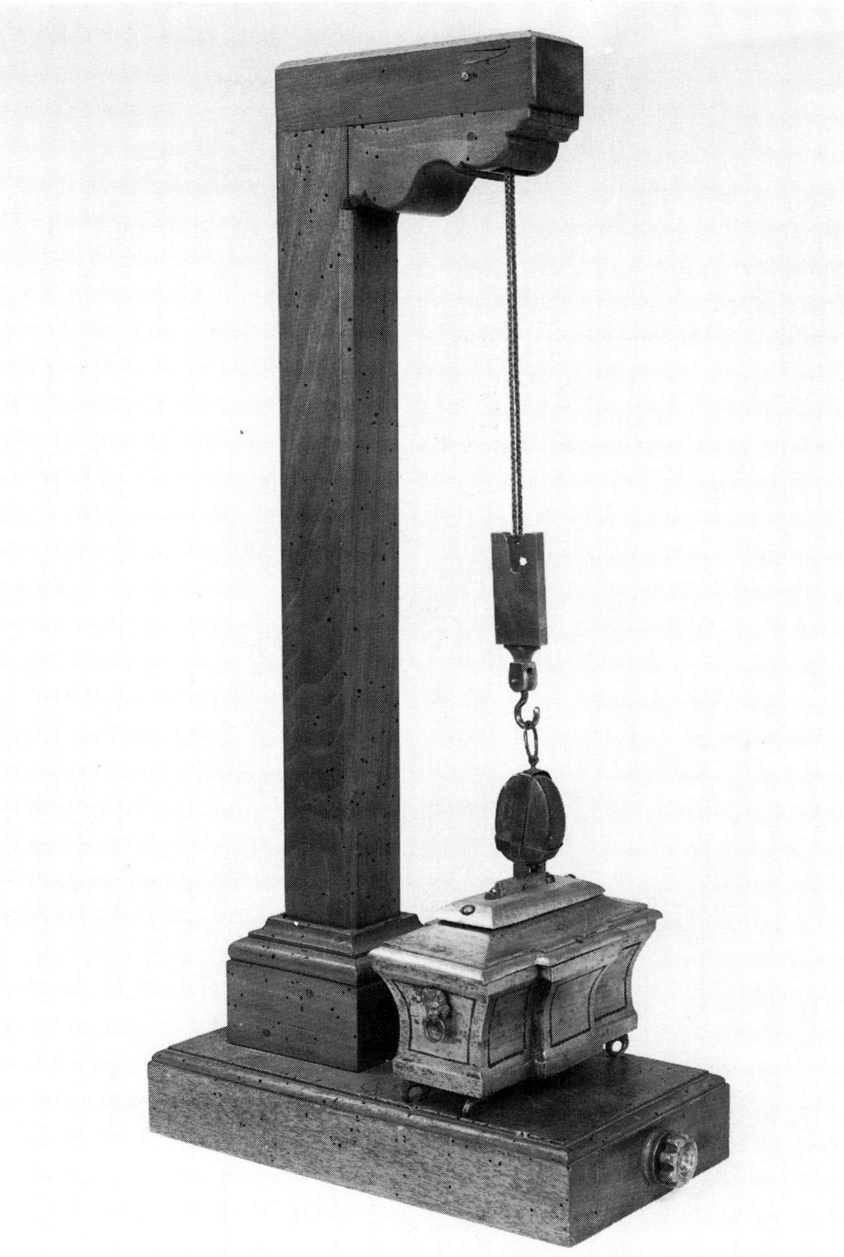

Figure 6. *Galileo's 'Moḥammad's tomb' lodestone,* c. *1608*
(Istituto e Museo di Storia della Scienza, Florence)

Figure 7. Large lodestones in the Uffizi Gallery in 1696

Figure 8. *Parigi's version of Stradanus' 'Flavio' in his study (Room 17, Uffizi Gallery)*

d'Alfonso Parigi in the late 1590s, depicting mathematical instruments, imaginary devices, and working machines.[92] An addition to this room was decorated almost a century later by Filippo Lucci, who included several pieces of electrical apparatus, reflecting the interest in that topic at the time. The differences in the imagery between the old and new decorations illustrate how science had developed during this period.[93]

The mystery of magnetism is represented by Parigi in a scene obviously derived from Stradanus' 'Flavio' engraving [Figure 8]. There are three armillary spheres, two on a shelf above the scholar's head, and another one behind him. Flavio sits at an ornate desk, facing the sea and a ship, and holding the smaller of two horizontally-pivoted needles. The larger one at the edge of his desk is roughly of the same size as the one depicted in Porta's plate. Both needles are pointing towards two lodestones. The smaller of these is floating in a large rectangular trough, and the larger one has been deposited on the shelf on which the name of 'Flavius' has been inscribed. Stradanus depicts no large lodestone in his engraving, so perhaps this one appeared in Florence sometime between the execution of Stradanus' engraving and Parigi's wall-painting, before it was eventually purchased by Cosimo II. It may not be too fanciful to assume that the same large lodestone is illustrated in the Meyer engraving ninety years later, as lodestones of this size were a rarity.

Stradanus' engraving became an enduring symbol of the scientific ideals of the Renaissance. However, I believe that politics, too, played a part in his choice of topic, the prominence he gave to Vespuccio and to Flavio, and indeed to the city of Florence.[94] This could only flatter his Medici patrons, always keen on aggrandisement and on commemorating their achievements. Parigi's image is much cruder than Stradanus', perhaps because of the medium in which he was working, but perhaps also because he was less concerned with scientific 'realism'. He seems to be expressing more the contemporary awe felt for such mysterious phenomena. In part it is the quest for the 'hidden cause' of such mysteries (as Du Bartas writes) that has given us our creative energy:

> I am not ignorant, that he, who seekes
> In *Romane* Robes to sute the *Sagest Greekes*,
> ... Hath sought to show with arguing subtiltie,
> The secret cause of this rare Sympathie.
> But say (*Lucretius*) what's the hidden cause
> That toward the North-Starre still the Needle drawes,
> Whose point is toucht with Load-stone? lose this knot,

---

in levare con la ruota le parti più terree et impure da essa Calamita, et armata poi la medesima d'una cassa d'ottone ridussi in brevi giorni l'istessa calamita tanta forzosa che non solamente tirava libre 25 di ferro, come haveva fatto per avanti, ma bensì fino a libre 55, ch'è più del doppio di quello tirava prima'.

92 G. L'E. Turner, 'Amorini scienziati', *Kos*, i, no. 10 (1984), pp. 17–66. Paolo Galluzzi, *Archimede e la storia delle matematiche nella Galleria degli Uffizi*, Florence: Arnaldo Lombardi, n.d., is based on a television script. See also M. Bacci, 'Le collezioni scientifiche', in G.C. Argan and L. Berti, eds, *Gli Uffizi. Storia e collezioni*, Florence: Giunti Martello, 1983, pp. 244–249, 255.

93 Hackmann *Op. cit.* (n. 35).

94 See, for instance, the prominent space occupied by Florence in pl. 1 (n. 23).

And still-green *Laurel* shall be still thy Lot:
Yet the more learned will I then confesse,
Then *Epicurus*, or *Empedocles*.[95]

Postscript

Paintings and engravings have often been reproduced in histories of science as illustrations subservient to the text, and with little knowledge of, or regard for, the intention of the original artist or illustrator. More recently, however, there has been increasing interest in these images of science, and they are beginning to receive serious study.[96] Scientists have used the skills of the artist as a means of communicating their discoveries.[97] Likewise, artists such as Stradanus, Holbein, Rembrandt, and a host of *genre* painters, have been inspired by science. Their work often reflects popular attitudes towards the science of their day.[98] Hopefully, this short piece will whet the appetite for this type of study.

[95] Snyder *Op. cit.* (n. 1), vol. 1, lines 962–972, p. 197.

[96] H. Michel, *Images des sciences. Les Anciens instruments scientifiques vus par les artistes de leur temps*, Rhode-St-Genèse: Albert de Visser, 1977; Martin Kemp, *The Science of Art. Optical Themes in Western Art from Brunelleschi to Seurat*, Yale University Press, 1990; Svetlana Alpers, *The Art of Describing*, London: Murray in association with the University of Chicago Press, 1983; and most recently John Cage, *Colour and Culture*, London: Thames & Hudson, 1993. To date there is little on the representation of scientific instruments in art, and on the use of instruments as visual metaphors in paintings. On instruments as applied to art, see Henri Michel, *Scientific Instruments in Art and History*, translated by R.E.W. and Francis Maddison, London, 1967.

[97] W.D. Hackmann, 'Natural Philosophy Textbook Illustrations 1600–1800', in R.G. Mazzolini, ed., *Non-Verbal Communications in Science prior to 1900*, Florence: Leo S. Olschki, 1993, pp. 169–196; D.M. Knight, 'Scientific Theory and Visual Language', in Allan Ellenius, ed., *The Natural Sciences and the Arts. Acta Universitatis Upsaliensis*, new series, xxii, 1985, pp. 107–124.

[98] Such studies have to be done with caution and with a knowledge of the methods of art-historical analysis. On this point, see C.R. Hill, 'Scientific Instruments: An Iconographic Note', in R.G. W. Anderson, J.A. Bennett, and W.F. Ryan, eds, *Making Instruments Count. Essays on Historical Scientific Instruments presented to Gerard L'Estrange Turner*, Aldershot: Variorum, 1993, pp. 88–98. We can learn a great deal about the methodology from such studies as Linda Seidel, *Jan van Eyck's Arnolfini Portrait. Stories of an Icon*, Cambridge University Press, 1993.

The Dependence of 'Hi-Tec' Science on Technology through the Rôle of Experimental Technique

Rom Harré

Much of our thinking about science and its relation to technology is determined by the idea that technology is applied science. I want to repudiate that view and defend the idea that science is applied technology. My argument will be directed towards a number of different cases to illustrate ways in which the manipulation of matter under the control of a skill, is prior to a more conventional scientific activity, such as classifying or explaining. I am going to be working with a notion of technology as what is required for the intentional pursuit of a material aim, i.e. some transformation of matter using a material skill. Science I take to be the intentional pursuit of a cognitive aim, the transformation of knowledge using cognitive skills. I want to show that this is an inadequate account of science. If the traditional way of thinking about the relation between technology and science were accepted the account of technology I have given is incomplete. In using material skills we draw upon scientific knowledge. There is supposed to be an input from scientific knowledge into the application of our material skills. That is the traditional view: technology is applied science. It argues that science is essentially autonomous from technology. I am going to try to argue that the matter should be conceived the other way round, that the account of science is inadequate, since I want to show that our capacity to use cognitive skills requires an input from material skills. Rather than having a one-way relationship from science to technology, which is the traditional picture, we should be seeing first of all an input from technology into science, leading to the perception of a dialectical relationship between them. It is a dialectic in which the technological 'end' is always in advance of the scientific. It is not that science fuels technology, it is that technology drives science.

I mentioned in the title the idea of high-technology science; I have to say what I take that to mean. I am going to assume an historical relativist idea of 'High

From *Learning, Language and Invention: Essays presented to Francis Maddison*, ed. W.D. Hackmann and A.J. Turner. Published by Variorum, Ashgate Publishing Ltd, Gower House, Croft Road, Aldershot, Hampshire, GU11 3HR, Great Britain.

Technology'. I choose the phrase 'hi-tec' to draw attention to the rôle of fashion in this matter. I want to indicate 'being on the boundary', on the frontiers of material practicality, so that for me 'hi-tec' is not a description of what we can do now, but rather indicates the boundaries of what was available at any given time. For instance, in the seventeenth century, glass-blowing was the hi-tec skill. Skill as a glass-blower was the essential component in what was possible or impossible in a scientific research programme. In the contemporary world there are other forms of high technology which are 'high' relative to our set of problems. So I mean something about the relationship between real skills and what is currently on offer, in the way of material objects, with which you can do experimentation. A hi-tec science is one in which both research programmes and leading concepts are dependently influenced by the available advanced technology of the era.

Another word about the notion of experiments is needed. A good deal of contemporary philosophy of science takes experimentation as a logical phenomenon: that one way or another it is an expression of the relationship between the general and the particular. An experimental programme is conceived within the framework of a transformation of the results into something which is explicable and logical. Experimentation is left as a neutral tool for producing statements and it is the relation of these statements (data recognition) with other statements (hypotheses) that is crucial.

I want to contrast this traditional idea with the conception of experimentation as a material practice. Experimentation should not be thought of as controlled by the need to realize such logical schemes, but rather by what is available as a material technology to carry out all sorts of different kinds of things. I tried to show in my experiments book[1] that there are not just one or two kinds of things you can do an experiment for; there are lots of possibilities, and fulfilling logical schemata is only one of them. For example there is the exploring of a field trying to find out what phenomena there actually are, say in the programme of Gilbert's *De magnete* (1600).

Then there is the problem of establishing whether a phenomenon is multiform or uniform (e.g. in Newton's optical experiments). There is trying to find out if a certain class of entity exists, and so on and so on. There is an enormous number of schemes to which an experimental programme is germane. I want to emphasize that the limits of the experimentation that is available determines which of these various possible experimentation programmes can be carried out. It is not that a scientific community presents itself with a set of logical problems, says 'Let's test this hypothesis' and then gets on with the experimentation. Rather it has available a set of experiment techniques and then it formulates the kind of research programmes that can be done with the help of that range of experimentation. It is an old platitude now, ten years after Knorr-Cetina's work[2] arguing that scientists' research programmes are more often the reformulation of failed programmes, to present experimentation as the rational pursuit of an experiment result *which you already have.* It is quite frequently the case that scientific

[1] R. Harré, *Varieties of Realism*, Oxford, 1986, chs 13 & 14.

[2] K. Knorr-Cetina, *The Manufacture of Knowledge*, Oxford, 1981.

research starts off by someone applying to some grant-giving body, formulating a research programme. He or she tries to carry out an experimental investigation which fails in various complicated ways, but nevertheless produces some results. Then one reformulates the research programme so the results already achieved are relevant to the new hypotheses.

The Relationship of Available Technology and the Existence of Phenomena

Let us start with Boyle's famous 'Spring of the Air'. We find Boyle and Hooke beset with technical problems. They try to investigate the spring of the air with a glass tube. They are doing it because Boyle is obsessed with theological problems. He has got into this not from physics but from theology. He wants to know if the absence of the vacuum in nature as we observe it is natural or whether it is somehow the result of a divine ukase. Do vacua disappear because air is springy or because of divine decree? So his concern is to solve the question of spring of the air. To do this he builds an air spring. He is beset with technical problems in particular the problem that the tube keeps breaking. In his book he advises experimenters to do such experiments in a big box because of the terrible problem of hunting around the floor for the mercury from broken glassware. As his glass technology improves so the phenomenon becomes more and more clearly defined. He has created the possibility of the phenomenon being represented by the development of his glass apparatus only if the glass is strong enough.

Another example is the vacuum pump. The idea of investigating an electrical discharge in tenuous gases depends crucially on the availability of vacuum pumps. So whether the phenomenon exist in laboratories or not, is not a matter of nature, it is a matter of whether vacuum pumps are able to produce low enough pressures.

These two cases, the air spring and the gas discharge, do have natural analogues. Gas pockets – in minerals of bubbles in volcanic lakes, for example – will obey Boyle's Law. But there are phenomena which exist nowhere but in the apparatus: electrical networks for example. They are only exemplified in the laboratory, and they depend upon the existence of accumulators, wires of constant resistance and so on.

I want to say more about the rôle of technical expertise in research programmes. One can take the story of Boyle's Law further, because at each stage advance and technique are inter-related. Amagat's idea of putting a tube down the shaft of a coal mine had obvious technical limitations. So he tried to construct apparatus in the laboratory to produce very high pressures. At high compression the volume of the molecules themselves become significant and this led Amagat to his formulation of a gas law: $P(v–b) = 7k$. There were all kinds of technical problems that needed to be solved to enable him to carry on a research programme designed to test how far Boyle's Law is universal. That is a simple kind of case and I want to go on to a more complex one.

The Relationship of Available Technology and the Domain of Phenomena

For convenience of reference I will roughly delineate three domains of beings relative to the possibility of human observation. In the first realm are those beings we can observe and identify without the help of sense-extending instruments. This is the domain of actual existence. The second realm we may imagine to contain beings which could exist, they are of plausible natural kinds, but we do not have the sense-extending instruments to reveal them. Many sciences also make reference to a third realm, beings which are beyond all possible experience. The ontological basis of the science is fixed not only by the choice of source models for theories, but also by the available technology and this manifests itself in the way that the boundaries between the three realms change under the influence of the invention of new instruments. These instruments are developed through technical achievements which are independent of the theories they are being used to test. For example, the invention of the bacterial theory of disease can be seen from an ontological point of view as the proposal of a new kind of being in realm two, organisms which are too small to be seen. The development of optical instruments owed something to the geometrical optics of Kepler and others, but a good deal to the practitioners of glass making and the skills of lens grinders. The transformation of the status of the bacterium from an unobservable organic entity to a researchable organism involved a movement of the boundary that separated the kinds typical of the first realm from those typical of the second. Bacteria became observable because microscopes became available. Similarly Kepler's investigations of the laws of luminosity, in particular the relationship we now call the inverse square law, enabled the construction of bolometers, which brought within the realm of observable facts the temperatures of stars.

High Energy Physics in the Technology of Experimentation

The relationship between high energy physics and experimental technique carries the dependence between the technical and the scientific to a very high degree. In order to follow the argument I need to expound some elementary points about the procedures of any example quantum field theory. A Lagrangean equation is written down describing the energy transactions of, let us say, a pair of particles. The final results of mathematical analysis of the interaction of two electrons is a series of amplitude terms, each with a probability, representing the set of possible states of the electrons. I have used, it will be noted, electron talk which already is tending towards a particle conception of the interaction so described, but the temptation to see the interaction in terms of particles must for the moment be suspended. Each amplitude term can be analysed into three components: in the case of a two-particle interaction, two of the components describe what I have just called particles, and the third component an interactional field term which links them. This term is said in quantum field theory to describe an intermediate vector particle. Research programmes designed to test such an analysis involve attempts by high energy physicists to find such 'intermediate vector particles' or at least the free species of a genus of particle which when they appear in interactions are called virtual. My interest is in looking into the question of why we leap so naturally into particle talk, why the concepts of high energy physics are called 'corpuscularian'.

One must remember that throughout contemporary physics the rules of Louis de Broglie allow any expression for a physical property cast in particle form to be transformed into a corresponding expression cast in wave form. So the choice of particle talk is at least in one sense arbitrary. However I want to argue that it is the techniques and technology of the experimental side of high energy physics that determine our choice between the two possible ways of describing physical phenomena. Experimental technology with which we are familiar is essentially a system of apparatus designed to produce *tracks*. There are tracks in cloud chambers, bubble chambers, made on photographic plates and created by a computer reconstruction of the firing of the arrays of detectors in more complex pieces of apparatus. We are invited to conceive of the phenomena in terms of thin longitudinal manifestations curved this way and that by magnetic and electric fields. It is very hard to resist the thought that where there is a track there is a particle. Looked at from this point of view I can express my hypothesis of the priority of technology in this field in the slogan: it is not particles that produce tracks, but tracks that produce particles.

But this is not all there is to it. There is a further, more profound, control, exerted by the kind of instruments we can make and detectors we have opted to create. On the frontiers of physics descriptions of nature characteristically take the form of dispositional statements. That is we ascribe to the underlying natural system that is interacting with our instruments, a disposition to produce the results that instruments detect. In classical physics there is a far going stability throughout the range of instruments we can create, so that it is easy to ignore the rôle of instruments in the formulation of dispositional properties. If essentially the same dispositions were manifested in all instruments, then the presence of the instrument could be conceptually divided out. Mass is a disposition of a material body to resist acceleration, and it does not much matter what particular piece of apparatus that resistance is manifested in.

In the psychology of perception in recent years another kind of dispositional descriptive concept has appeared, which J. J. Gibson called the 'affordance'.[3] An affordance is a disposition in which the manifestation is specifically designed in terms of some human artefact, or human activity or human practice. So, for example, we can speak of scissors 'affording' cutting. Cutting is a human practice which is incorporated in the very concept of a pair of scissors, yet that concept is applied to a metal object which exists in some sense independently of human beings. In Gibson's famous example, some ice in winter 'affords' walking. It is easy to see that the way in which I have emphasized the role of track making equipment might lead one to think of the kinds of descriptions that we make in physics of the unknown basic natural matter to be thought of as affordances. When we analyse an amplitude term and express this analysis in a Feynman diagram [Figure 1] we are choosing a mode of expression which

[3] J. J. Gibson, *The Ecological Approach to Visual Perception*, Boston: Howard Mifilin, 1979.

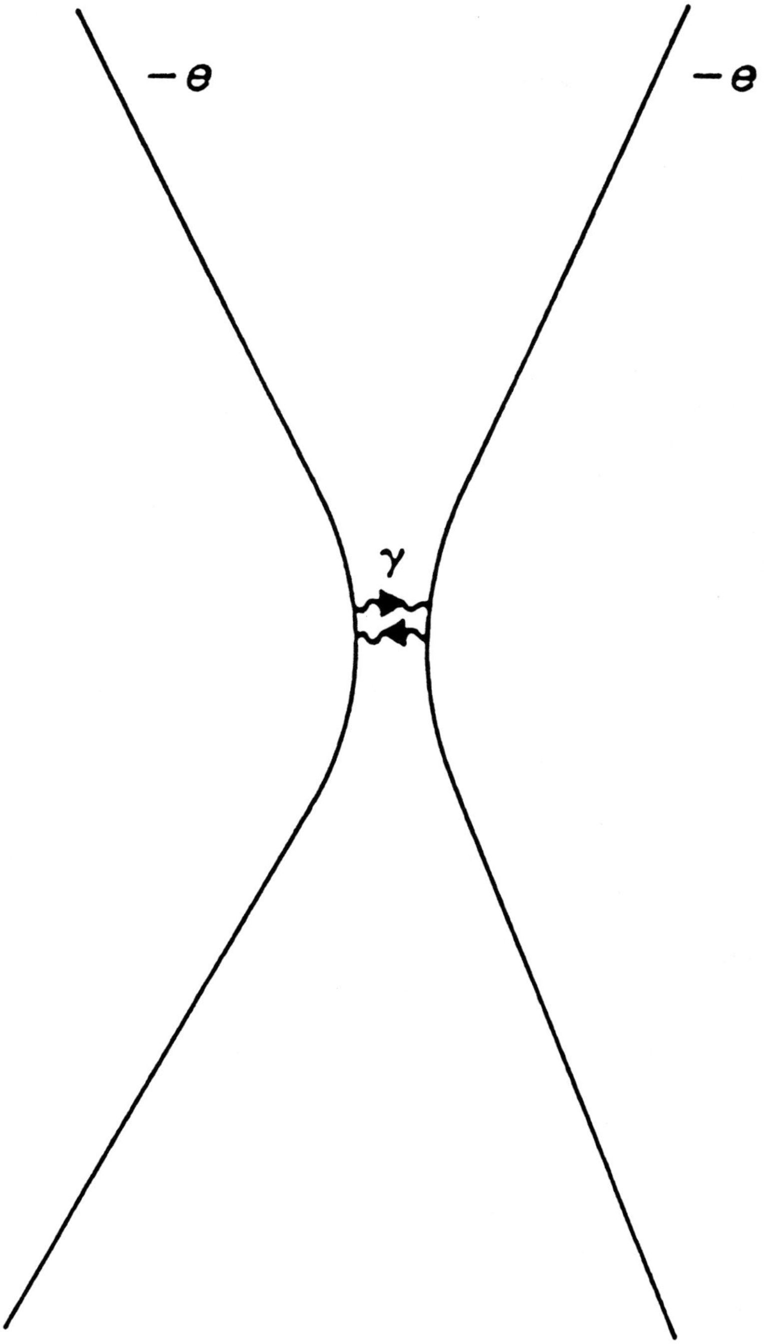

Figure 1. Feynman diagram

is determined by the kind of apparatus which it is convenient for us to construct. To put this colloquially the properties that we ascribe to nature are those affording particles. The Feynman diagram represents in a very schematic way an interpretation of the mathematics in terms of particle affordances. It is the apparatus of track producing which so to say forms the iconic representation of electrons as lines in the figure. Having made it our interpretation of the amplitude term in this way for two components, we are naturally tempted, and rightly so, to treat the third component similarly and to create the virtual particle or intermediate vector boson.

On the basis of this way of parsing amplitudes, quantum field theory led to research programmes for the detection of particles which were determined by our capacity to build instruments which made tracks. So the project of finding intermediate vector bosons became the project of finding the W and Z particles. It is clear, I hope, that it is the availability of the technology of track making that drives conceptual interpretations towards the use of particle concepts, consequently to a certain kind of affordance. According to quantum mechanics, if nature can afford particles it can also afford waves. And I owe to Sarah Foster the thought that there is another kind of apparatus, an apparatus that makes clicks. Just as it is not particles that make tracks, but tracks particles, it is not waves that make clicks, but clicks that make waves. In working out the statistical properties of large numbers of clicks, each representing the collapse of the wave packet at a particular Geiger counter, we get a wave description of nature and a wave affordance is ascribed to the ultimate matter. A question of whether the world consists really of particles or really of waves is quite misplaced; we can construct click-making apparatus and track-making apparatus and the rules of quantum mechanics tell us we cannot use them together. Nature can afford particles and waves, i.e. manifest tracks and clicks; the only common description is that of the mathematical formulation which we broke up in the analysis above that I call 'parsing the amplitude'. Now this is not a counsel of despair, it is simply a recognition of the fact that our interaction with nature is controlled by the availability of technology and it is in terms of that technology that the very concepts that we apply to nature are ultimately determined. Another culture, arriving from another galaxy might very well have developed apparatus of a different kind and the ultimate natural matter might be persuaded to display a range of affordances for them too. The limit of affordances are only the limits of the technology available to us. The history of science shows us that there is no satisfactory way of deciding in advance what those limits might be.

The Basic Dialectical Relation
Technique 1 makes Science 1 possible
Science 1 makes Technique 2 possible
Technique 2 makes Science 2 possible

At each level of techniques, 1, 2 etc. a skill element is also required in order to understand the relationship between the technique and the science. This can be illustrated with the transformation wrought in chemistry by Berzelius' amazing training of his own manual dexterity to enable him to carry out chemical analyses of a precision hitherto undreamt of. The interposition of skill brought into the story of the dialectic relation and its effect is illustrated by the way Berzelius' transformation of chemistry altered the very fundamental structure of chemical theory itself. At that time Prout's hypothesis was still relatively popular: that the relative weights of the chemical elements were integral multiples of the weight of hydrogen and that differences from whole number ratios were to be put down to errors in experimentation. So great was Berzelius' skill that he was able to produce chemical results which convinced the community that Prout's hypothesis must be abandoned. Berzelius' results were awkward but 'correct'. So Berzelius' technique coupled with his extraordinary dexterity made a step in the science of chemistry itself, and it was made wholly in terms of technique. One final qualification of this dialectic table: I owe to Mary Tyles the point that it is not just a science of the previous era that makes the technology of the science of the present era possible. Skill is one additional ingredient, but the level of industrial development of a community is another. For instance the possibility of constructing an apparatus that could contain a magnetic bottle and so could make fusion possible, is determined by the level of industrial skill amongst the engineers who are entrusted with the task of making tokomaks.

In summary then, I have been trying to show that it makes sense to reject the traditional idea of technology as applied science. Rather we should think of science as applied technology.[4] Applied to what? Applied to the task of understanding nature by revealing the kinds of beings of which the natural world is intimately and one hopes ultimately composed. To illustrate this I have taken two kinds of examples. Those in which the apparatus made phenomena possible and those in which techniques not only affect phenomena, but also, and importantly, the concepts by means of which those phenomena are described.

[4] See also a recent article by S. Lelas, 1993, 'Science as Technology', *British Journal for the Philosophy of Science*, xliv, 423–442.

From Philo to Al-Jazarī

D.R. Hill

There are a number of technical treatises from Hellenistic and Islamic times that provide us with the best available information about early mechanical engineering. Of the machines that they describe some – water-clocks, water-raising machines, specialized machines – served utilitarian purposes, while others – trick vessels, automata, fountains – were designed to give amusement and aesthetic pleasure to the members of courtly circles. In general it is convenient to use the term 'fine technology' for the machines described in the treatises. (The expression also embraces instrumentation, but this subject is not included in the scope of the present article.) The range of mechanisms, components and techniques incorporated in the treatises is wide. Much of this array was to enter the vocabulary of European engineering and to be of importance in the genesis of the Industrial Revolution. Moreover, the best of the treatises provide us with information about techniques of manufacture and construction that are to be found in no other pre-Renaissance sources.[1]

The pre-history of fine technology begins with such devices as the simple outflow clepsydra, which was known in Egypt and Babylonia before 1500 BC An important development was the inflow clepsydra, whose time and date of origin are unknown. It may have been a Greek invention. At first this consisted of two vessels, an outflow clepsydra that discharged into a receiver having an hour-scale marked on its interior. Clearly, this was no more accurate than the simple outflow type, but the next innovation made it a much better timekeeper. The reservoir vessel was provided with an overflow pipe and a constant supply of water, flowing in at a greater rate than the outflow from the orifice. A constant head was therefore maintained in the reservoir.

[1] Donald Hill, 'Arabic Mechanical Engineering: Survey of the Historical Sources', *Arabic sciences and philosophy*, i, 1991, pp. 167–186.

From *Learning, Language and Invention: Essays presented to Francis Maddison*, ed. W.D. Hackmann and A.J. Turner. Published by Variorum, Ashgate Publishing Ltd, Gower House, Croft Road, Aldershot, Hampshire, GU11 3HR, Great Britain.

The water supply could be from a large vessel, a stream or the city mains.[2] Both Plato (d. 347) and Aristotle (d. 322) are said to have had 'alarm clocks'.[3]

The earliest types of automata were probably the articulated figures such as those from the Egyptian tombs from the Twelfth Dynasty onwards. The next stage of development is also found in ancient Egypt: talking statues worked by means of a speaking trumpet concealed in hollows leading down from the mouths. The beginnings of sophisticated, mechanized constructions, however, are to be found in Hellenistic Egypt.[4]

The earliest Hellenistic engineer about whom we have any information is Ctesibius, who is known only by passages from the writings of Vitruvius.[5] According to Vitruvius he constructed two inflow clepsydras, in one of which a float in the receiver rose at constant speed. A rod soldered to the top of the float carried a figure whose finger moved over an hour-scale. In the other clock the rod was toothed and engaged a pinion which transmitted the movement to various automata. Ctesibius is also credited with the invention of the water-organ and the force-pump.[6] Two celestial globes are said to have been constructed by Archimedes.[7] There is also a work attributed to him on the construction of water-clocks, which will be discussed in connection with monumental Islamic water-clocks. The bulk of our information on fine technology, however, comes from four sources, two Greek and two Arabic. These, in their modern editions, are:

Philo of Byzantium, *c.* 200 BC, *Le Livre des appareils pneumatiques et des machines hydrauliques par Philon de Byzance,* Arabic text with French trans. by Carra de Vaux, *Paris Académie des Inscriptions et Belles Lettres,* 38, 1903, part 1.

Hero of Alexandris, fl. mid-first century AD, *Heron Alexandrinis Opera,* vol. 1, *Pneumatica and Automata,* edited Greek text with German trans., G. Schmidt, Leipzig, 1899.

Banū Mūsà, *The Book of Ingenious Devices,* trans. and annotated by Donald R. Hill, Reidel, Dordrecht, 1979.

Banū Mūsà, *Kitāb al-ḥiyal,* Arabic text of the above, ed. A. Y. Hassan, Institute for the History of Arabic Science, Aleppo University, 1981.

---

2 Donald Hill, *A History of Engineering in Classical and Medieval Times,* London, 1984, p. 225.

3 A. J. Turner, *The Time Museum, Catalogue of the Collection,* vol. 1, *Time Measuring Instruments,* Part 3, 'Water-clocks, Sand-glasses, Fire-clocks', Rockford, Ill., 1984, p. 6.

4 One is preserved in the Louvre, one, a large limestone bust of the god Re-Harmarkhis of Lower Egypt, is in the Cairo Museum. Jointed and talking figures were not confined to Egypt, but probably occurred in early civilizations and are widespread. See Derek de Solla Price, *Science since Babylon,* New Haven and London, 1975, pp. 50–51.

5 Vitruvius, *De Architectura,* trans. F. Granger, London, 1934, vol. 2, pp. 255–267.

6 *Ibid.*, pp. 311–314.

7 Emilie Savage-Smith, *Islamicate Celestial Globes,* Washington, 1985, p. 5.

Al-Jazarī, ibn al-Razzāz, *The Book of Knowledge of Ingenious Mechanical Devices*, annotated and trans. by Donald R. Hill, Reidel, Dordrecht, 1974

Al-jami' bain al-'ilm wal-'amal al-nāfi' fi ṣinā' at al-ḥiyal, Arabic text of the above, ed. A. Y. Hassan, Institute for the History of Arabic Science, Aleppo University, 1979.

These are the only major treatises on fine technology to have come down to us.[8] It might be argued that there are other works of comparable importance awaiting discovery, but if such works had been written they would surely have been mentioned by other writers, particularly scientists and the compilers of biographical dictionaries. But no other comparable works are referred to. Philo is mentioned by Vitruvius[9] and by Hero.[10] The *Pneumatics* exists only in an Arabic translation probably dating from the ninth century. None of Philo's treatises is known to have circulated in Europe in the Middle Ages, but in 1615 Solomon de Caus listed him among his predecessors.[11]

Hero was also known to the Arabs.[12] His *Mechanics* has survived in an Arabic translation made by Qūsta b. Lūgā (d. 912). But although some of his works were known in the Middle Ages only one Greek manuscript from the late fifteenth century is known to have survived. From the sixteenth century onwards, however, nearly one hundred copies of the *Pneumatics* are extant. In most of these the *Automata* is also included. Hero's influence on Renaissance writers on mechanics, hydraulics and pneumatics was considerable.[13]

All the major Arabic biographical dictionaries dealing with scholars, from Ibn al-Nadīm (d. 995) onwards, devote entries to the Banū Mūsà and to no other writers on machines.[14] The great historian Ibn Khaldūn (d. 1406) refers to their machine book in glowing terms,[15] and al-Jazarī mentions them as the most illustrious of his predecessors in the field of fine technology.

Al-Jazarī lived too late to be included in the *Fihrist* of Ibn al-Nadīm and later biographers do not mention him, either because they usually copied from the *Fihrist* or because his reputation did not spread beyond the limits of the Jazīra until some time after his death. He is, however, mentioned by the Egyptian scholar al-Qalqashandī

[8] This does not preclude the possibility of shorter works having been written. There are, for example, short works on musical automata, extant in Arabic but probably from Greek originals. There is also a chapter on perpetual motion machines to be found in several Arabic manuscripts. See note 1 above.

[9] Vitruvius, *De Architectura* (n. 5), vol. 2, p. 75.

[10] *Heron Alexandrinis Opera*, vol. 1, *Pneumatica and Automata*, edited Greek text with German trans., G. Schmidt, Leipzig, 1899, pp. 405–407.

[11] The following works are valuable for a study of the Hellenistic engineers: A.G. Drachmann, *The Mechanical Technology of Greek and Roman Antiquity*, Madison, Wisconsin and Hafner, London, 1963; by the same author, 'Ktesibos, Philon and Heron: a Study in Ancient Pneumatics', *Acta historica scientiarum et medicinalium*, ed. Biblioteca Universitatis Hauensis, Copenhagen, 4, 1948, pp. 1–107; Albert de Rochas, *La Science des philosophes et l'art des thaumaturges dans l'Antiquité*, 2nd edn, Paris, n.d.,. 1930.

[12] E. Wiedemann, *Aufsätze zur arabischen Wissenschafisgeschichte*, Hildesheim, 1970, vol. 1, pp. 69–70.

[13] Turner *Op. cit.* (n. 3), pp. 31–33.

[14] The Banū Mūsà, trans. Hill, p. 3.

[15] Ibn Khaldūn, *Muqaddimah*, trans. F. Rosenthal, 2nd edn, London, 1967, vol. 3, p. 132.

(d. 1418) as pre-eminent in the science of machines,[16] a pre-eminence that is emphasized by his having been given the soubriquet *Badī' al-Zamān* ('The wonder of the age'). At least 17 manuscripts of his work, from the thirteenth century to the eighteenth, also testify to the esteem in which he was held in the world of Islam.

There were other shorter treatises, some of which embody ideas that are not found elsewhere. These will be referred to as occasions arise, to ensure that this survey includes all the important techniques, but in general attention will be concentrated on the works of the four major writers.

Certain techniques were common to all writers on fine technology. Above all, nearly all their devices rely upon water or other liquids as their prime movers. The water might be used directly so that its pressure, or flow, or both, produced the desired results. Very frequently, however, the water was used to produce pneumatic results – it is significant that the main treatises of Philo and Hero are called *Pneumatica.* Pneumatic effects are caused by differences in aerostatic pressure, usually when the pressure of the atmosphere (1 atm = 101.325 kN/m^2 or 14.7 lbs/in^2) is greater than that in an adjacent container. For example, if a pitcher with a narrow spout is filled with liquid and then made airtight, nothing can be poured from the spout once a tiny drop has emerged, reducing the internal pressure to slightly less than 1 atm. The pressure of the surrounding air prevents any further discharge. If, however, there is a concealed hole in the handle leading to the body of the vessel, then once the hole is uncovered the internal and external pressures are equal and outpouring can occur. Siphons work on the same principle. When water reaches the top of the bend, the pressure there drops below atmospheric. Provided the outer leg is longer than the inner, discharge will occur until the water drops below the opening of the inner leg. Another device that makes use of the pneumatic effect is the occluded orifice. The orifice is let into an otherwise airtight vessel. The hydraulic system can be so arranged that the orifice is opened and closed at very short intervals, so producing a kind of feedback control.

All the early writers believed that pneumatic effects were caused by the impossibility of a vacuum, but this did not prevent them from using these effects in a variety of sophisticated ways. Provided the desired results could be produced by the engineer, using his empirical knowledge, the actual final velocity of the fluids was unimportant. It was not usually necessary to be able to quantify pneumatic phenomena. With notable exceptions, which will be discussed in the appropriate places, liquid flows were slow.

The liquid flows were themselves often the end-product of a particular device. For example three differently coloured liquids might be poured into a vessel and issue from its tap in the reverse order from which they were poured in.

The general formula for flow of liquid in pipes is known as 'Bernoulli's Equation' and is given by:

[16] Al-Qalqashandī, *Subḥ al-a'sha,* ed. M. A. R. Ibrahim, 14 vols, Cairo, 1913–20, vol. 1, p. 477.

$$\frac{p}{w} + \frac{V^2}{2}g + h = A\ constant$$

where p = pressure, w = specific (unit) weight, V = velocity, g = the gravitational constant and h = height. The early engineers did not, of course, know this equation, nor did they understand the concepts of pressure and gravitational constant. They knew that water under gravity flows downwards and this knowledge was sufficient in the many cases where the rate of flow was unimportant. When, as with water-clocks, a precise rate of flow had to be maintained, empirical methods were resorted to.

The following basic hydraulic components were used by all those concerned with fine technology, from Philo onwards. When each writer is discussed the additional components and techniques that he introduced will be described separately.

Pipes	usually made by bending copper sheet round a wooden former and soldering along the seams. Pipes were occasionally made from brass or lead.
Siphons	bent-tube and concentric types.
Pipettes	
Vessels	jars, pitchers, cups, basins.
Funnels	
Orifices	plain and occluded.
Partitioned vessels	
Taps	with single and multiple borings.
Clack valves	
Floats	
Human and animal figures	

Philo of Byzantium

Nothing is known of Philo's life. By his own account he wrote a large number of works dealing with different branches of mechanics, of which only two have survived: two books of a compilation on war machines called *Belopoiika*[17] and the *Pneumatics.* In the former he mentions, among other works he had written, treatises on levers and on the void. Philo is mentioned by Vitruvius[18] as having written on machinery. Hero criticizes Philo's work in his *Automata*[19] saying that his constructions were too cumbersome. This is the only evidence we have of Philo having written about automata.

[17] Carra de Vaux, p. 37.

[18] See n. 9 above.

[19] See n. 10 above.

Philo was known to the Arabs. He is mentioned, for example, by the Egyptian historian al-Kindī (d. 961), who says that he built water-wheels, mills and ingenious devices.[20] He was, however, known earlier than this. His *Pneumatics* was translated into Arabic, probably in the ninth century. This text was edited and translated into French by Carra de Vaux, using MS. 3713 of Santa Sophia as the foundation of the work, with other Arabic texts and a Latin fragment as supporting documents.

It should be mentioned that the drawings which accompany the text in the version of Carra de Vaux are somewhat unsatisfactory. They are derived from the text and the original drawings, but appear as Renaissance-style illustrations in perspective, with the internal parts shown. This sometimes makes the exact layout of the equipment difficult to follow, although in general these drawings are a useful adjunct to the text. The Greek letters are marked on the various components in accordance with Philo's instructions. Nevertheless, it would have been preferable if the original drawings had also been shown. After all, the adapted drawings are themselves translations.

There are 65 devices described in the *Pneumatics.* The first 16 of these are demonstrations of the principles of pneumatics as understood at the time. Examples include siphons – bent-tube and concentric – and pipettes. Although the theory expounded by Philo is faulty, his descriptions of the operation of the components are correct. Of the 65 devices, 41 are operated by pneumatic means, two by steam and the remainder by hydraulic/mechanical power. All except the last five devices can be classified as trick vessels of various kinds, some of them including human or animal figures, or both.

A quadruple lamp that replenishes itself is fairly typical of the pneumatic devices (No. 20). Four lamps are grouped symmetrically around an air tight vessel filled with oil. A pipe runs vertically up the centre of this vessel and from its protruding lower section a pipe is led out into one of the lamps – say, A. Pipes lead out from the centre of the reservoir, one over the centre of each lamp. When the lamps are lit and the fire consumes the oil, and the end of the pipe in A is uncovered, oil then runs into all the lamps from the reservoir until the pipe in lamp A is again covered – and so on. This is an example of feedback control using an occluded orifice.

The hydraulic devices include a pitcher from which only a measured quantity can be drawn each time the tap is opened (No. 28). There is also an automatic lavabo (No. 31) which Philo says is 'like the clocks'. The machine incorporates a tipping-spoon, which is clearly a precursor of the tipping-buckets used to such effect by al-Jazarī.

Heat was used to produce steam in only two of Philo's models, Nos 57 and 58. The first directs steam on to burning incense to produce a draught. The second has steam forced through pipes in the bodies of model birds in order to simulate a whistling sound.

There are several taps with two or more borings. Using such a component attached to a vessel divided into two compartments, turning the tap to one side produces water, turning it to the other side produces wine. Another tap, more complex than this, is

[20] Wiedemann, *Aufsätze* (n. 12), vol. 1, pp. 71–73.

able to throw up a vertical jet of water that turns a small paddle-wheel located in the extended shank of the tap. The vocabulary is the most interesting part of this chapter. Philo calls the wheel a *nā'ūra*, i.e. a noria, and later says that the jet rotates the mill (*raḥā*). This is a clear indication that water-mills were in use, at the latest, by 200 BC.

The final five of Philo's devices are water-wheels. The first is a tympanum in reverse; when water is directed into its compartments it whistles (No. 61). No. 62 is a miniature noria with a similar purpose. No. 63 is also a small noria, this time intended to produce a stream of water for ablutions. No. 64 is a manually operated suction pump and the last machine is a water driven *sāqiya*. There has been considerable discussion as to the possibility of these five chapters being later interpolations, but there seem to be no valid reasons for excluding them from Philo's own work.

Later Philo's work can be evaluated in the light of the writings of his successors. For the moment it is worthy of note that he used no transmission systems – neither gears nor pulleys.

Hero's works on mathematics, mechanics, war machines and surveying are not within the scope of the present chapter, which is concerned only with the *Pneumatics* and the *Automata*. The *Pneumatics* may originally have been intended as notes for pupils, since after the initial didactic chapters there is no order and the chapters appear haphazardly. Nevertheless, the chapters themselves are not haphazard. All are written in the same concise style and according to a fixed plan, beginning with a description of the apparatus, then a description of how it works and finally an explanation (if necessary).[21] They are very clear and it might be possible from the descriptions to reconstruct the devices. To attempt this, however, one would be severely hampered by the lack of dimensions.

The *Pneumatics* is divided into two books, containing in all 80 devices. Many of these are copied more or less exactly from similar constructions in Philo's work. It will therefore be profitable to discuss only those devices and components in Hero's work that are not found in Philo's.

After describing the bent-tube and concentric siphons, Hero describes (I 4 & 5) a siphon that passes through a float to which it is soldered with watertight joints. As the water-level falls or rises the discharge from the siphon remains unaltered, since the distance from the water-level to the end of the longer leg of the siphon is constant [Figure 4]. Three double siphons are described in II 14. These have a bent-tube siphon over the top of the tube and an inverted concentric siphon at the bottom. These siphons were used to produce artificial airlocks. Hero simply says that they could be used in a variety of applications, but gives no examples. As will be seen, double concentric siphons were frequently used to great effect by the Banū Mūsà.

In I11 the manufacture of a clack-valve is described. Philo used the same component but does not describe it, presumably assuming that his readers were familiar with its construction.

[21] A. G. Drachmann, *Hero of Alexandria*, in *Dictionary of Scientific Biography*, New York, 1972, vol. 6, pp. 310–313.

In I 16 pulley systems are described for the first time in the *Pneumatics*; far more complex systems are described in the *Automata* (see below).

Hero described the use of gears in his *Barulkos*, a machine describing how a load of 1000 talents can be lifted by a force of five talents, that is, the power of a single man. The machine consists of a number of parallel meshing gears, but it is only a theoretical solution; parallel-toothed wheels were not known to have been used in antiquity for the transmission of heavy torque.[22] In the *Pneumatics* (II 32) Hero has a pair of gears meshing at right angles.

Model I 41 contains a large conical valve. This is a 'bath-plug' type of valve that is lifted in and out of its seat by a chain. This is the only example in Hero's work of this important component.

Of the complete machines described by Hero, the force-pump (I 28) and the water-organ (I 42) were probably copied from a lost work of Ctesibius.[23] The pump is a twin-cylinder, manually operated machine. The cylinders are vertical and stand directly in the water; there are plate-valves on the entry ports.

The water-organ is supplied with air from a pump from which the air is forced into a vessel containing water. From the vessel a pipe leads vertically into an air-chest. There are eight pipes fitted vertically into the air-chest; these can be opened and closed from a keyboard. A second organ is wind operated. A vertical wind-wheel (*anemouria*) with paddle-like vanes is mounted on an axle which carries four cams at 90° to one another. As the *anemouria* is turned by the wind the cams bear down in turn on one end of a rocking arm. This is pin-jointed at the other end to the connecting rod of a vertical piston that moves in a cylinder. A pipe leads from the underside of the cylinder, goes horizontally through a plinth then turns vertically and enters the air-chest which rests on top of the plinth. Eight vertical pipes are fitted to the air-chest. There is no keyboard (I 43).

In models I 38 and 39 temple doors are made to open by the action of heat. In I 38 for example, the temple doors rest on a hollow plinth, into which their hinges are extended as circular axles, resting in bearings at the bottom of the plinth. A pulley system is rigged around these axles: on one end of it is a counterweight, on the other end a container is suspended. A bent tube connects this container to a sphere half filled with water. A pipe leads from the sphere vertically into an altar. All joints are leak-proof. When a fire is lit on the altar the water in the sphere is heated and some of it flows through the bent tube into the vessel, which becomes heavier than the counterweight. The pulley system therefore opens the doors. If the fire is allowed to go out the water returns to the globe and the doors close.

There are several lamps in the *Pneumatics.* A self-trimming lamp is described in I 34. As the level of the oil in the lamps falls a pair of pinions are turned and one of these engages a toothed, curved rack on the end of which is the wick. The wick is

---

[22] *Ibid.*, p. 11.

[23] See nn. 5, 6 and 10 above. Vitrivius' mentions of the writings of Ctesibius in general and of the organ and the force-pump in particular lend credence to the likelihood of Hero having known these machines.

therefore gradually extruded. Three other lamps are described (II 22, 23, 24), all of which can be replenished without the intervention of an operator.

Of those devices of Hero's that use heat; the small steam-driven machine called the Äolipile has always intrigued historians of technology. On top of a boiler with a flat lid two arms in the shape of inverted L's were erected. A metal sphere was suspended between these. One of them turned in a bearing fixed to the side of the sphere, but the other was a pipe that entered the sphere and the boiler. Protruding from the sphere were two L-shaped pipes, the two outer legs facing in opposite directions. The fire was lit under the boiler. When the water boiled the steam entered the sphere and emerged from the two pipes, causing a turning moment that made the sphere revolve. Although this is a toy steam turbine, it did not, of course, inspire the invention of a practical machine. Nor was this to be expected: the other techniques and materials needed to construct a working steam engine lay far in the future.

The *Automata* has attracted far less attention than it deserves. Whereas Hero's *Pneumatics* is in a direct line of descent from Philo's *Pneumatics* to the Banū Mūsà's *Book of Ingenious Devices*, the *Automata* is in a category of its own. Admittedly, Hero speaks of the ancients having made automatic theatres and indeed he criticizes Philo's designs as being too cumbersome and as having omitted important features,[24] but we have no survivals of any writings on automatic theatres before Hero. In any case, it would seem that his designs were considerably in advance of those of his predecessors. Apart from the intrinsic interest of the constructions themselves, Hero's introductory remarks are noteworthy, since they give an insight into the motives of those who made automata. 'The automatic theatre' says Hero 'enjoyed great popularity among the ancients [*proteros*], first because a variety of skills were developed from it, and secondly because the action of the play was a cause of wonder'. Also, he continues, the manufacture of automata demands a knowledge of every branch of mechanics. It does not appear that any further enquiry into the purposes of fine technology needs to be added to these remarks of Hero.

There were two types of automatic theatre: moving and static. In the first type the whole theatre moved on wheels in front of the audience, various actions were performed, and the theatre then returned to its original position. In the static type the theatre itself did not move, but the scenes were somewhat more complicated than in the moving type.

The moving theatre in Hero's version consisted of a plinth about 46 cm long by 31 cm wide by 23 cm high. The wheels were almost hidden at the base of this plinth. At each corner of the plinth was a pilaster about 62 cm high and 15.5 cm wide, supporting an architrave some 7.7 cm high. (It is unusual for Hero to give measurements. He did so in this case, he says, so that onlookers could see that the theatre was too small to conceal a human operator!) The architrave was decked over and on the deck was a little circular temple, with six columns and a dome surmounted by a spike. On the spike stood Nike with outspread wings, holding a wreath. A Bacchus-figure

[24] See n. 10 above.

stood in the centre of the temple holding a thyrsus in the left hand and a goblet in the right. At his feet sat a small panther and close by was an altar. At each column stood a Bacchante.

When the theatre stopped the altar fire ignited, milk or water flowed from Bacchus' Thyrsus, wine flowed from the goblet, the Bacchantes danced, drum rolls and cymbal strikes were heard. When the noise died down both the Nike and Bacchus turned outwards. All these actions were repeated, then the theatre returned to its starting point.

The motor for all the movements consisted of a heavy weight in a circular chamber filled with millet or mustard seed. The motor was started by pulling a string that moved a shutter from an orifice at the bottom of the chamber, whereupon the seed began to trickle into a container below the weight chamber and the weight began gradually to descend. Cords from the top of the weight led to pulley systems that caused the movements of the theatre and the actions of the automata.

These systems were very ingenious. To take but one example, the forward and reverse movements of the theatre. Between the two front wheels of the theatre's plinth was a wooden cylinder, i.e. a very thick axle. A cord from the weight passed over a pulley then down to this cylinder. A length of this cord was wound around the cylinder in one direction, then a slack section was tacked to the cylinder with some wax, then the rest of the cord was coiled round the cylinder in the reverse direction. The theatre moved forward as the cord uncoiled, remained static as the slack was taken up, then went into reverse when the cord uncoiled in the opposite direction. More complex actions are described: for instance Hero tells us how the theatre can be made to move in a circle or round the sides of a rectangle.

The static theatre was similar in many ways. Dry sand instead of seeds was used, since lightness was unimportant. The movements of the actors were more prolonged and varied than in the static version.

Hero's specifications are a curious mixture of the meticulous and the sketchy. He says that the surface upon which the theatre rolls must be perfectly smooth and level. He gives very careful instructions for preparing the materials. The cords must be pre-stretched so that they do not stretch further while in use. The wheels and axles are to be carefully made and balanced and they are to be oiled before they are used. *This is the only mention of lubrication in any of the treatises on fine technology.*

On the other hand, apart from the overall measurements of the theatre, no dimensions at all are given. There are also no clear directions as to how the various parts are to be assembled. We are not told, for example, how the weight-chamber is located in relation to all the moving parts. It would therefore be very difficult to make a reconstruction of the theatre. It would seem that for a complex construction the early engineers found it difficult to give clear instructions for manufacture. Clarity was not assisted by the rudimentary state of technical drawing. Even so, the *Automata* is a remarkable achievement. It is doubtful whether there is a more sophisticated use of pulleys before modern times. And of course the moving theatre was the first 'horseless carriage', albeit a miniature one.

With the Banū Mūsà we return to the mainstream of fine technology, but with important innovations. These three brothers were named, in order of seniority, Muḥammad, Aḥmad and al-Ḥasan. Their father, Mūsà b. Shākir, was a noted astronomer who was a close companion of the future Caliph al-Ma'mūn when the latter was living in Khurasan. After Mūsà's death the brothers were entrusted to the care of al-Ma'mūn (d. 833) and became members of the courtly circles of this Caliph and his successors in Baghdad. They took part in the turbulent palace politics of the time, undertook major public works, and extended their patronage to a number of eminent scientists and translators. They also made journeys to Byzantium and were instrumental in bringing back to Baghdad copies of scientific works, which were then translated from Greek or Syriac into Arabic. They were also eminent scientists in their own right and composed a number of works on astronomy and mathematics, one of which, *On the Measurement of Plane and Spherical Figures*, was translated into Latin by Gerard of Cremona in the twelfth century. There are entries on the Banū Mūsà in all the major Arabic biographical dictionaries and their political activities are mentioned by the great historian al-Ṭabarī (d. 923), who was alive when they were prominent in public affairs.[25] Apart from the mathematical work, only two of the Banū Mūsà's works have survived. One of these is a short treatise on a musical automaton, describing a very advanced machine that embodies a number of delicate mechanisms and control systems. Unfortunately, space does not permit any discussion of this machine here. It can be studied in two scholarly publications, one an edition of the Arabic text, the other an English translation.[26]

Their major surviving work is *The Book of Ingenious Devices*, containing descriptions of 100 devices. It was probably mainly the work of Aḥmad and it was widely appreciated in medieval Islam. Ibn Khaldūn says of it: 'There exists a book that mentions every astonishing remarkable and nice mechanical contrivance. It is often difficult to understand, because the geometrical proofs occurring in it are difficult. People have copies of it. They ascribe it to the Banū Shākir'.[27]

The Banū Mūsà, unlike Philo and Hero, give no exposition of the general principles of pneumatics, nor do they describe a number of experiments that demonstrate those principles. The first model describes a concentric siphon, but thereafter there are descriptions of various devices, each accompanied by an explanation and an illustration. The illustrations are sectionalized elevants of the complete devices and are in general a useful aid to understanding the construction and operation of the devices. As with their Greek predecessors, the various components on the illustrations are marked with identifying letters and these are referred to in the text.

Models 2–74 are trick vessels producing a bewildering variety of effects. To take but three:

---

[25] The Banū Mūsà, trans. Hill, pp. 3–6.

[26] Arabic text, ed. L. Cheikho, *al-Machriq*, ix, Beirut, 1906, p. 9; English trans. by H. G. Farmer, The *Organ of the Ancients*, London, 1931, pp. 80–85.

[27] See n. 12 above.

Model 17 A boiler with an outlet tap. Hot water issues from the tap only so long as cold water is poured in at the top of the boiler.

Model 43 A jar with a tap, into which three different liquids can be poured without mixing. When the tap is opened the liquids discharge in the sequence in which they were poured in.

Model 62 A jar with an outlet pipe from which, after the jar has been filled with wine, wine discharges as long as water is poured in; then water discharges as long as wine is poured in.

Models 74 to 87 are self-replenishing vessels.
Models 88 to 94 are alternating fountains.
Model 99 is a bellows for removing foul air from wells.
Model 100 is a small clamshell grab.

It would clearly be impossible to describe the construction of even a selection of the trick vessels in detail. Instead, it may give some insight into the methods of the Banū Mūsà if the components that they used frequently are described. In the first place, of course, they used components that appear in the *Pneumatics* of Philo and Hero. These include the bent-tube and concentric siphons [Figures 1 and 2] (not the floating siphon, which is unique to Hero), occluded orifices, human and animal figures, pipes and vessels of various kinds. The new components are shown in Figures 3, 5, 6, 7, 8.

Figure 3 shows a component very frequently used by Banū Mūsà, a double concentric siphon. Apart from a rudimentary drawing in Hero's *Pneumatics*, this device appears in no other treatise on fine technology. It acts like a switch. When liquid is introduced it behaves like a normal siphon. When filling is interrupted the level in the upper tank sinks to position A and some air enters the upper siphon. A state of equilibrium is reached such that $x = y$. Inpouring cannot be resumed through this siphon unless $H > a + c$. When inpouring is restarted the liquid will take an alternative route, through another siphon or a pipe, if such a route has been provided.

Figure 5 shows a simple but effective way of separating two liquids when they are poured into a vessel. The first liquid (say) is poured in forcefully and runs from the bent end of the funnel into tank c. The second liquid is poured in gently and trickles into tank d. The liquids can be evacuated from the tanks through valves, siphons or by some other means.

Figures 6 to 8 show different applications of a component that was used frequently and with great confidence by the Banū Mūsà – the conical valve. This type of valve occurs in the pseudo-Archimedes treatise as part of the feedback control system in a water-clock and it was used for the same purpose by Riḍwān and al-Jazarī (see below). Nowhere else, however, do we find it used so effectively and in such a wide variety of applications. The valves were made of cast bronze: first the plug and the seat were cast together in the same mould, then ground together to a watertight fit with emery powder. They were used by the Banū Mūsà mainly as switching devices. In Figure 6 water runs into tank k, in which there is a float a. This float is connected by rods, as

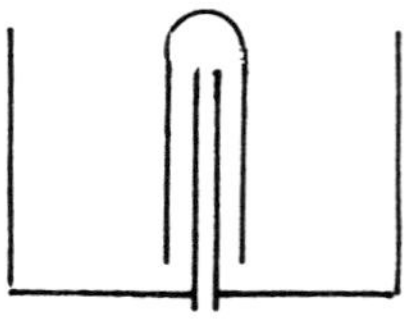

Figure 1. *Concentric siphon*

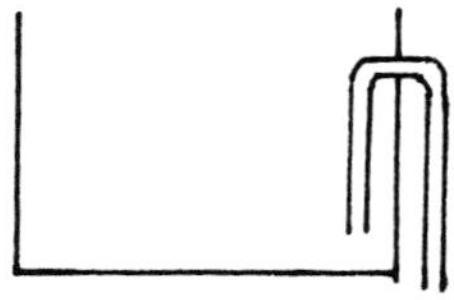

Figure 2. *Bent-tube siphon*

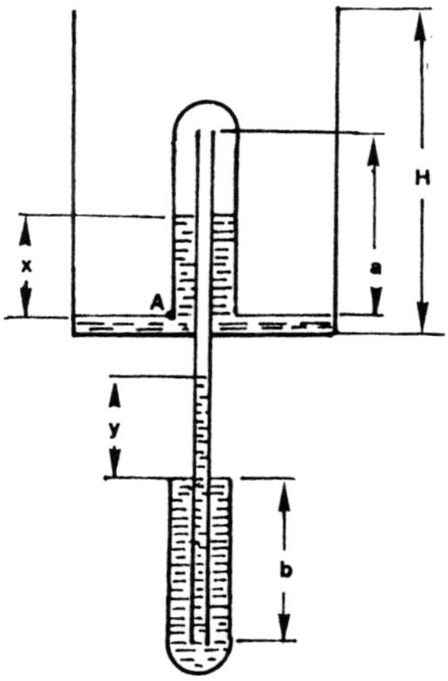

Figure 3. *Double concentric siphon*

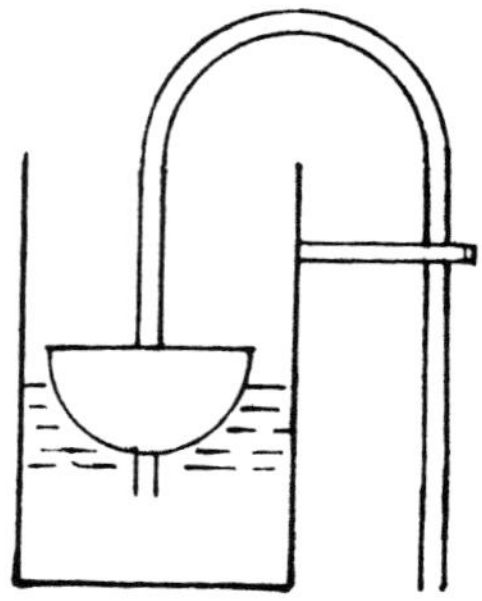

Figure 4. *Floating siphon*

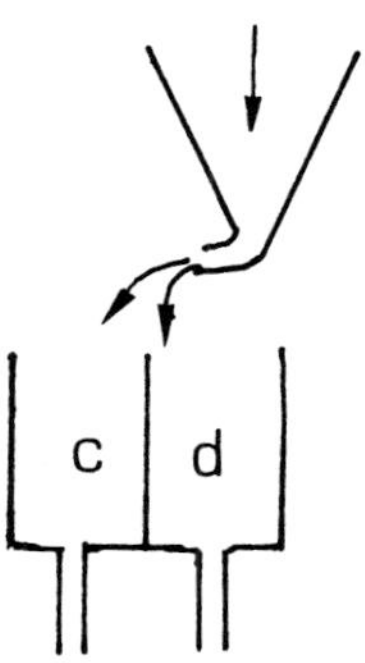

Figure 5. *Method of separating two liquids*

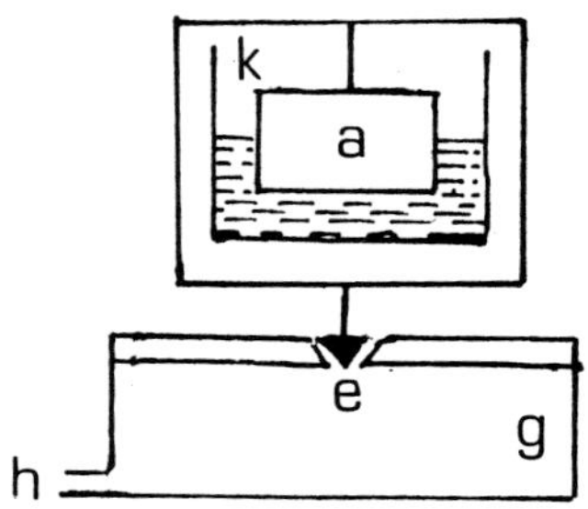

Figure 6. *Hydraulic switches for conical valves*

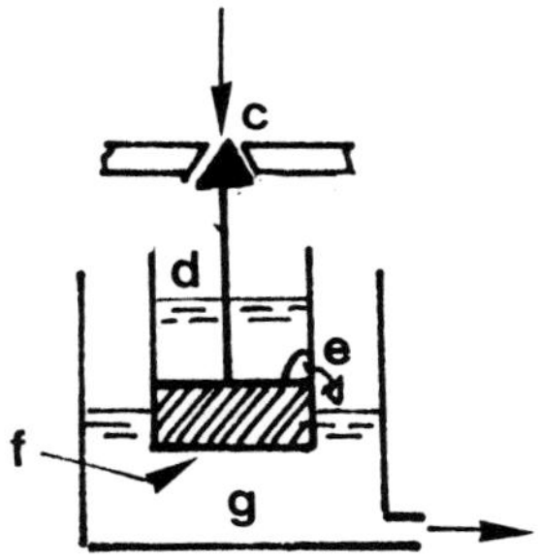

Figure 7. *Hydraulic switches for conical valves*

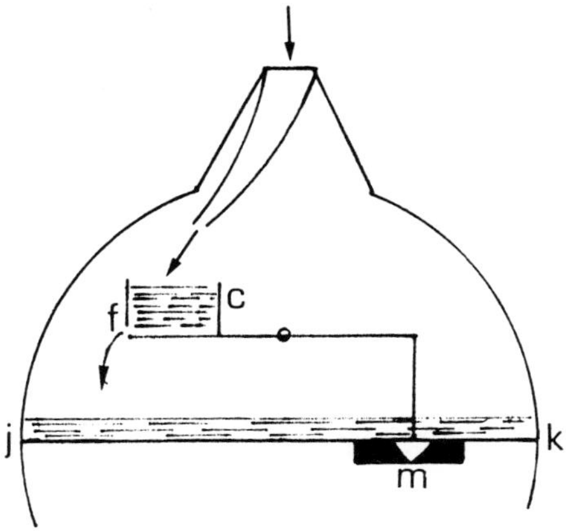

Figure 8. *Hydraulic switches for conical valves*

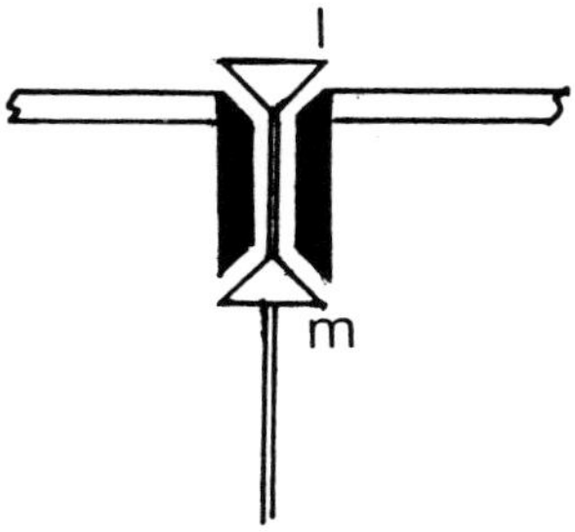

Figure 9. *Double-acting conical valve*

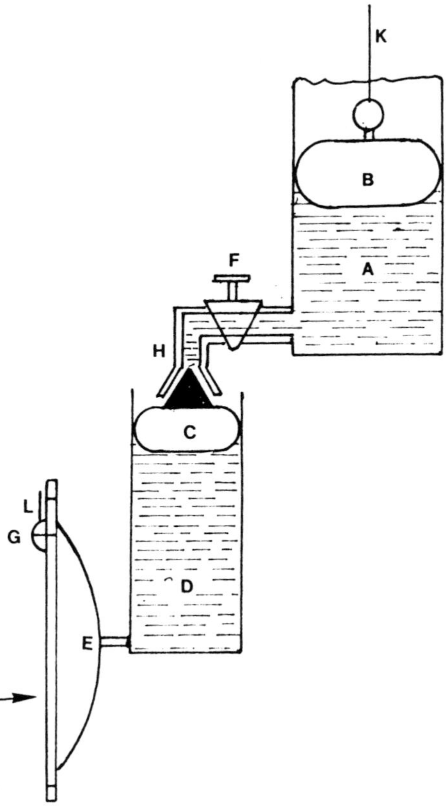

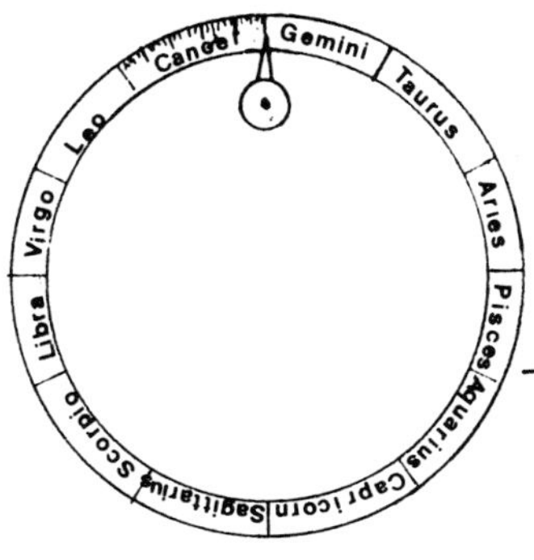

Figure 10. *Water-machinery for al-Jazarī's first clock*

shown, to a conical valve in a tank g, under tank k. When the water in tank k reaches a certain level, valve e opens and the water runs into tank g and issues from outlet h. In Figure 7 the water runs through the open valve c into tank d, which is on the top of float f, and through hole e into tank g. When filling is stopped tank d evacuates into tank g. Float f rises and closes valve c and no more water can be poured in. Figure 8 shows a valve with the reverse effect of the valve in Figure 7. When poured in the water runs into tank c which fills, the water leaking from hole f on to the partition jk of the vessel. When pouring is stopped, tank c evacuates, allowing the lever arm to tilt to the right and open valve m. The water therefore runs into the main body of the vessel.

The additional components used by the Banū Mūsà, especially the conical valves, allow them to produce a greater variety of effects in their trick vessels than either Philo or Hero. There is also a range of devices, Models 75–87, that have no counterpart in the Greek works. These are all self-replenishing vessels. In the simplest of these the vessel is located outside a large cylindrical container and is connected to a tank on the floor of this container by a pipe, also close to floor level. The top compartment of the container is a reservoir from which a pipe leads down towards the tank. There is a tap in this pipe and this is connected by a rod to the float in the tank. When water is removed from the outer vessel, the level in the inner tank also falls, the tap opens, and the inner tank and the outer vessel are replenished to their original level. The tap then closes (Model 75).

Model 77 is a more sophisticated version. In this case the rod from the float carries two conical valve plugs on its upper end, separated by a short length of rod, which passes through a hole in the floor of the reservoir. This hole is fitted with two valve seats, one just below to fit valve plug m, the other inside the reservoir to fit valve plug l [Figure 9]. When a moderate amount of water is taken from the outer vessel the lower valve opens and the tank and the outer vessel are replenished, as in Model 75. If, however, a large amount of water is withdrawn, valve l drops into its seat and no further replenishment can occur. This is an early example of a fail-safe mechanism and also of a valve rod with twin valves.

Models 88 to 94 are alternating fountains; for example, the fountain-head emits a single jet for a while, then a spray then a jet again and so on. The most interesting feature is that one of these fountains incorporates a miniature horizontal water-wheel driven by axial flow. When a component occurs in a miniature form it is almost always the case that a full-size practical version was already in use. The earliest known axial-flow wheels in Europe are the so-called 'tub-wheels' introduced in the sixteenth century.[28]

Models 95 to 98 are lamps, the first two being self-filling types. This type of lamp had been known for centuries – Philo says that there are many types of self-filling lamps (p. 143). Lamp No. 97 of the Banū Mūsà combines the self-replenishing motif with the self-trimming variety, as in Hero's *Pneumatics* I 34. Indeed, so close is the

[28] Norman Smith, *Man and Water*, London, 1976, p. 165.

Banū Mūsà's design to Hero's that it is certain that it is a direct copy. Model 98 is a 'hurricane' lamp with a shield that always turns into the wind to protect the flame. This lamp together with the last two devices – the bellows and the clamshell grab – were probably developed by the Banū Mūsà to meet certain needs in their civil engineering works.

The Banū Mūsà do not include in their work anything similar to the water-wheels of Philo or the pump and organs of Hero. These machines must have been well-known in their time and it is likely that they did not feel that it was appropriate to include them in the *Book of Ingenious Devices.* The end results of most of their devices were in many ways more sophisticated than those of Philo or Hero, but they were equally trivial. Their mastery of control systems and delicate mechanisms, their use of conical valves and switching mechanisms and of small variations in aerostatic and hydrostatic pressures were, however, unique in their time and indeed in the ensuing centuries. Not until relatively recent times, with pneumatic instrumentation, were their achievements in these fields surpassed.

In the interval between the Banū Mūsà's *Book of Ingenious Devices* written about 850 and al-Jazarī's great machine book, completed in 1204 or 1206, there were only two treatises on fine technology of any significance. The first was written in Muslim Spain in the eleventh century by a certain Ibn Jalaf or Ibn Khalaf al-Murādī. It consists of brief descriptions of water-clocks and automata-machines. Although the unique manuscript is badly defaced, the important aspects of the machines can be discerned. First is their ruggedness, with counterweights up to 3 kg and gear-wheels up to 72 cm in diameter. Secondly is the use of complex gear-trains, including segmental gears, i.e. gears with teeth on only part of their circumference in order to allow intermittent working. Although manually operated gear-trains had been known in Hellenistic times, al-Murādī's complex gears are the first to have carried heavy torque. As such they are important in the history of clockwork.[29]

The second treatise was written in 1203 by Riḍwān b. al-Sa'ātī (son of the clockmaker), describing repairs he made to the large water-clock built by his father Muḥammad in Damascus in the reign of Nūr al-Dīn b. Zenkī (1154–1174). This was a large clock with a working face about 4.23 m wide by 2.78 m high. It was very similar to al-Jazarī's first clock but not so well designed; it suffered at least two breakdowns. Also, Riḍwān was a doctor, not an engineer and hence his descriptions often lack clarity. His lack of expertise, however, is occasionally an advantage, since he sometimes describes the manufacture of components in detail, whereas an engineer would take them for granted.[30]

We know nothing of the life of al-Jazarī except what he tells us in the introduction to his book, namely that when he completed his book he had been for 25 years in the service of the Artuqid princes of Diyar Bakr. The book is divided into six categories:

---

[29] Donald Hill, *Arabic Water-clocks*, Aleppo, 1981, pp. 36–46.

[30] *Ibid.*, pp. 69–88.

I. Water-clocks and candle-clocks; 10 chapters
II. Trick vessels; 10 chapters
III. Water dispensers and phlebotomy measuring devices; 10 chapters
IV. Fountains and musical automata; 10 chapters
V. Water-raising machines; 5 chapters
VI. Miscellaneous; 5 chapters

Category I occupies about 40 per cent of the entire work. The first of the water-clocks was also the most elaborate and indeed this clock alone occupies about 12 per cent of the book. A description of the operation and construction of the most important parts of this machine will serve to illustrate, in large part, the unique features of al-Jazarī's methods and skills.

The working face of the clock was a screen of bronze or wood about 1.35 m wide by 2.25 m high, with another 0.75 m for the semidiameter of the zodiac circle. Below the top of the screen was a row of 12 double-leaved doors and below this was a row of 12 single-leaved doors. In front of this row was a small crescent moon. Beneath this was an inverted semicircle of 12 glass roundels and at either side of these was the figure of a falcon below which was a vase with a cymbal in it. Finally, on a platform at the bottom of the screen were the figures of five musicians – two trumpeters, two drummers and a cymbalist. The screen was in the centre of the front wall of a roofless wooden 'house' about 3 m square which contained the water-machinery.

Throughout the day and the night the zodiac circle moved at constant speed at the top of the clock, one sign rising above the 'horizon' – the top of the screen – as another set. The crescent moon moved, also at constant speed, in front of the lower row of doors. At every hour one of the upper doors opened to reveal a standing figure, the lower door rotated to show a different colour and the falcons leant forward, spread their wings and dropped balls from their beaks on to the cymbals. One roundel had become illuminated. At the end of the day all the upper doors were open, all the lower doors had changed colour and all roundels were illuminated. The musicians played at the sixth, ninth and twelfth hours. They were operated by the discharge of water from the orifice.

All the other mechanisms were operated by the water-machinery, attributed by both Riḍwān and al-Jazarī, probably correctly, to Archimedes. This machinery incorporated a feedback control system [Figure 10]. The drive was from the steady descent of the heavy float B in the reservoir A (only the lower part is shown). The cord k went through pulley systems that activated the time-recording automata. A tube made of cast bronze, with a tap F, led out from the bottom of the reservoir, its end H bent down to form the seat of a conical valve. The plug of the valve was soldered to the top of the small float C in float-chamber D. The outlet pipe from the float-chamber entered the flow regulator (see below). When the tap was opened the water entered the float-chamber through the conical valve. The water-level in D rose and closed the valve momentarily, the water flowed through the outlet pipe, the valve opened momentarily, and so on. A virtually constant head was maintained in the float-chamber and the rate of discharge from the reservoir was therefore constant.

The purpose of the flow regulator was to produce the unequal hours, which were used in early clocks. The unequal hours are obtained by dividing the hours of daylight (or darkness) by 12 to give 'hours' that vary in length from day to day. This was done by varying the head of the orifice G below the water-level in the float chamber by rotating it, a piece of drilled onyx, inside an annulus graduated into the 12 zodiacal signs, each subdivided into 30 degrees. The pointer L, on the same radius as the orifice, was set to the correct degree for that day. Al-Jazarī graduated the annulus by painstaking trial-and-error methods, having ascertained that earlier flow regulators, including that of Archimedes, were inaccurate, since they simply divided the annulus into 12 equal divisions of 30° each.

Only one of al-Jazarī's complete machines is of significance in the history of machine technology. This is a water-driven piston pump with horizontally opposed cylinders and true suction pipes. In Greek and Roman pumps the cylinders stood vertically in the water (Category V, Chapter 5).

Throughout the book, however, there are components and techniques that appear for the first time in any treatise. The tipping-bucket occurs in the second clock and in a number of other devices. This was a vessel like 'half a boat' from which stub axles protruded at the top, close to the broad end. These rested in bearings. Water dripped into the vessel from an orifice and after a set time the bucket tipped and discharged all its contents before returning to the horizontal. The sudden tipping of the bucket activated various mechanisms. In one of al-Jazarī's water-raising devices (Category V, Chapter 4) there is a crank as an integral part of the machine. This is the first known use of a non-manually operated crank. Another mechanism used by al-Jazarī was the submersible bowl (*ṭarjahār*); this had an orifice in its underside which caused it to submerge in a pre-determined time. This was used in two of the water-clocks (Category I, Chapters 3 and 4), both of which operated on the closed-loop principle. The *ṭarjahār* was used before this to time the allocation of irrigation water.[31]

A number of techniques that were to be of importance in the history of machine design first appear in the pages of al-Jazarī's book. This does not mean, of course, that they were all invented by him; earlier craftsmen did not usually commit their results to writing. All the same, it is noteworthy that al-Jazarī almost invariably used the best technique available, whether it was his own invention or not. Among these practices were the accurate calibration of orifices; the static balancing of large pulley wheels by fixing small lead weights to their perimeters, much as is done nowadays; the lamination of timber to prevent warping; the use of paper models to establish designs. When manufacturing large vessels from beaten copper, circular wooden templates were used to ensure that the vessel was of uniform circular cross-section along its entire length. In

[31] Al-Muqaddasī, *Ahsān al-taqāsim*, ed. M. J. de Goeje, Leiden, 1906, p. 357. Al-Muqaddasī (late tenth century) refers to the use of the submersible bowl, which he calls the *ṭarjahara*, for allocation of irrigation water in northern Iran.

the construction of a large ornamental door for the royal palace at Amid (Category VI, Chapter 1) al-Jazarī describes how he cast some of the brass sections first used in Europe at the end of the fifteenth century.[32]

The machines, components and techniques used in al-Jazarī's book would alone be sufficient to guarantee it a foremost place in the history of engineering writing. When to these features his method of presentation is added, however, it becomes no exaggeration to say that this is the most important engineering document, from any cultural area, before the Renaissance. He was writing with the avowed intention of enabling later craftsmen to reconstruct his machines, and in this he was successful.[33] His descriptions, whether of large machines or small devices, always follow the same pattern. First a description of the outside appearance of the machine, then its operation, then a step-by-step description of its manufacture and assembly. Finally, a repetition of its operation. Illustrations include general assemblies of the complete machines, with detailed drawings of various components and sub-assemblies. Although these semi-perspective drawings may at first look strange to our eyes, they are in fact an essential aid to understanding the instructions in the text. As with the other fine technology treatises, they all carry the letters of the alphabet on the various components. And – most importantly – each device is fully dimensioned, not on the illustrations themselves, as in modern practice, but by referring to the identifying letters. Of the earlier writers Philo gives only a few dimensions, Hero and the Banū Mūsà scarcely any. Riḍwān does give some measurements but in a haphazard manner. Indeed, few European treatises before modern times give such accurate specifications and dimensions for the benefit of constructors.

It would not be very rewarding to attempt to fit the four major writers into a progressive sequence. The treatises of all the writers have merits peculiar to themselves: Philo's pioneering pneumatic devices, Hero's experiments with steam and the remarkable pulley systems of the *Automata*, the Banū Mūsà's use of conical valves in switching mechanisms. Al-Jazarī, in some aspects of his work, was clearly drawing upon the work of his predecessors, known and unknown, but his whole book carries the imprint of a brilliant engineering mind. The greatest tribute we can pay to these writers is to study their works and appreciate the wealth of engineering knowledge that they incorporate.

---

[32] Cyril Stanley Smith, 'The Early History of Casting, Molds and the Science of Solidification', in *Metal Transformations*, ed. W. W. Mullins and M. C. Shaw, New York, 1968, pp. 23 and 27.

[33] For example, al-Jazarī's first clock was reconstructed full-size by P. N. Haward in the Science Museum, London for the 1976 World of Islam Festival. It was built according to al-Jazarī's instructions and worked perfectly. Mr Haward also built a phlebotomy-measuring device for the same occasion and the Museum built a quarter-scale model of the piston pump. These worked equally well, as did a full-size reconstruction of the elephant clock built by Mr Haward in 1993 for the Musée de l'Horlogerie in Le Locle.

Patient Grisilde: Sun and Moon in *The Clerk's Tale*

John North

The Canterbury Tales and the Wife of Bath

Even those unfamiliar with the work of the late-fourteenth-century English poet Geoffrey Chaucer will be aware that his *Canterbury Tales* were supposedly stories told by a group of pilgrims. The pilgrims were on their way from Southwark, then a small place to the south of London, to the shrine of Thomas Becket at Canterbury. The use of such a framework for a collection of tales is of course ancient: the *Thousand and One Nights* provides another good example, and there are many more. The closest parallel to Chaucer's in the fourteenth century was with Boccaccio's *Decameron*, although it is not known whether Chaucer actually knew that work. It was a clever idea to use a pilgrimage as a framework, for what characterized a pilgrimage was the great diversity of human beings taking part in it – rich and poor, young and old, good and bad, intelligent and otherwise – and male and female. The first to tell his tale in Chaucer's scheme was a knight, a soldier high-born and eloquent. The second was the Miller, a drunken, plain-spoken, boorish man at quite the other end of the spectrum of courtesy. And so it went on, with the Reeve, the Cook, the Man of Law. At last there came the first shot in the battle of the sexes, with the tale told by the Wife of Bath, one of Chaucer's best-known characters.

The Wife had been married five times, had gone on many pilgrimages, was highly sociable, but above all was characterized by a high degree of sexuality and general independence of spirit. She begins by stating her philosophy: virginity is all very well for those who aspire to a state of religious perfection, but it is not for her. Before she starts her tale she tells us her own history, and is so convincing that one of the pilgrims (the Pardoner), who was planning to marry, begins to have second thoughts. She tells in particular of her marriage to her fifth and last husband, a man twenty years younger than she, who had been an Oxford scholar. He had caused her to fly into a rage by

From *Learning, Language and Invention: Essays presented to Francis Maddison*, ed. W.D. Hackmann and A.J. Turner. Published by Variorum, Ashgate Publishing Ltd, Gower House, Croft Road, Aldershot, Hampshire, GU11 3HR, Great Britain.

reading to her a series of well-known stories of wicked women who caused their husbands suffering or death. She explains how, after exchanging blows, in the course of which he thought he had killed her, she finally got mastery of him.

In short, the Wife was a very liberated lady from a period of history when that meant much more than it does now. Her story too concerns the theme of woman's emancipation. A knight has been condemned to death for raping a maiden. He will be reprieved if he can find the answer to the question *What is it that women most desire?* She must inevitably have given offence to many in her audience, in particular to those who sincerely believed in the Church's doctrines of marriage as a sacrament and virginity as a state of perfection. The Clerk was one of these people, and it is not surprising that he later hits back at this lusty lady who pays lip-service to the ideals of the Christian church, but nothing more.

There are other tales in what is sometimes called 'the marriage group' in the *Canterbury Tales*[1] and one of them, the Clerk's, is in a sense an answer to the Wife of Bath. The Clerk, like her fifth and last husband – she was hoping for a sixth – was an Oxford scholar who had long before studied logic at Oxford. It will soon become clear that he had a smattering of astronomy too.

The Clerk's Tale, and its Sources

The Clerk's love of philosophy required him to spend his limited riches on books. His demeanour was respectful and his ideas were profound.[2] It is not easy to explain why Chaucer made him render an account of what to most modern readers seems to have been no more than a husband's wanton mental cruelty to his own wife. He was plainly no sadist. Perhaps Chaucer thought a large measure of logical expertise would enable the Clerk, in the end, to reconcile his audience to his rebarbative tale.

As the Clerk himself tells us, he relied on one of Petrarch's narratives, *De insigni obedientia et fide uxoris.* This was written in 1373 and revised in 1374. Chaucer used the revision, and also an undated French version, *Le livre Griseldis.*[3] As the Clerk continues, his tale approaches rather nearer to the French. The differences of content are not great, but in the French the language portraying the sufferings of the wife, Grisilde, seems rather more humane than Petrarch's. The tale is callous, as we shall see, and this is no doubt to be explained in part by its origins in folk-tradition. Boccaccio had adapted it, twenty years before Petrarch, at the end of the *Decameron*, but only with Petrarch did it take on a high moral and religious tone. This was not lost on Chaucer, despite the heightened level of emotion in the Clerk's delivery, achieved in large measure by Chaucer's much greater use of direct speech than either of his sources, and by a much better portrayal of the sentiment of a mother for her child.

---

[1] This description does not find universal acceptance.

[2] *General Prologue*, 285–308. References to Chaucer are to *The Riverside Chaucer*, 3rd edn, ed. L.D. Benson, Boston, Mass., 1987.

[3] W.F. Bryan and G. Dempster, *Sources and Analogues of Chaucer's Canterbury Tales*, Chicago, 1941, pp. 288–331.

After an abbreviated rendering of Petrarch's introduction, but like him setting the tale in Saluzzo in northern Italy, the Clerk tells his tale. He tells of Walter, the young Marquis of Saluzzo, and of Grisilde, the virtuous daughter of Janicula, an old and very poor man living in a nearby hamlet. Walter's subjects, worried lest he die without an heir, beg him to take a wife, and this he does on the condition that he be allowed to make his own choice. Walter sets the day of the wedding, and rides with a great retinue to their dwelling, where he meets Grisilde returning with water from the well. It is at this point in the Clerk's account that the first significant additions to the sources are made. (I shall consider only Petrarch and the French.) Grisilde, innocent of the fact that the commotion is on her account,

> To fecchen water at a welle is went ... (276).

Petrarch's word is *fons*, which could have denoted a well, but more probably a spring, and the source was 'distant' (*e longinquo*); in the French her 'croche de l'eaue' came from an unspecified distant place ('de bien loing'). As she was about to cross her own threshold,

> The markys cam and gan hire for to calle;
> And she set doun hir water pot anon,
> Biside the thresshfold, in an oxes stalle ... (289–91)

None of this is in either Petrarch or the French source, and there have been various speculations as to why, for example, Chaucer introduced here the idea that Grisilde's home was an ox's stall. Of course it symbolized her lowly origins. Was it also, for example, conceived as a *religious* allegory, with Grisilde as the Virgin and the stall the lowly place of Christ's birth? While Chaucer might have welcomed such an association in the hearer's mind, it is possible that he had a more strategic purpose in view, and that, in at least one layer of allegorical meaning, the zodiacal sign of *Taurus* was Grisilde's home. This idea will need to be carefully defended, but if I am right it will eventually prove to throw some new light on Chaucer's attitudes to the chief characters in the story.

Called by the Marquis, Grisilde fell to her knees and so remained until she had heard his will. Asked where was her father, Grisilde

> Answerde, 'Lorde, he is al redy heere'.
> And in she gooth withouten lenger lette,
> And to the markys she hir fader fette. (299–301)

Walter takes the old man by the hand to one side, and astounds him by asking for Grisilde's hand in marriage. The Marquis promises to take her for his wife 'unto hir lyves end'. The old man shakes with astonishment, but as liege-man to Walter has no real choice in the matter.

Grisilde meanwhile, awed by the presence of such a high-born guest, and his retinue outside, looked on 'with ful pale face' – a Chaucerian addition (340). Walter now proposed marriage, giving her an opportunity to think the matter over if she wished.

He warned her, however, that if she accepted she would have to be ready to obey him, no matter what his wishes, whether they cause her pleasure or pain. She was to swear perpetual obedience. She did so, of course. She would never willingly disobey him in thought or act. Her fate was sealed. He left the chamber first, declared his choice to the people, and commanded them to honour and love her. He then commanded that she be stripped of her old clothes and dressed in new, crowned and covered with jewels. He married her with a ring, and set her on her way to his palace, on a snow-white horse, and led by the people. They spent the day 'in revel, til the Sonne gan descende' (392), another Chaucerian addition, as is a second reference to the ox's stall in which the new Marquise had been born and raised. News of her virtue spread, and she earned a reputation as a peace-maker sent to the town by heaven. In due course, she gave birth to a daughter. She would rather it had been a son, but she was not barren, and no doubt the pleasure of bearing a son would follow.

We have now reached the end of Part II of the tale, and my own small contribution to an understanding of it really goes no further than here. I shall suggest that the relatively slight additions that Chaucer has made up to this point are capable of a simple astronomical interpretation. Since it is not easy to set out the argument briefly, to avoid losing the thread of the story as a whole I shall continue with the tale, and content myself for the time being with the remark that *Grisilde, 'with ful pale face', was represented by the Moon, while Walter, whose clothes she wears, whom she obeys, from whom she derives her light and her glory, was represented by the Sun.*

Even while the baby was sucking at her mother's breast, Walter was obsessed by the idea of testing, 'assaying', his wife. The Clerk expresses a distaste for the idea of placing someone needlessly in anguish and dread, and no doubt here we have a sign of Chaucer's own unease at the cruel turn the story is about to take. Walter tells Grisilde that people are complaining about her low origins. To appease them, the child must go. He does not say that the child will be killed, only that it shall be disposed of as the people wish it to be. Grisilde, apparently unmoved and unaggrieved, answered that she and the child were his to do with as he wished. They were his *property.* Showing sign of neither pleasure nor displeasure, Walter left, and sent to her a villainous executioner of the law, who took away the child with a little speech of a kind used throughout all of human history in such situations, to the effect that 'painful though this my duty is, we servants must obey'. The mother bids her child a painful farewell, and begs the man to give the child a decent burial. In fact, unbeknown to her, Walter has arranged for the child to be brought up by his sister, secretly near Bologna.

Four years later, Grisilde bears a much-wanted son, but yet again – although he waits two more years before doing so – Walter is overcome by a desire to test his wife's obedience. The same procedure is followed as before, and the child is taken from its mother. Chaucer is of course here dealing with a situation that tests the reader's patience as surely as Walter tested Grisilde's. How could she have been so meek in the face of the apparent murder of her children? Chaucer cannot change the story, but he does add words that hold the reader's sympathy for her, and he tells of the hatred of the people for her husband's cruelty. The cruelty, however, does not end here. When the daughter is twelve years old, Walter has some forged letters sent from Rome, allowing him to divorce his wife and remarry. A public announcement is made that he is divorcing Grisilde to please the people. Asked to return to her father's house, needless to say his wife agrees, and returns in a simple smock, stripped of her richer apparel.

The children are fetched from Bologna, and Walter sends for Grisilde, who is asked to prepare a room for the new bride – in reality their daughter. Of course she agrees, and not realizing the identity of the girl, helps to prepare her for the wedding. The people are fickle, and are delighted with Walter's choice of bride. As for Grisilde, she pleads with him only to refrain from tormenting his new bride as he had done her. It is at this point – the closest she comes to reproaching him – that he confesses all. She is his only wife. She has survived these tests with unqualified devotion to him. Following a moving and tearful reunion with her children and her husband after these twelve years of tribulation, she is stripped once again of her rough clothes, dressed in jewels and cloth of gold, and allowed to experience the happiness that she could have had all along, had Walter been of a different disposition.

Before returning to the moral side of the story, we should notice here that we are dealing with a variety of story-tellers, each of whom has his own point of view. There is the pre-Chaucerian material, there is the Clerk who speaks to us, and in the background there is always Chaucer. Indeed in this tale the poet comes into view in an unusual way, for there is an 'envoy' by Chaucer tacked on the end of the tale and – while still in the voice of the Clerk – it provides some material additional to Petrarch's. As we have seen, Chaucer also made a number of slight changes in the course of the tale, changes that serve to humanize a very inhuman and improbable story. Did he – as he so often did elsewhere – introduce an element of astronomical allegory through those very changes to Part II of *The Clerk's Tale*?

Janicula, Grisilde, and the Ox's Stall

On the basis of its style, the work as a whole is widely thought to have been written at an early date in relation to the *Canterbury Tales* as a whole. Of course it might have been written early, and modified when it was incorporated into that collection, but if so, the seams in the work are hardly visible. It seems more probable that an astronomical thread was woven into it from the start, but that it was a very slender thread. The evidence from his other works suggests that Chaucer began to introduce astronomy into his poetry in a serious way around the year 1382. Joining together these very slight pieces of additional evidence, we can say that if the tale was indeed astronomically tinted from the time of its composition, then it is likely to have been put together a few years either way of 1382. But why should the additions be read astronomically? The chief reason is that they seem to be reminiscent of the style of *The Complaint of Mars*, an allegory the astronomical character of which cannot be seriously questioned. *The Clerk's Tale* is admittedly less explicit. There are no planetary deities, but there is so much talk of places and of crossing their limits, of 'paleys' and 'halle', 'chambre' and 'house', 'dore' and 'thresshfold', not to mention the ox's 'stalle', that we should at least consider the possibility of a closer similarity of style.[4] Had Chaucer left us only a single work, *The Clerk's Tale*, we should have had no grounds firm enough to lead to a solution, but fortunately we have several other works of his that can help us to assess the light touches of astronomical language.

If it is a planetary allegory, then certainly the identities of Walter and Grisilde are not immediately obvious. They are clearly not Mars and Venus – the tale is not lascivious enough for that astrological combination, which fitted so well with the Wife of Bath. Jupiter and Venus? Most certainly not. Walter could not have been conceived as having the magnanimity or religious qualities of Jupiter, and Venus is not appropriate to a tale of unwavering female obedience. Mars and the Moon? As a combination, this could have carried the notion of the nobleman's cruelty had it not been a reluctant and rational cruelty. Walter was neither a martial character nor an assassin, and it is not possible to make sense of this combination. Saturn would have been totally inappropriate, but what of Mercury for the changeable Walter? Again, Chaucer would not have made a ruler and an autocrat into a subject of Mercury. Arguing in this most general way, we are driven to consider the Sun and Moon; but for the time being the whole question must be left in suspense.

The real starting-point is the ox's stall. If his motives were astronomical then Chaucer surely wanted to represent *Taurus*, or a part of that sign. Does this mean that Grisilde was Venus after all – since Taurus was her home – or that she was represented by a planet with a chamber (term or face), in Taurus? At this stage we ought to reserve judgement, but one thing we can feel strongly about is the identity of her father, Janicula. If he is associated with any planet at all, then it must be with *Saturn*. There is plenty of corroboration in the text for the idea that he is a feeble old man. Philology notwithstanding, perhaps *Janicula* was a name that made Chaucer think of January, and so of Saturn, lord of the sign of Capricorn, where the Sun is found in January.

Now Saturn has a chamber in Taurus – in fact he has a face (the last, running from 20° to 30° of Taurus, that is, from 50° to 60° of the ecliptic), and he also has a term within that face. The term runs from 52° to 57° of the ecliptic. The planet Saturn moves so very slowly that it was in these chambers *only at one period* during Chaucer's entire writing career. Janicula, be it noted, spends most of the tale in or near his home.

If we are to use Saturn as a guide to other possibilities, then we must decide on the limits placed on its whereabouts. Was it only with the 30° *sign* of Taurus? Or the 10° face? Or the 6° term? It was in the sign between June 1380 and April 1383, an encouraging result, but as yet not very precise.

Saturn's behaviour when in its face and term within the sign – this in 1382 – is singularly interesting, however. By virtue of its retrogradation, it *stays in their neigh-*

---

[4] A solution I found in 1967, when – with the encouragement of Francis Maddison – I was writing a series of articles on some of Chaucer's other works, was not very satisfactory, and I abandoned it. A letter from Dr Susan Innes, in June 1991, led me to look at the matter again. She noted such ostensibly astrological language as 'obeisance', 'reverence', 'benigne', 'estaat', 'dignitee', 'degree', and 'adversitee'.

bourhood for several weeks, and in that period *there is one date, 3 or 4 May, that presents itself in a most natural manner.* Readers of my *Chaucer's Universe* will recognize 3 May as a date Chaucer favoured for days of disaster or near-disaster.

Its occurrence in *The Clerk's Tale* may now be plausibly explained. When we perform a series of calculations using the Alfonsine planetary tables, we find that the Alfonsine Saturn would have been held to behave in the following way – and differences between this and the real Saturn were minimal.

The planet passed into its face on 19 April, and into its term in the early hours of 4 May 1382. We have as yet no particular reason for assuming one time of day rather than another in specifying Saturn's positions, but since Chaucer mentions undren as the time for which the wedding was planned (260–1), we might as well take that hour for our first approximations.[5] For want of anything better I shall accept my old 10 a.m. reading for undren, although the matter is certainly not critical in the case of ponderous Saturn. On 3 May 1382, Saturn was only 6 minutes of arc outside its term, and so might possibly have been thought within it.[6] I prefer not to rest the case with even so close a miss, for we can do better by assuming that the *face* rather than the *term* is the chamber of Janicula. This will be later supported by the circumstance that we need the face to accommodate the three main planetary characters simultaneously.

During all of this time, Saturn was close to the Sun. The strict (Alfonsine) conjunction of the two occurred on the evening of 4 May. We need not interpret these things for the moment. After the Sun had overtaken Saturn in the zodiac, the planet advanced, until by 8 July it left the sign. After a time it turned back, and 122 days after leaving, re-entered Taurus, which it left finally only on 30 March 1383. It was not to return to any part of Taurus in Chaucer's lifetime.[7] The planet must certainly have given an unusual impression in 1382 of a fondness for its chamber in Taurus.

Walter and Grisilde, Sun and Moon

We are now faced with a familiar problem, and one that Chaucer faced constantly: it is not easy to devise a scheme that would have the characters move around as quickly as the action in the tale requires, if they are to be strictly tied to the slow-moving planets. In *The Complaint of Mars*, which the language of the material he added to the *Clerk's Tale* seems to resemble most nearly, Chaucer openly accepts a more measured pace for the planets' movements in and out of their chambers; but here he cannot, for he is tied to his sources.

---

[5] Petrarch has only *hora prandii*, and in the French no hour is mentioned. Later in the work (981) Chaucer translates Petrarch's *hora tercia* (in the French *tierce du jour*) as 'abouten undren'. The third hour in question was of course reckoned in seasonal, 'unequal' hours.

[6] The Sun was in Saturn's face from the evening of 2 May onwards, and did not enter the term until the evening of 4 May.

[7] On 9 June 1382 it left its term; by 8 July it left Taurus. It re-entered its face 120 days after leaving the sign, returned to its term on 13 December 1382, and stayed there until eventually reversing its motion again. It left its term for the last time on 24 February 1383, and quit Taurus (and its face) completely on 30 March 1383.

On the face of things, the 21 verses that seem to be meant to cover the day of the wedding (lines 246, 'The day of wedding came...' to 392, '...til the Sun gan descende') look more like a clock than a calendar. To be sure, they are in sevens, which one could perhaps interpret as weeks, but a London calendar in Chaucer's time would have given the length of the 'vulgar day' in the first few days of May as something close to 21 hours.[8] The placing of the two lines quoted was of course something he could control relatively easily, and we are reminded here of his handling of the *Parliament of Fowls.*

In searching for a pair of planets for the leading characters, notions like 'commanding' and 'obeying' make one think of the commanding and obeying *signs,* but here there is nothing of promise.[9] The planets, apart from Saturn, are equally unpromising, but the *Sun and Moon* do have an interesting relationship with Saturn and Taurus generally. This must be set out at some length, because the overall pattern of the behaviour of the Sun, the Moon, and Saturn does resemble that of the action on the day of the wedding in several respects.

The year 1382 is taken to have been rendered probable by the Saturn–Taurus link, an allegory involving Janicula and the ox's stall. Now of course there is of necessity a meeting of the Sun and Moon within any period of more than a month, but one of their meetings in 1382 turns out to have been of a very unusual sort: when they met on 14 April in that year, it was *at the precise degree of the exaltation of the Moon, inside Taurus.* In London, at undren (to take a time for which we have as yet no clear justification) they were virtually at mid-degree.[10]

If Chaucer regarded this as representing one of the various meetings of Walter and Grisilde on the day of the wedding, then the action on the wedding day in the poem is unlikely to have corresponded to a single day in the history of the heavens. This conviction is reinforced by the fact that there were more correspondences occurring, later in the planetary calendar. On *May,* at undren, London, the Sun was at 50;45° and the Moon at 290;32°, *a near-perfect trine aspect.* The perfection of this arrangement, taken together with the proximity of the Sun to Saturn (51;54°), the date, and especially the named time of day, make it probable that the hour of undren was chosen with some deliberation. As mentioned earlier, the strict conjunction of the Sun and Moon on 14 April had also occurred, in the middle of the degree of exaltation, *almost precisely at undren.* Since it occurred as the first of the notable astronomical events relating to Sun and Moon, this conjunction was perhaps seen as the occasion of her kneeling to Walter, and hearing his will.[11] Chaucer's own reference to undren was to identify the time at which the action reported on the wedding day began. Both of

---

[8] Nicholas of Lynn's figures for Oxford are nearer to 21 hours 30 minutes, but they are on the high side. The onset of twilight is a somewhat arbitrary affair.

[9] See my *Chaucer's Universe,* 2nd edn, Oxford, 1990, p. 194; Taurus obeys Leo, for instance, and that might be relevant, but note that various conventions exist. No planets show any connection with the term or face of Saturn whilst Saturn is there. (So much for those who hold that such relationships are easy to find.)

[10] They are at longitudes 32;29° (Sun) and 32;26° (Moon), on Alfonsine reckoning.

[11] I reckon conjunction to have been ten minutes or so before undren, using Oxford tables, and six minutes after, using London tables.

the events referred to here, which form a part of this action, fit with the hour of undren in allegorical time.

There is another place in the sky that appears to correspond to a place in the poem. This is the perfectly placed well, *puteus*, just outside Taurus (in the penultimate degree of Aries). Although the usual translation for this astrological term is 'pit', the Latin word means 'well'; and we have already seen that one of the Chaucerian additions was to make Grisilde fetch water from a well (276). Walter first called her as she was about to cross the threshold of her dwelling, and she set down the jug just inside it, before kneeling to him (288–92). Although the Moon was not actually near either *puteus* or exaltation on 3 or 4 May 1382, it was at both on 14 April, at times spaced by an interval of some six or ten hours, depending on how we interpret 'being in a degree'. It has to be remembered that, on our reading, this interval is in *long-term* allegorical time, and not in the real time of the wedding day.

Ladies and Other People

Another property wonderfully suited to the Sun, at what seemed to have been a very special time on 3 May, was the fact that Walter had lords and ladies in his company (267ff). We know from other of Chaucer's poems that when we see a reference to groups of *ladies* we should consider the possibility that they are analogues of stars. When we look to the position of the Sun, as here conjectured, the stars we find are the *Pleiades*, exactly as in the *Nun's Priest's Tale*, where they represented hens. The position quoted earlier for the Sun was Taurus 50;45°, and if it was in that neighbourhood, then it would have been counted adjacent to the Pleiades.[12] The Pleiades could have been the *ladies*. Chaucer would surely not have made them lords, but there were several other stars at the end of the sign that might have served as lords. Taurus, home of the music-loving Venus, was a place highly appropriate to 'many a soun of sondry melodye' (271) such as that made by the revellers.

There might have been further intentional astrological colouring in the form of Walter's beginning to *call* Grisilde (289). This would mean the Sun approaching the Moon. The characters speak to one another and answer (296–9). Saturn, throughout the period of interest, was outside (although within a whisker of) his own chamber (term) but securely within his tower (if we may use that word for face). Saturn's face in Taurus may thus be taken to be Janicula's 'chambre', in which Walter suggests to him that 'I and thou and she / Have a collacioun' (324–5).

These interpretations are not without their difficulties. After kneeling to Walter, Grisilde goes ahead of him and fetches her father. The Moon naturally outruns the Sun, but how can Grisilde be said to *fetch* her father (301)? The Moon, unlike the planets, cannot reverse along its path. The Sun can meet with Saturn, though (5 May), after the Moon has met with the planet. Walter is said to have taken Janicula by the hand (302), and the language perhaps amounts to saying that Saturn and the Sun were within near (platic) conjunction, and not strictly conjoined.

---

[12] I earlier gave their position (*Chaucer's Universe*, pp. 462–3, as quoted in a contemporary source), as between 21;22° and 22;52°.

We have to bring Grisilde into their presence again, of course, but the Moon has gone on ahead, and will come back through the same entrance as before, having circled the sky completely. If we follow the movements of the Sun for about ten days following 3 May, we find that it moves from the 21st degree of Taurus (3 May) to the 60th (12 May), and thence out of Taurus, at 10.28 p.m., London time on 12 May. This movement covers the third and last *face* of Saturn. It gives the Moon time to return. Saturn, the Sun and the Moon are never in perfect conjunction in the face of Saturn, but the Moon is at the entrance to the chamber as the Sun is at its end, with Saturn at 53;12°. It does no violence to language to count this as a tripartite *collacioun.*

The event occurs at about 1.55 a.m., London time, on 13 May. The Sun leaves the chamber then, as explained, and the Moon leaves late on 13 May. In the story, it is now Walter who goes out of the door before Grisilde (cf. 366–7), and this explicit reversal of roles, fitting perfectly the astronomical facts, should help our confidence a little.

Having left the home of Janicula, Walter addresses 'the people' (368ff). When the Sun is leaving Taurus it is among the Hyades. We now think of the Hyades as female, but to the Romans they were little pigs, and I do not find it hard to accept that they could have corresponded to the people at the door. There is in any case a whole bevy of easily visible stars in the vicinity of the Hyades in addition to the half-dozen or so that were traditionally given names.

The tale is now beginning to look very much more Complaint-of-Mars-like. The espousal takes place (386–7), and presumably this event corresponds to a strict Sun–Moon conjunction in allegorical time. According to an Alfonsine calculation, that occurred *in the very first degree of Gemini*, at about 10 p.m. [*sic*] on 13 May 1382. Not a very special place, perhaps, but of course it had to be *outside* the 'ox's stall', Taurus.

Before the wedding, Grisilde had been robed, and now, after it, she is sent on to Walter's palace. And what, in the allegory, is that? The most obvious choice is the Sun's domicile, the sign of Leo. There is another possibility, and this gives the whole work a very attractive symmetry. If we simply take the third conjunction in a row, of the Sun and Moon, then we come to 8.10 a.m. on 12 June 1383, and the place is the penultimate degree of Gemini. The luminaries that represent the two leading characters are united there, in a *face of the Sun*, his palace on this hypothesis, at the very end of spring at this period of history. The Sun is about to reach its highest point in the sky, and Grisilde's trials are about to begin. At this point, Chaucer's little astronomical allegory seems to end.

A Celestial Crown

Were the crown and the ring mentioned in connection with Grisilde's wedding, and the snow-white horse that took her to Walter's palace, all in the heavens, by any chance?[13]

[13] There is no difficulty in having the Moon move through all the constellations (and *a fortiori* through Corona Borealis, and Pegasus, for instance), since the Sun needs more time to get to his palace, whether Leo or the end of Gemini, than does the Moon.

These elements are not new to Chaucer. Corona Borealis would be an obvious choice for the crown in the heavens corresponding to that placed on Grisilde's head by the ladies (381), but there are other distinctly preferable ways of finding a crown in the allegory. If we accept that his was a Sun-Moon allegory, with clothing of gold and brooches of gold and azure plainly intimating the fact, then it is conceivable that Chaucer had this well-known and mildly apposite text from the Book of Revelation in mind when he was writing:

> And there appeared a great wonder in heaven; a woman clothed with the Sun, and the Moon under her feet, and upon her head a crown of twelve stars. And she being with child cried, travailing in birth, and pained to be delivered.[14]

There are other biblical references that come to mind, introducing woman and childbirth. There was the medieval tradition that the constellation of Taurus represented the ox that stood by the ass in the manger at the nativity of Christ. Whatever the truth of the matter, our interpretation would be agreeably strengthened if we could find a crown in the heavens near the end of Taurus or the beginning of Gemini.

Now one of the outstanding puzzles in Chaucerian astronomical reference is his statement at the end of *The Legend of Ariadne* (in the *Legend of Good Women*) that after Ariadne ('Adryane') was deserted by Theseus – and note the similarity of theme – and after she died in child-bed, the gods placed the stones of her crown *in Taurus*:

> And in the sign of Taurus men may se
> The stones of hire corone shyne clere. (2222–4)[15]

Chaucer certainly used Ovid's *Metamorphoses*, viii.182, although there the crown was set between the constellations of Ophiuchus and Hercules, both well away from Taurus.[16] It is now customary to follow Skeat's explanation, which is that Corona Borealis – as we know the crown – came to the meridian with the zodiacal sign of Scorpio, and by virtue of this fact was conspicuous when *the Sun* was in the sign of Taurus. It is hard to believe that Chaucer used the language he did, with this meaning, but it is equally hard to accept that he did not know Ariadne's crown, which was always included in large standard catalogues of stars, such as that in MS Brussels 10119. In that manuscript it is illustrated with a drawing of a crown over which are the words *forma corone septentrionalis et stelle eius, et dicitur Adriagnes.*

If we accept that Chaucer simply erred in this work of his early astronomical period, then we can see various possible aids to misunderstanding. In the Brussels manuscript, when the Hyades are listed, a gloss gives their names:

> Hyades stelle 7, nomina earum Ambrosia, Cydora [*sic*, for Eudora], Pheidele [usu. Pedile], Coronis, Polixo, Phito [usu. an alternative to Polixo], Thyene. Alio nomine Plyades [*sic*].[17]

---

[14] *Revelation*, 12:1–2 (quoted from the King James version).

[15] Ariadne is also mentioned in *The House of Fame*, pp. 407–11, but not with a crown.

[16] Strictly 'the Serpent-holder' and the 'Kneeler' (*qui medius Nixique genu et Anguemque tenentis*).

[17] F. 63v.

This makes one wonder whether for a time Chaucer really did believe that Ariadne's crown was in Taurus. His memory might even have played a trick on him, had he half-recalled the lines of Virgil, in which farmers cultivating cereals were advised, before planting, first to 'let the daughters of Atlas [the Pleiades] pass from view in the morning, and then let the Cretan star of the fiery Crown [Corona Borealis] abscond'.[18] Here Pleiades and Hyades are confusingly juxtaposed.

One of the oddities about the Hyades is that, although given seven or fewer names in mythology by different authorities, the individual *stars* are not usually given separate names. This is so even in our large star catalogues, deriving from Ptolemy's, where five of the Hyades are listed and described simply as 'the five stars in the face [of the bull]'. The brightest of them, 'the reddish one on the southern eye', was usually glossed with its Arabic name, Aldebaran. We cannot readily tag any individual star with the name 'Coronis', and yet it is true to say that the longitude Chaucer would have given to the star *epsilon Tauri*, the last on Ptolemy's list of Hyades, was almost exactly that shared by the Sun and Moon at their conjunction late on 13 May.

A Chronological Summary

How are we to present the curious chronology here derived? Within a single day, whether 12 June, or 13 May, or 3 May, or 14 April, or some other, it is impossible to compress so much action without a completely different type of interpretation. We need a convention that is unusual even for Chaucer, and to introduce a more leisurely interpretation than the literal one. The descriptions – one literal, the other allegorical – are overlaid, rather than consecutive, as they might have been. There seems to be no interpretation that precisely matches the calendar by line or by verse, but here is a summary of the reconstruction offered above. Stanza-numbers come first, counting only within the twenty-one that cover this part of the tale, and line-numbers second:

Stanza/line	*Event in the Tale*	*Date*	*Celestial counterpart*
1/246	The day of wedding came	–	–
4/267ff	Walter with retinue	3–4 May?	Sun with Pleiades etc.
5/276	Grisilde fetches water	14 April	Moon at *puteus*
7/290	Grisilde sets down water	14 April	Moon within Taurus
7/292	Grisilde kneels to Walter	14 April	Sun-Moon conjunction
8/300–101	Grisilde goes for father	14 April	Moon leaves Sun
9/302	Walter greets Janicula	3 May	Sun in face of Saturn
12/325	Walter negotiates	3–13 May	Moon has circled sky, re-enters face
18/368	Walter addresses the people	13 May	Sun with Hyades
19/374–83	Ladies dress Grisilde	13 May	Moon with Pleiades
21/386	Marriage	13 May	Sun-Moon conjunction?
21/389	Grisilde sets out to palace	13 May	Moon leaves Sun

[18] *Georgics*, I.221–2.

–	–	12 June	Moon and Sun conjoin at end of Gemini
–	–	10 August	Moon and Sun conjoin in Leo
21/392	sunset on wedding day	–	–

The only problem here is a minor one. Why is there mention of Walter's retinue at a point that seems to be out of sequence, that is, coming before the Sun meets up with the Pleiades or Hyades? The only answer I can give is that the common people and his retinue are mentioned several times at points corresponding to the Sun's presence in Taurus or at its door into Gemini (267–73, 275, 368ff), that the sign as a whole conjured up the image of numerous people, and that in any case the original texts have the information at this point.

Giving some sort of support to the general scheme, we have not only another occurrence of the date of 3 May that Chaucer used at least three times elsewhere, but of 13 May, which he used in *The Squire's Tale* and *Troilus and Criseyde.*

Weddings and the Astrologer

Despite its many human associations, my presentation up to this point has been for the most part purely astronomical, and we can hardly leave the astronomical scene without considering its potential astrological counterpart. Might there not have been a broad hint of an *interrogation* or *election* in Walter's expressed wish that the three of them have a 'collacioun' in Janicula's chamber?

> For I wol *axe* if it hire wille be
> To be my wyf, and reule hire after me. (326–7)

Albumasar on elections would read the trine aspect of the Sun and Moon as signifying that it was opportune to write to kings and princes. This is to say very little to us about the probability of the scheme, since kings appear constantly in elections involving the Sun. Nevertheless, that simple fact is a consequence of a standard astrological doctrine, to the effect that (as the same author has it in his *Flowers*):

> Sol signifieth grete lordes and richemen and honorable men... (Trinity Coll., Cambridge, MS O.5.26, f. 47r)

Astrologers judged questions relating to marriage, or the most propitious times for a wedding, on the relationship between the Sun and Moon, either at the nativities, or (in the case of an election) at some later date. Alkabucius' *Introductorie* has 'the partye of weddinges' taken by day and night from Saturn to Venus, and cast from the ascendent.[19] Zael's book of elections is a standard source that, like Albumasar, made the fifth and seventh mundane houses the crucial places in an election on wedlock.

[19] Col. 102. See my *Op. cit.* (n. *9), Chaucer's Universe,* Appendix 3.

The standard warnings issued by Zael[20] are tacitly heeded in the *Clerk's Tale*, but that is to say little. He goes on to speak of improving the stability of a wife, where the relevant sign is Libra, but her equilibrium seems to be intended here. On the other hand, for the wife's 'firmation', by which something like *obedience* is meant, Taurus is a favoured sign for the Moon to be in,[21] and this gives us slight but positive support.

Whilst it seemed at first that this tale was impossible of solution, on the grounds that there is no hard astronomical information included in it, it now seems that there is indeed one invaluable datum. This is the almost invisible lynchpin of the solution, the reference to the time of day, undren. We cannot be precise here, but if that time is materially changed, the tightness-of-fit of the calculated positions is lost, and this is a point we should not undervalue.

The Theme of the Tale and the Envoy

If we are right to relate Walter to the Sun, and Grisilde to the Moon, does the commanding-and-obeying motif have no bearing on the case? This might not have been so much a case of technical astrology as a simple expression of the common idea that the wife's destiny to be ruled by her husband bears comparison with the Moon's destiny to be illuminated, and outshone by the light of the Sun. In preparation for the marriage, Walter had gems set in *gold and azure* for Grisilde's sake (254–5), that is, in the colours of the Sun and the sky. After her downfall, her final enlightenment, and her restoration, she is dressed in *cloth of gold* (1117).

The obvious reading is very probably the correct one: Walter was insisting that she wear his colour, his apparel, not her own. If so, it is the husband who is of gold, and Grisilde who reflects his glory. It would surely have been too modern a piece of irony to have reversed these roles. On the other hand one could ask whether Walter was not superficially *changeable*, and so resembling the Moon. The Clerk himself (998ff) has no such perspective. There is throughout the poem an interplay of twin notions of *richesse* and *gentilesse*, and though Grisilde may be transformed artificially by her husband into a woman of great estate, she has an intrinsic virtue that can survive being stripped of her husband's raiment, just as the Moon's virtues survive the withdrawal of the Sun's rays. The imperious character of the Marquis would have been seen as a prerogative of a subject of the Sun, whose house was Leo.

Changeability was not an appropriate epithet to apply to a man whose merest whim was law, by virtue of his having been a noble and a man. The Clerk actually uses the

[20] 'Whan thu wilnest eleccioun of wedlok, be war lest Luna be in the 12, and be war lest hit be in signes that be not profitable to this, whiche beth Aries, Cancer, Libra and Capricornus. And be war of signes in whiche were evele and Cauda.' (Trinity Coll., Cambridge, MS O.5.26, f. 61^{r-v}.)

[21] 'Whan thu wolt make ferme or stable a woman, be this whan Luna were joyned to fortunes and were in a mevable signe and than of hem better is the signe of Libre. And be war in firmacioun or in fastnynge, lest Luna be in a fixe signe. In coniunccioun sothly, that is whan eny man entreth to his wif and useth hir, be war lest the Mone be in a mevable signe neither in a comune, be this whan Luna were in a fixe signe and better than othere beth Leo and Taurus...' (loc. cit.).

metaphor of the waxing and waning Moon to underline the fickle character of the *populace*, when Grisilde's fortunes are at their lowest. The point to note here is that it is the *Clerk's* perspective on things that makes Grisilde a mere reflection of her lord's glory. The Clerk's audience was allowed its own opinion.

The Clerk, having told a moral tale of obedience to properly constituted authorities, ends his story in two different ways. The main story is followed by a simple statement by the Clerk that he has no intention of trying to make wives follow Grisilde's example. We should, though, all be as constant in adversity as we are able, as was Grisilde. This seems to break down into three principles: first, that we should accept the act of an oppressor, because this is decreed by God. This is a case of 'turning the other cheek', and it is good training for the endurance of suffering when there is no human agency visible. Second, that loyalty to an ideal may be a virtue in itself. Grisilde was constant to her marriage vows, come what may. But third, by implication, there seems to be the idea that we should accept oppression all the more readily if the oppressor is in authority over us, because in doing so we may thereby learn an important religious lesson about receiving whatever God may send.[22] This is where we may part company with the Clerk, and where many a woman of the fourteenth century would have done so. What Grisilde did might have been good for her immortal soul, but it is not clear that it was good for Walter's.

After his quasi-theological ending, the Clerk begs leave to say one thing more. It is not easy to find Grisildes today, he tells us. If you put them to the test, what you may think is gold turns out to be allayed with brass.[23] The metaphor is particularly interesting, since gold is the metal of the Sun. In reading the tale with the solar allegory in mind, we might have been excused for wondering time and again whether the word used of Walter's trial of his wife, namely 'assay', was invested with that metallurgical meaning. It proves to have been so used.

In view of the imperfection of the modern woman, the Clerk seems to say, he will sing us a refrain, out of affection for the Wife of Bath. This 'Envoy' to the Clerk's tale (usually called *Lenvoy de Chaucer* only because it is not Petrarch's) is written in an ironical vein more mature and skilful than the rest, and is generally presumed to have been grafted on to the tale, a piece of his earlier and less mature work, when the time came to incorporate it in *The Canterbury Tales* as a whole. It is as though Chaucer could not permit even one of his characters to be so insensitive to the fate of the married woman as to allow the story to go unqualified. This is uncharacteristic of Chaucer's own outlook. So what does the Clerk finally say?

He releases some of the tension that has built up during the telling of the story, and allows his audience to laugh, but he now uses a mocking tone, advising wives not to allow their tongues to be nailed. He takes it as read that women are, in matters of native humility, talkativeness, shrewishness, and essential impatience, made differently from men. Act as did Echo, who was never silent, and who always answered back.

---

[22] Cf. 1142–51.

[23] Lines 1166–9.

Do this for the common good. Slender wives may be feeble in battle but, faced with men's insults, should be fierce as an Indian tiger and noisy as a mill. Do not fear your husband. If he wears armour, pierce it with your crabby eloquence. Show that you mistrust him, and so bind him with a feeling of guilt. If you are beautiful, show yourself off in public. If you are ugly, spend money and so find friends. Either way, let your husband be the one who worries, weeps, wrings his hands, and wails. This charter of liberation is as extreme in its way as the behaviour of Walter to Grisilde. By comparison with that tailpiece, a few final words added by the Host seem refreshing in their honesty. In a stanza that was ultimately – as it seems – rejected by Chaucer, or moved elsewhere, the Host swears that he would sacrifice a barrel of ale in exchange for a wife who had heard the tale of Walter and Grisilde.

The Clerk's cynical advice comes at the end of a tale in which the central character has behaved quite otherwise. Grisilde was not Echo, who answered back, but the Moon, whose 'ful pale face' reflected the glory of her lord and master, the Sun. The astronomical point is a trifling one, but it does seem to show to us Chaucer trying out skills that over the following decade he would constantly strengthen and refine.

L'astrolabe sphérique dans l'Occident latin

Emmanuel Poulle

L'astrolabe sphérique est un peu le mal-aimé des historiens de l'instrumentation astronomique médiévale. Alors qu'il figure assez régulièrement parmi les instruments énumérés comme constituant la panoplie de l'astronome,[1] il a été peu étudié, peu recensé.[2] Il est probable, d'ailleurs, qu'un traitement aussi injuste est la conséquence de la situation documentaire de cet instrument: son absence dans les collections et les musées contribue à le confiner dans un rôle subalterne, qu'accentue la place au contraire largement prépondérante tenue par son homonyme planisphérique.

A vrai dire, jusqu'en 1962, l'astrolabe sphérique pouvait passer pour un instrument quasi mythique, puisque son existence n'était attestée que par une littérature technique relativement peu diffusée, et qu'on n'en connaissait aucun exemplaire. L'identification d'un astrolabe sphérique arabe du XVe siècle, par Francis Maddison, suivie de l'entrée de cet instrument, cette année-là, dans les collections du Museum of the History of Science, à Oxford, ont donc représenté un événement considérable dans notre connaissance des instruments anciens.[3] Quelques années plus tard, en 1976, un autre

[1] Voici par exemple une énumération d'instruments astronomiques figurant dans le *De spera concava* de William Batecombe: 'Armille quidem, rectangulus, albion, turketus et spera solida radices omnium stellarum verificant; almanak pluresque tabule calculate varios motus planetarum declarant; astrolabium, quadrans, horologium et saphea et temporum et horarum momenta certificant', Paris, nouv acq. lat. 625, f. 85^v. Autre liste en tête du traité du turquet de Regiomontanus: 'Qui astronomicis student exercitiis duplex instrumentorum genus versare soliti sunt; habent enim genus quoddam portabile quod videlicet contrectari potest ubicunque collibitum fuerit, nunc quidem sub divo nunc autem in loco quolibet occluso ad quem stellarum radii penetrare nequeunt, quale est opus albionis, sphaerae solidae, aequatorii sapheaeque et astrolabii vulgaris quadrantisque horarii et cylindri ac caetera hujuscemodi, regulae demum Hipparchi quibus ipse diametros luminarium permensus est ...', Regiomontanus, *Scripta* ..., Nuremberg, 1544. Sur l'homonymie entre l'astrolabe sphérique et la sphère solide, v. ci-après la n. 6.

[2] Le seul travail d'ensemble est: Richard Lorch, 'The sphera solida and related instruments', *Centaurus*, xxiv, 1980, pp. 153–161.

From *Learning, Language and Invention: Essays presented to Francis Maddison*, ed. W.D. Hackmann and A.J. Turner. Published by Variorum, Ashgate Publishing Ltd, Gower House, Croft Road, Aldershot, Hampshire, GU11 3HR, Great Britain.

astrolabe sphérique arabe, sans doute aussi de la fin du Moyen Age, malheureusement incomplet de son araignée, a été signalé dans une collection privée.[4]

On explique aisément la rareté, aujourd'hui, de l'astrolabe sphérique dans le matériel survivant, surtout au regard de la prolifération des astrolabes planisphériques, comme étant à l'image de la situation ancienne et du peu d'enthousiasme qu'ont dû manifester les praticiens du Moyen Age pour un instrument qui présentait sur la concurrence deux handicaps sérieux: la complexité de sa réalisation, génératrice de frais importants, et les inconvénients qu'il présentait lors des transports, tous deux résultant de sa forme sphérique.

Il n'est pas inutile de rappeler la définition de l'astrolabe sphérique, définition d'ailleurs très proche de celle de l'astrolabe planisphérique: il offre une double représentation des deux sphères, la sphère fixe de référence, et la sphère céleste mobile, constituée par le zodiaque et les étoiles, représentation telle que la rotation de la seconde devant la première puisse figurer le mouvement diurne. Alors que l'astrolabe planisphérique réalise cette figuration par le recours à la projection stéréographique, qui ramène les deux sphères sur des disques superposés, l'astrolabe sphérique en reste à une représentation du monde sous forme de deux sphères concentriques, la sphère extérieure étant réalisée schématiquement de façon à laisser voir les indications portées sur l'autre qui lui est intérieure. La présence ou l'absence de deux autres caractéristiques, le système de visée et le calendrier zodiacal, tenues pour participer de la définition de l'astrolabe planisphérique, est finalement de peu d'importance.

Cette définition a l'avantage de bien distinguer, en principe, l'astrolabe sphérique d'autres instruments également 'en relief', de forme sphérique: le globe céleste, qui n'a qu'un embryon de sphère fixe de référence, ou la sphère armillaire, où la représentation de la sphère céleste mobile se borne à la prise en compte du zodiaque, mais pas des étoiles. Mais, dans la pratique, il est certain qu'il n'est pas facile de faire le départ entre l'astrolabe sphérique et des 'sphères' qui pourraient à bon droit revendiquer d'y être rattachées.[5] D'autant que l'ambiguïté est entretenue par le flottement du vocabulaire dans les traités latins, qui emploient volontiers des termes comme 'spera solida', 'spera materialis' ou 'spera volubilis' pour désigner l'astrolabe sphérique.[6]

[3] Francis Maddison, 'A 15th-century Islamic spherical astrolabe', *Physis*, iv, 1962, pp. 101–109.

[4] Ernesto Canobbio, 'An important fragment of a west Islamic spherical astrolabe', *Annali dell'Istituto e Museo di Storia della Scienza di Firenze*, i, 1976, pp. 37–41.

[5] Que penser, par exemple, de la '*Compositio sphere materialis quam translucidam aut armillarem vocant*' ('*In superficie aliqua plana describe primo circulum* ...', Tübingen Ms.64, f. 2–12^v, du XVIe siècle), ou d'un bref texte anonyme du XVe siècle ('*Nota ad componendum globum in primis composita armilla* ...', Prague 433, f. 90^v), qui traitent d'un matériel applicable seulement au mouvement du soleil et à ses effets cosmographiques? Ou de la *Compositio sphere volubilis magistri Conradi Norici* (1509)) ('*Scire debetur quod hic non intendo ordinare spheram solidam nec spheram projectam in plano, sed spheram volubilem* ...', Vienne 5280, f. 27^v-29)? Le commentaire '*Quia presens scientia est introductoria ad astronomiam* ...', (Vienne 5145, f. 72–94^v), en revanche, ne nous concerne pas. L'absence de *globum* rend plus facile l'identification des sphères armillaires, dont il existe plusieurs traités de construction.

[6] Je ne suis pas d'accord pour suivre la distinction que fait Lorch *Op. cit.* (n. 2), entre la sphère solide et l'astrolabe sphérique: les deux expressions sont synonymes; c'est ainsi, par exemple, que le manuscrit

Sur les deux astrolabes arabes conservés, le programme de la double représentation des sphères est réalisé en inscrivant les cercles et graduations de la sphère céleste fixe de référence sur la sphère intérieure, et ceux de la sphère céleste mobile sur la sphère extérieure; le découpage et l'aspect de la seule sphère céleste qui soit conservée, celle de l'astrolabe sphérique d'Oxford, est morphologiquement assez semblable à celui d'une araignée d'astrolabe planisphérique, dont elle tient exactement le rôle.

Il semble que la littérature technique arabe propose de façon préférentielle la même disposition: la brève description d'al-Marrakushi, publiée en traduction française par Sédillot,[7] le traité d'Isaac ben Sid, connu par la version castillane insérée dans les *Libros del saber* d'Alphonse X de Castille,[8] ou le texte anonyme traduit en latin sans doute au XIe siècle sous le nom d'horologium secundum alkoram[9] inscrivent la sphère fixe de référence sur la sphère intérieure, et la sphère céleste mobile, c'est-à-dire le zodiaque et les étoiles, sur la sphère extérieure.

Mais une autre organisation est concevable: mettre la sphère céleste mobile à l'intérieur de la sphère fixe de référence. C'est la solution à laquelle se rallie Qusta ben Luqa, dont le texte, en soixante-cinq chapitres, existe en arabe,[10] en castillan,[11] en hébreu[12] et en latin. Il existe deux versions latines: l'une est anonyme ('Dixit Quasti filius Luce: Augeat Deus valorem principis... Spera est vas quoddam optime consolidatum in seipso et est infra quemdam circulum ...'),[13] l'autre ('Spera est corpus rotundum cujus centrum...')[14] fut faite par Stephanus Arlandi, médecin de Barcelone, qui l'a, en 1301, traduit de la version hébraïque de Profacius. La version de Stephanus Arlandi comporte un bref 66e chapitre sur la construction de l'instrument, tandis que la version anonyme est dotée d'un préambule.

Vienne 5412, f. 238, donne au traité de Jean de Harlebeke le titre: *Incipit tractatus de spera solida sive de astrolabio sperico compositus anno Domini 1303.*

[7] Texte analysé par L.-Am. Sédillot, 'Mémoire sur les instruments astronomiques des Arabes', *Mémoires présentés par divers savants à l'Académie royale des inscriptions et belles-lettres*, 1re série, i, 1844, pp. 1–229, aux pp. 142–148.

[8] Manuel Rico y Sinobas, éd. *Libros del saber del rey D. Alfonso X de Castilla*, 2, Madrid, 1863, pp. 113–222.

[9] 'Quando queris scire horas diei...', éd. J. Millàs Vallicrosa, *Assaig d'historia de les idees fisiques i matemàtiques a la Catalunya medieval*, Barcelone: Estudis universitaris catalans, série monogràfica, 1, 1931, pp. 288–290.

[10] Traduction partielle (quatorze chapitres sur soixante-cinq, les autres étant seulement présentés par leur titre) par W.H. Worrell, 'Qusta ibn Luqa on the use of the celestial globe', *Isis*, xxxv, 1944, pp. 285–293.

[11] M. Rico y Sinobas, ed. *Op. cit.* (n. 8), vol. 1, Madrid, 1863, pp. 153–205. La version castillane ajoute quatre chapitres de construction.

[12] *Histoire littéraire de la France*, vol. 27, 1877, p. 603. La version en hébreu, due à Profacius, daterait de 1256.

[13] TK 457 et Carmody 132. Munich, Clm 10662, f. 118–127v: 'Liber in opere spere volubilis quem composuit Quasti filius Luce ad Abjelhassin servum Dei filium Johannis'; Vienne 5273, f. 139–155v.

[14] TK 1523 et 1524. Wolfenbuttel 2816, f. 116–121v: 'Incipit practica spere solide cum utilitatibus suis, et fuit liber iste translatus de arabico in hebraycum in Montepessulano per Prophacium Judeum et postmodum de hebrayco in latinum per magistrum Stephanum Arlandi phisicum Barchinone anno Christi 1301 mense augusto, qui eum in certis capitulis reduxit'. Sur Stephanus Arlandi, v. E. Wickersheimer, *Dictionnaire biographique des médecins en France au Moyen Age*, Paris, 1936, pp. 135–136, et D. Jacquart, *Dictionnaire biographique..., supplément*, Genève, 1979, pp. 67–68.

Peut-être parce que ces traductions latines furent, plus que le texte fort obscur de l'*horologium secundum alkoram*, à l'origine de l'introduction de l'astrolabe sphérique dans l'Occident latin, c'est la combinaison de Qusta ben Luqa qui, semble-t-il, fut préférée par l'ensemble de la littérature latine sur l'astrolabe sphérique (à l'exception de l'*horologium secundum alkoram* cité ci-dessus).

Le texte latin sur l'astrolabe sphérique qui eut le plus de diffusion est assurément un traité de construction et d'usages, daté à l'occasion de 1303 et dont l'auteur est donné par plusieurs manuscrits comme Jean de Harlebeke: 'Totius astrologice speculationis radix et fundamentum ...'.[15] Il en existe de nombreux manuscrits. Mais on rencontre d'autres textes qui, pour avoir eu moins de succès, n'en attestent pas moins l'intérêt marqué par le monde latin pour l'astrolabe sphérique.[16]

La disposition suivie par Qusta ben Luqa et par les textes latins offre assurément des avantages pratiques considérables, en ce qu'elle simplifie sensiblement la confection de la sphère céleste mobile, qu'il n'est plus nécessaire de découper comme une araignée dans une tôle plus épaisse que celle qui sert à façonner un globe plein, et qu'elle améliore quelque peu les prestations escomptées de l'instrument puisqu'on peut alors sans inconvénient multiplier les étoiles et les regrouper selon les constellations. En revanche, la situation présente un notable inconvénient: il n'est plus possible de figurer les almucantarats et les azimuts de la sphère de référence.

Privée de ses almucantarats et de ses azimuts, la sphère fixe de référence se réduit donc à l'horizon et au méridien, c'est-à-dire au programme minimum auquel se tiennent ordinairement les globes célestes: c'est à bon droit que se pose le problème de la distinction à établir entre le globe céleste et l'astrolabe sphérique de type à sphère céleste intérieure. Je propose de l'établir sur des bases plus intellectuelles que matérielles, en retenant que, au niveau des principes, la référence à l'astrolabe sphérique traduirait l'effort fait pour tirer le maximum des possibilités de la rotation d'une sphère céleste devant un horizon et un méridien, tandis qu'une rotation produisant seulement la contemplation du mouvement du monde définirait alors l'instrument comme un simple globe céleste.

C'est peut-être en ayant conscience de cette ambiguïté qu'un auteur anonyme latin a entrepris de pallier les inconvénients propres à l'astrolabe sphérique occidental en y apportant un aménagement destiné à y introduire les almucantarats et les azimuts défaillants. Son texte ('Licet Costabenluce in suo tractatu de spera ...'),[17] que l'on

---

[15] TK 1576 et 298. Impr. : Venise, 1518. La date de 1303 figure par exemple dans les manuscrits Paris lat. 7294, f. 30ᵛ, et Vienne 5412, f. 238. L'autre nom alternativement parfois proposé pour l'auteur de ce texte, Accursius de Parme, doit être celui d'un copiste (cf. Lorch, *Op. cit.* (n. 6), p. 155).

[16] Voici quelques références. Construction, '*Habeas ergo primo armillam meridianam...*': Bruxelles 10117–26, f. 139–142 (*Tractatus compendiosus pro spera solida componenda tradendus intelligentibus et apprehendentibus sub brevibus verbis*). Construction, '*Ut vero faciamus exemplum per speram corporalem secundum quod jam firmavimus ...*', Paris, lat. 7281, f. 247–248 (sans doute partie d'un traité *De sphera celesti* de Jean de Troyes, '*Centrum est cujus ab initio circumductio...*'). Construction et usages, '*Volens spheram componere solidam prout globum habeat rotundum...*', Vienne 5228, f. 39–45ᵛ). Construction, '*Formentur primo duo circuli equales ...*', Munich, Clm 19689, f. 269ᵛ–270.

[17] Paris, Bibl. nat., lat. 10266, f. 45–49ᵛ. Cf. E. Poulle, *La bibliothèque scientifique d'un imprimeur humaniste au XVe siècle, catalogue des manuscrits d'Arnaud de Bruxelles à la Bibliothèque nationale de Paris*, Genève: Travaux d'humanisme et renaissance 1963, 56, p. 63.

trouvera ci-après en annexe, présente l'intérêt supplémentaire de s'ouvrir par une critique extrèmement vive du traité de Qusta ben Luqa, dont, avec virulence, il énumère les usages abusifs, dénonce les erreurs, souligne les insuffisances. Je ne puis affirmer avec certitude à laquelle des deux versions latines de Qusta ben Luqa il est fait référence: apparemment, les citations qui en sont données renverraient plus volontiers à la version anonyme 'Spera est vas quoddam ...', mais je n'ai pu retrouver dans celle-ci l'emploi, qui traduit effectivement une incroyable bévue, qui y serait fait de *aux* pour désigner la hauteur méridienne, emploi qui est en revanche attesté dans la version de Stephanus Arlandi.

Après cette volée de bois vert, l'auteur propose d'ajouter à l'astrolabe sphérique, outre un calendrier zodiacal et un système de visée qui font trop partie de la 'culture' astrolabique pour que leur absence ne soit pas regrettée,[18] un *membrum* en forme de triangle sphérique réalisé sur trois quarts de grands cercles, dont un des côtés est gradué en quatre-vingt-dix degrés; ce triangle peut glisser sur la sphère céleste en passant sous l'armille qui représente le méridien et en prenant appui sur l'armille qui traduit l'horizon, le point 90 de la graduation étant au sommet opposé au côté qui coulisse sur l'horizon: quand le triangle pivote autour de son sommet en prenant appui sur l'horizon, chaque point de la graduation du triangle décrit un cercle parallèle à l'horizon, donc un almucantarat virtuel.[19]

Par opposition à l'exposé jugé pléthorique de Qusta ben Luqa, les usages de notre auteur latin se veulent en petit nombre et font explicitement référence au modèle fourni par la littérature technique sur l'astrolabe planisphérique, promu au rang de norme de ce qu'on peut attendre d'un astrolabe, quel que soit son type. Le texte visé pour cette référence est le traité de Jean de Séville, qui ne peut être mieux identifié: plusieurs traductions de l'arabe lui sont imputées, la plus banale étant celle du traité dit de Messahalla;[20] mais celle-ci est constamment alléguée sous le nom de son prétendu auteur arabe et non sous celui de son traducteur.

Je ne connais le texte 'Licet Costabenluce' que par un manuscrit tardif (et d'ailleurs peu sûr: j'ai dû le corriger à plusieurs reprises pour le rendre intelligible), provenant de la bibliothèque d'Arnaud de Bruxelles, à la fin du XVe siècle. La seule référence qui pourrait être utilisable pour le dater, la mention d'un *episcopus Consanus* qui serait le promoteur ou le parrain de l'entreprise, reste incompréhensible; je serais tenté d'y

---

[18] Notre auteur anonyme latin n'est évidemment pas seul à prévoir un système de visée et un calendrier zodiacal, dont il est assez simple d'équiper un astrolabe sphérique, encore que le système de visée donne facilement lieu à des constructions peu convaincantes; par exemple, dans le traité de Jean de Harlebeke, le système de visée dépend d'une suspension de l'instrument réalisée par quatre cordes qui ne sont pas sans évoquer un enscensoir.

[19] Le traité d'al-Marrakushi comportait aussi un triangle sphérique pour compléter l'araignée et la pourvoir d'un réseau de cercles de hauteur (cf. L.-A. Sédillot *Op. cit.* (n. 7), p. 148). Jean de Harlebeke propose aussi un dispositif pour doter la sphère fixe de référence d'almucantarats virtuels.

[20] Sur le véritable auteur de ce texte, v. P. Kunitzsch, 'On the authenticity of the treatise on the composition and use of the astrolabe ascribed to Messahalla', *Archives internationales d'histoire des sciences*, xxxi, 1981, pp. 42–62.

voir l'intervention de Nicolas de Cues s'il n'eût pas été plus logique de désigner celui-ci par son titre de cardinal.

Edition d'un traité anonyme latin de l'astrolabe sphérique (Paris, lat. 10266, f. 45–49v)

Licet Costabenluce* in suo tractatu de spera pluribus utilitatibus docuit uti quam Johannes Hispalensis in tractatu de astrolabio, dico quod omnes utilitates quas adinvenit ultra utilitates quas docet Johannes aut sunt parum utiles et non necessarie ad judicia astrologie aut sunt manifeste omni astrologo; et, si necessarie, doctrina earum non est bona, nec etiam multarum aliarum que sunt de numero utilitatum Johannis; nec etiam docet omnes illas quas docet idem Johannes; et quamvis fecerit 65 capitula non sunt 65 utilitates: nam aliqua sunt per modum narrationis et aliqua alia significantia rem unam. Primum, secundum et tertium sunt per modum narrationis nisi quod in fine tertii docet scire arcum diurnum et nocturnum diei ignoti in gradibus orizontis, quare erravit vel fuit vicium scriptoris (1). Quartum, quintum, septimum et octavum debet scire omnis astrologus. Sextum, nonum, decimum, undecimum et duodecimum paterent per finem tertii dum tamen doctrina esset bona; et omnis utilitas que precedit 9m est pro die ignoto, et in nono dixit: 'primo scias gradum solis' (2), quem non docet scire per speram, et sic utilitas noni capituli et alie que per gradum solis sciuntur non possunt sciri per solam speram. Item capitulo tertio decimo docet scire quot hore equales sunt transacte ex die per ascendentem, et quarto decimo quot inequales, et, cum ascendens sit ignotus, erunt hore equales et inequales ignote (3); sed presupposito quod sciatur ascendens per aliquod aliud instrumentum, dico quod quicquid scitur per ascendens nescitur per solam speram. Item capitulo 15 docet scire ascendentem per horas equales, et sexto decimo per inequales (4), sed hore equales et inequales per suam doctrinam non possunt sciri nisi prius sciatur ascendens et cum sciatur frustra indagatur, quare ista duo capitula frustra ponuntur. Item 24 et 23 significant rem unam (5). Item 40 dicit sic: 'horam qua ascendit luna vel aliqua stellarum errantium in quacumque civitate et nocte potes scire elevando polum articum ab orizonte secundum latitudinem civitatis; postea scias gradum lune vel alicujus planete' (6); tunc dico quod hora qua ascendit planeta aut est nota aut est ignota; si nota, frustra indagatur; si ignota, impossibile est scire gradum planete; et [idem] (7) peccatum accidit sibi capitulo sequenti de occasu (8): in luna posset contingere error propter velocitatem sui motus, sed in aliis est insensibilis quia sufficit scire gradum pro illo die vel nocte, et est sciendum quod hec doctrina habet locum tam de die quam de nocte (9). Item 53 docet scire altitudinem solis cum spera et jubet ponere [stilum] (10) supra gradum solis in zodiaco firmiter cum modico cere (11), et non dicit quod illud signum sit oblongum et equale in grossicie et erectum supra superficiem spere, que omnia hec accidentia sunt necessaria, aliter continget error. Item capitulo 58 docet scire locum lune vel alicujus planete de nocte (12); dico etiam quod hec doctrina habet locum dum planeta caret latitudine vel non caret et invenitur et in capite cancri vel capricorni; sed si fuerit in capite arietis vel libre continget error; nam si caput arietis fuerit in linea medii celi et etiam planeta inveniatur in eadem linea et ejus latitudo sit septentrionalis erit planeta in gradu vel gradibus arietis, et si fuerit meridionalis erit in gradibus piscium; et in capite libre fit [econverso] (13) idest si latitudo erit septentrionalis erit in gradibus virginis, et si meridionalis erit in gradu vel gradibus libre; quod satis patet cuique ymaginanti talem statum seu situm spere; et secundum quod plus distabit a capite arietis vel libre minor continget error, et quia in hoc contingit error, et in sequenti continget sed insensibilis. Item capitulo 60 docet prescire per unum diem vel unum diem cum dimidio vel duos fere si luna debet eclipsari vel non (14), et dicit quod si luna nullam habuerit latitudinem [15a] (15) nocte eclipsabitur illo mense, et si habuerit majorem uno gradu et 4 (16) minutis non eclipsabitur, et si minorem eclipsabitur; dico ego quod in omnibus tribus terminis est possibile errorem contingere: cum enim media oppositio fiat post 14 dies 18 horas et 22 minuta novi autem quandoque citius quandoque tardius, et sic est bene possibile inter [15am] (15) noctem et veram oppositionem esse duo dies fere, et est possibile quod in duobus diebus moveatur luna 26 gradibus et 33 minutis fere et sic in 13a nocte nullam habuerit latitudinem; in hora vere oppositionis poterit habere latitudinem fere duorum graduum quare non eclipsabitur, quod est contra ipsum; et si in [15a] (15) nocte habuerit minorem latitudinem uno gradu et 4 (16) minutis et sit modicum ante caput vel cauda draconis adhuc poterit esse latitudo tanta in hora oppositionis quod non poterit eclipsari; et si post caput vel caudam, multo minus, quod est contra tertium; et si latitudo fuerit minor uno gradu et 4 (16) minutis et precedat caput vel caudam, est bene possibile quod in hora vere oppositionis sit latitudo nulla vel tam modica quod poterit luna eclipsari, quia semidiameter umbre terre subceditur 40 minutis et duobus tertiis minute et semidiameter lune succenditur 15 minutis et medietate et 6e minuti sicut patet 14o capitulo 5e dictionis libri Almagesti; et hoc patet quod, si in hoc oppositionis latitudo minor hiis duabus quantitatibus aggregatis que sunt 56 minuta et 20 secunda, erit possibile lunam eclipsari, quod est contra secundum. Item idem peccatum accidit sibi capitulo sequenti de eclipsi solari. Item altitudinem meridianam nominavit augem (17); sed omnes astrologi, compositores tabularum ac actores theorice planetarum et ceteri punctum in ecentrico vel in epiciclo qui maxime removetur a centro terre vel a centro equantis nominaverunt augem, et etiam quendam arcum nominaverunt augem in 2a significatione, sed nunquam nominaverunt altitudinem meridianam alicujus stelle augem nec altitudo meridiana plus removetur a centro terre quam alia altitudo, quare videtur ignorare et a vocabulis aliorum astrologorum deviare.

Hec sunt utilitates que habentur per speram solidam reverendi in Christo patris ac domini domini episcopi Consani, que breviter sunt omnes que habentur per astrolabium Johannis Hyspalensis, hoc addito: quia habet locum pro omnibus regionibus mundi sine additione et diminutione et etiam quia spera est integra, poterit haberi noticia cujuslibet stelle ut per eam possit judicari cum fuerit necesse, quod non potest per astrolabium. Et est sciendum quod, de mandato prefati reverendi in Christo patris, preter facturam communium sperarum in hac spera aliqua addidi et mutavi et ad hoc ut plures habeantur utilitates et cum spera sola sine alio instrumento.

Ut autem habeatur verus locus solis sicut in dorso astrolabii, addidi in superficie spere circulum mensium et situavi eum prope tropicum hiemalem versus polum antarticum ubi sunt pauce stelle, et designavi supra polum zodiaci; verumtamen divisi eum supra circulum ecentricum et prope ipsum designavi alium quasi contingentem undique in quo sunt gradus signorum designati. Item addidi unam alidadam cujus polum posui in medio arcus orizontis orientalis et est hec alidada reflexa ita quod, quando ipsa movetur, una extremitas ejus tangit circulum meridianum in cujus una quartarum sunt gradus altitudinis, in alia vero puncta umbre; et in medio istarum duarum quartarum est una armilla suspensoria. Et hic meridianus est fixus in orizonte, et pars que est super orizontem non contingit superficiem spere ymo est aliquantulum remota ad hoc ut possit duci supra superficiem spere unum aliud membrum factum ad modum trianguli continentis octavam partem superficiei spere cujus unumquodque latus est equale quarte parti circuli magni superficiei spere et alterum trium laterum divisum est in 90 partes equales, per quod membrum multe utilitates habebuntur ut infra continetur. Et etiam ut idem membrum possit duci super omnem partem superficiei spere existentem supra orizontem, non posui axem in polo artico et dico quod sufficit solummodo in antartico quia spera non potest moveri ante nec retro a destris nec a sinistris propter contactum orizontis sui sive propter sui molem deorsum propter contactum sue sedis. Et feci axem qui intrat polum existentem sub orizonte artificiose currentem per unam quartam circuli meridiani existentem sub orizonte. Et quia hec omnia supradicta et alia cuique intuenti patebunt, nunc de tali narratione facture sufficiat et ad utilitates revertamur, et primo dicamus de vero loco solis.

[1]. De vero loco solis. Signum et gradum in quo erit sol sic reperies. Quere in circulo mensium diem in quo eris et in ejus directo in circulo sequenti versus zodiacum invenies gradum in quo erit sol, in cujus directo invenies nomen signi in zodiaco.

[2]. Sequitur de altitudine solis vel stelle ut per eam possit haberi figura celi. Altitudinem solis sic accipies. Et primo ad hoc quod spera cum suis circunstanciis sit equeponderans, pone triangulum versus occidens ita quod medium unius lateris contingat medium arcus orizontis occidentalis, contingente concava superficie trianguli conexam superficiem spere, et immoto triangulo tene speram per armillam suspensoriam cum manu dextera et dirige eam versus solem et eleva alidadam quousque radii solares transeant ambo foramina duarum tabularum; tunc vide in quarta meridiani divisa in 90 partes equales quotum gradum extremitas alidade tetigerit, quia ille erit gradus altitudinis. Et si puncta umbre volueris invenire tene armillam manu sinistra et fac ut superius dictum est, et extremitas alidade ostendet in alia quarta puncta umbre. Et si ceciderit extremitas alidade supra medietatem illius quarte versus orizontem, erunt puncta umbre recte; et si versus cenit capitum, erunt puncta umbre verse. Eodem modo facies cum volueris scire altitudinem stellarum vel aliorum corporum in gradibus vel in punctis, aspiciendo per ambo foramina duarum tabularum, et sic poteris judicare de altitudinibus turrium, profunditatibus puteorum, dictanciis locorum et hujusmodi ceteris, operando secundum doctrinam practice geometrie.

[3]. De declinatione solis ab equinoctiali. Quanta est declinatio ab equinoctiali sic scies. Latus trianguli quod subtenditur angulo in quo scribitur 90 pone taliter quod sit in contactu superficiei orizontis et superficiei spere, et cum unumquodque latus trianguli sit equale quadranti circuli magni superficiei spere, angulus in quo scribitur 90 cadet in cenit, et sic immoto hoc angulo a cenit duc triangulum quousque angulus in quo incipit numerus sit supra lineam seu notam in orizonte que est in directo circuli meridiani, tunc manifestum est quod latus divisum erit in linea medii celi; et immoto hoc triangulo volve speram quousque gradus solis sit in contactu lateris divisi, et vide quot gradus erunt inter gradum solis et equinoctialem, quia tanta erit declinatio solis. Et si propter motum spere gradus solis non obviet lateri diviso, volve triangulum in partem oppositam. Item si gradus solis obviet lateri diviso et non obviet equinoctialis eidem lateri, vide quot gradus erunt inter gradum solis et cenit, et immota spera volve triangulum in partem oppositam sicut prius et vide quot gradus erunt inter equinoctialem et cenit, quos aggregabit cum primis et aggregatum erit declinatio solis.

Vel sic. Angulum cui subtenditur latus divisum pone in contactu medii arcus orizontis orientalis vel occidentalis ubi equinoctialis secat orizontem, et immoto hoc angulo duc triangulum a septentrione ad meridiem vel econverso quousque movente spera gradus solis et equinoctialis obvient lateri diviso, quod, cum triangulus contineat octavam partem superficiei spere, erit latus divisum in linea medii celi; tunc volve speram etc.

Vel aliter. Pone duos polos spere in contactu orizontis et immotis polis volve speram quousque gradus solis sit in contactu orizontis orientalis vel occidentalis et numera gradus orizontis qui sunt inter gradum solis et equinoctialem, quia tanta erit declinatio solis. Cum enim fuerit triangulus in dispositione supradicta, scilicet quod angulus in quo scribitur 90 sit in puncto cenit capitum et hoc in angulo fixo, circumducatur triangulus contingente superficie ejus superficiem spere, manifestum est quod divisiones seu note lateris divisi describent circulos equidistantes orizonti qui in tractatu astrolabii nominantur almucantarath; et etiam cum latus divisum cadat supra divisiones orizontis et supra cenit, denotabit azimuth, quapropter in sequentibus nominentur almucantarath et azimuth.

[4]. De latitudine regionis, et primo quid sit latitudo. Latitudo regionis est distancia equinoctialis a cenith capitum computato in linea meridiana que est equaliter elevationi poli supra orizontem. Cum igitur volueris scire latitudinem regionis in qua eris, accipe altitudinem solis existentis in linea meridiana secundum doctrinam secundi hujus; et si linea meridiana fuerit tibi ignota, accipe altitudinem solis ipso ascendente frequenter quousque incipiat declinari, et major altitudo quam invenisti est altitudo meridiana. Item secundum doctrinam 3ii vide quanta est declinatio solis, quam adde altitudini solis invente si fuerit sol in signis meridionalibus, et si in septentrionalibus subtrahe, et quod remanserit erit altitudo capitis arietis et libre, quam subtrahe a 90 et quod remanserit erit latitudo vel elevatio poli supra orizontem ipsius regionis.

[5]. De aptatione spere seu elevatione poli secundum latitudinem regionis cujuslibet. Cum scita fuerit latitudo cujuslibet regionis et secundum hanc latitudinem volentes

aptare speram, idest elevare polum supra orizontem ut per eam possint haberi multe alie utilitates, pone latus trianguli divisum in linea meridiana versus polum articum sicut dictum est in 30; deinde duc axem qui est in polo antartico quousque polus articus cadat supra tot gradus in latere diviso quot sunt in latitudine regionis computando ab orizonte versus cenit, vel pone ipsum in parte opposita et duc axem quousque equinoctialis cadat supra tot gradus quot sunt in latitudine regionis, computando a cenit versus orizontem, vel quot sunt in altitudine capitis arietis et libre, computando ab orizonte versus cenit.

Vel sic. Accipe latitudinem meridianam sicut dictum est in precedenti et volve circa polos quousque gradus solis sit in linea meridiana; deinde duc axem quousque gradus solis cadat supra tot gradus quot sunt in altitudine meridiana; et sicut fecisti per gradum solis facies per aliquam stellam fixam ponendo centrum stelle in linea meridiana. Et tunc fige axem artificiose cum uno conio ita quod non recedat spera a tali situ, quia aptata erit spera pro illa regione ita quod si volvatur ab oriente ad occidens describet supra orizontem equales angulos et similes arcus angulis et arcubus spere celestis descriptis supra orizontem ipsius regionis. Et sciendum quod hec ultima doctrina valet pro hac utilitate et pro precedenti.

[6]. De quatuor plagis mundi inveniendis. Si quatuor plagas mundi scire volueris, accipe altitudinem solis et intrantibus radiis solaribus foramina duarum tabularum alidade depone speram supra planum et, licet superius dictum sit quod alidada est reflexa, tamen linea protracta a foramine unius tabule ad foramen alterius describit circulum cujus superficies est equedistans superficiei circuli meridiani spere. Quapropter cum deposita fuerit spera supra planum intrantibus etc., meridianus erit equedistans azimuth in quo erit sol et sic vera linea meridiana tantum distabit a circulo meridiano versus occidens si fuerit ante meridiem, vel versus oriens si fuerit post, quantum azimuth solis a linea meridiana. Et si volueris scire quanta est hec distancia pone gradum solis supra similem altitudinem in almucantarath hoc modo: quot gradus sunt in altitudine solis inventa, tot computa in latere trianguli ab orizonte versus cenit, et volve speram ab oriente versus occidens et triangulum circa cenit quousque gradus solis et gradus altitudinis simul concurrant. Tunc vide in orizonte quot gradus sunt inter meridianum et locum quem tangit extremitas lateris divisi quia tot erunt inter veram lineam meridianam et circulum meridianum spere versus occidens si fuerit ante meridiem, vel econverso.

Vel fac sic, quod est brevius. Pone gradum solis supra similem altitudinem in almucantarath versus occidens si fuerit ante meridiem, vel econverso, et latus divisum ostendet in orizonte plagam meridionalem, et nadir ejus septentrionalem; orientalem et occidentalem invenies computando 90 gradus in utramque partem a vera linea meridiana inventa. Et sicut scivisti per solem scies per aliquam stellam fixam ponendo centrum corporis stelle supra similem altitudinem almucantarath. Et si supra idem planum volueris designare has lineas inventas, fac unum perpendiculum oblongum cujus extremitas inferior sit acuta; tunc pone filum perpendiculi in directo note invente in orizonte et dimitte descendere perpendiculum quousque extremitas inferior tangat planum et fac ibi notam; et fac sic in aliis; deinde amove speram ab illo plano et ab unaquaque nota ad aliam sibi oppositam protrahe lineam rectam et habebis quod queris.

[7]. De latitudine ortus vel occasus solis. Si scire volueris latitudinem ortus vel occasus solis, idest in quo loco orizontis obliqui oriatur vel occidat sol quolibet die, volve speram quousque gradus solis sit in contactu orizontis orientalis vel occidentalis, et computa quot gradus sunt in orizonte inter gradum solis et locum in quo equinoctialis secat orizontem, quia tanta erit latitudo ortus vel occasus solis. Quod si hoc scire volueris supra terram, pone speram supra planum taliter quod circulus meridianus spere sit in directo linee meridiane invente supra planum, quod examinari poterit cum perpendiculo. Item cum eodem perpendiculo poterunt designari note ortus et occasus sicut superius dictum est.

Vel sic. In sectione duarum linearum protractarum in plano pone unum pedem circini et secundum quantitatem linearum describe circulum cujus unamquamque quartarum divide in 90 partes ad similitudinem orizontis spere, et similem latitudinem latitudini quam invenisti in spera quere in circulo illius plani, quia versus illam partem orizontis orietur sol vel occidet. Et hoc idem poteris scire de qualibet stella fixa situata in spera, ponendo centrum corporis stelle in contactu orizontis orientalis vel occidentalis etc.

[8]. De altitudine meridiana. Altitudinem meridianam cujuslibet diei sic scies. Pone latus trianguli divisum in linea medii celi sicut in 3o dictum est. Deinde volve speram quousque gradus solis sit in contactu ipsius lateris, et vide in hoc latere quot gradus erunt inter orizontem et gradum solis, quia tanta erit altitudo meridiana.

[9]. De gradu ascendentis inveniendo. Cum volueris scire gradum zodiaci qui est in ascendente, accipe altitudinem solis et gradum solis pone supra similem altitudinem in almucantarath; tunc vide quis gradus zodiaci est in contactu orizontis orientalis, quia ille erit ascendens in hora acceptionis altitudinis solis. Et hoc idem de nocte scies per aliquam stellam fixam situatam in spera, accipiendo altitudinem ipsius et ponendo notam ejus supra similem altitudinem in almucantarath, et gradus zodiaci inventus in orizonte orientali erit ascendens.

[10]. De horis equalibus. Si volueris scire quot hore equales transierunt de die quavis et in quavis hora diei, pone gradum solis in orizonte orientali et ubi idem orizon tangit equinoctialem ibi fac notam. Deinde accepta altitudine solis pone gradum solis in almucantarath cum eadem altitudine, et ubi tunc orizon orientalis tangit equinoctialem fac secundam notam. Postea accipe distanciam sive quantitatem graduum qui sunt inter primam notam et secundam, quos gradus divide per 15, et quod provenerit erunt hore, et quod remansit erunt partes hore.

Hoc idem sic scire poteris de nocte: pone nadir solis supra orizontem orientalem et ubi idem orizon tangit equinoctialem fac notam; deinde accepta altitudine alicujus stelle fixe signate in spera pone ostensorium solis supra almucantarath cum eadem altitudine, et ubi tunc orizon orientalis tangit equinoctialem fac notam; postea accipe distanciam et operare ut supra.

[11]. De quantitate arcus diurni. Si volueris scire quantitatem arcus diurni et quot horas equales continet quilibet dies, pone gradum solis in orizonte orientali et ubi idem orizon tangit equinoctialem ibi fac notam. Deinde move speram versus lineam meridiei donec gradus solis perveniat ad orizontem occidentalem, et ubi orizon orientalis tangit equinoctialem ibi fac secundam notam. Tunc accipe distanciam sive quantitatem graduum qui sunt inter utramque notam, quos divide per 15 et exibunt hore equales; et si aliquid superfuerit erit pars hore. Et si volueris scire quantitatem arcus nocturni accipe gradus inventos inter utramque notam, quos minue de 360 et quod remanserit divide per 15 et habebis horas noctis. Vel minue horas diei de 24 et restabunt hore noctis.

[12]. De quantitatibus hore inequalis cujuslibet diei. Si volueris scire quot gradus continet quelibet hora inequalis, divide arcum diurnum per 12 et secundum illud quod supervenerit tot gradus continet quelibet hora inequalis. Et similiter si arcum nocturnum diviseris per 12: secundum illud quod supervenerit tot gradus continet quelibet hora noctis.

[13]. De horis inequalibus preteritis et futuris inveniendis. Si volueris scire quot hore inequales transiverunt de die quavis et in quavis hora diei, scito quot hore equales transiverunt de die, gradus earum aggregatos divide per quantitatem unius hore inequalis diei et quot provenerit erunt hore inequales. Et si hoc idem de nocte volueris scire, scito quot hore equales noctis transiverunt, gradus eorum aggregatos divide per quantitatem unius hore inequalis noctis, et quot provenerit erunt hore inequales.

Alias utilitates scies intrando canones tractatus astrolabii sequentes precedentem, verumtamen ubi invenies 'ubi almuri tangit limbum' intellige 'ubi orizon orientalis tangit equinoctialem'; et ubi invenies 'move rethe' intellige 'move speram' prout in 11o et 12o hujus. Item ubi invenies 'in gradibus dorsi' intellige 'in gradibus circuli meridiani qui sunt in quarta divisa in 90 partes equales, in cujus alia quarta sunt puncta umbre', prout est in secundo hujus.

[14]. De equatione duodecim domorum. Item duodecim domus equantur per alium modum quam per modum tractatus astrolabii, et equantur hoc modo. Gradum qui est ascendens pone in orizonte orientali qui erit initium prime domus, et ejus nadir erit initium 7e domus; deinde pone triangulum taliter quod alterum duorum laterum indivisorum sit in contactu orizontis occidentalis superficiei spere et angulus cui subtenditur latus divisum sit in contactu linee seu note que est in directo circuli meridiani; tunc manifestum est quod unum duorum laterum indivisorum erit in linea medii celi, et sic gradus zodiaci qui erit in contactu ejus erit initium 10e domus, et ejus nadir erit initium 4e domus. Hoc facto, immoto angulo cui subtenditur latus divisum, move triangulum ab occidente versus oriens per 30 gradus quos computabis in latere diviso a puncto qui erat in cenit versus occidens, et sic pone 30m gradum in directo claviculi qui est in cenit, et scias quod gradus zodiaci qui est in contactu lateris indivisi versus occidens est initium 8e domus, et nadir ejus est initium 2e domus; et ille qui est in contactu lateris indivisi versus oriens est initium 11e domus, et ejus nadir est initium

5e domus. Et iterum move triangulum per alios 30 gradus versus oriens et gradus zodiaci qui erit in contactu lateris indivisi versus occidens erit initium 9e domus, et ejus nadir erit initium 3e domus; et ille qui erit in contactu lateris indivisi versus oriens erit initium 12e domus, et ejus nadir erit initium 6e domus. Et si in contactu alicujus duorum laterum indivisorum non esset aliquis gradus zodiaci, angulum cui subtenditur latus divisum pone in partem oppositam et operare ut supra. Et est sciendum quod iste sunt domus de quibus loquitur Albumazar in Introductorio suo tractatu 6, differentia 26.

Et est sciendum quod omnibus utilitatibus poteris uti a regione cujus latitudo est 0, scilicet ubi equinoctialis est ad cenit, usque ad regionem cujus latitudo est 66 gradus et 27 minuta; ultra vero non est possibile uti omnibus utilitatibus et omni tempore quia plus quam 24 horas equales stat sol supra terram una vice, et alia vice plus quam per totidem horas sub terra. Et si hoc volueris experiri, apta speram secundum latitudinem regionis datam, et invenies proprietates linearum que sunt ab equinoctiali usque ad polum articum de quibus loquitur Ptholomeus Almagesti dictione 2a capitulo 6o. Et si volueris scire ab equinoctiali usque ad polum antarticum, eleva corpus spere a sua sede et pone polum articum in axe inferius et antarticum superius, et orizon orientalis vocetur occidentalis et econverso; et moveatur spera a manu dextra versus sinistram, idest ab orizonte in quo scribitur 'orizon occidentalis', et invenies sicut in alia parte; et sic spera habet locum pro universo mundo sine addito, quod dictum fuit in principio.

* Pour apprécier la pertinence des critiques faites au traité de Qusta ben Luqa, les passages visés par ces critiques seront donnés en notes, dans l'une et l'autre version. La version de Stephanus Arlandi (d'après le manuscrit de Wolfenbuttel 2816) sera désignée par A, la version anonyme (d'après le manuscrit de Munich, Clm 10662) par B.

1. Titre du ch. 3: *A.* De figure ipsius (= spere); *B.* In sciencia figure celi et dispositionis ejus et motus ejus. Le reproche ne porte pas sur la détermination de l'arc diurne, effectivement mentionnée en fin de chapitre, mais sur la présence de cette détermination dans un chapitre où elle n'est pas annoncée.
2. Ch. 9: *A.* Ad sciendum diversitatem duorum dierum in una civitate. Diversitatem duorum dierum in qualibet civitate scire poteris si polum articum eleves secundum altitudinem civitatis illius. Scias primo locum solis in utroque duorum dierum quorum diversitatem vis scire ...; *B.* In scientia diversitatis que est inter quoscumque dies volueris in quacumque villa volueris. Cum volueris hoc, eleva polum septentrionalem super orizonto [*sic*] in mensura latitudinis ville in qua scire volueris; et scito gradum solis in diebus quorum volueris scire diversitatem...
3. Titre du ch. 13: *A.* Ad sciendum horas que preterierunt ex die per gradum ascendentis; *B.* In scientia ejus quod transit de die de horis equalibus cum fuerit ascendens notus. Titre du ch. 14: *A.* Ad sciendum quid transactum est ex horis naturalibus alicujus diei per gradum ascendentis; *B.* In scientia ejus quod preteriit de die de horis inequalibus quando fuerit ascendens notus.

4. Titre du ch. 15: *A.* Ad sciendum ascendens per horas equales diei transactas; *B.* In extractione ascendentis quando erit quod preteriit de die de horis equalibus notum. Titre du ch. 16: *A.* Ad sciendum ascendens per horas diei naturales transactas; *B.* In extractione ascendentis quando erit quod preteriit de die de horis inequalibus notum.
5. Titre du ch. 23: *A.* Ad sciendum quantum distat stella fixa a zenith capitis nostri; *B.* In extractione cujuscumque stelle volueris de stellis signatis super spera a cenith capitum hominum cujuscumque ville volueris. Titre du ch. 24: *A.* Ad sciendum altitudinem altiorem alicujus stelle fixe in hoc instrumento posite in quolibet orizonte; *B.* In scientia maxime altitudinis cujuscumque stelle fixe de stellis signatis in spera in quacumque villa volueris.
6. Ch. 40: *A.* Ad sciendum ortum planetarum. Scias verum locum illius planete in zodiaco tam in longitudine quam in latitudine, similiter partem latitudinis, quem gradum pone sub meridiano; postea computa a gradu meridionalis circuli a loco illius gradus quanta est latitudo stelle; quo facto fac signum in spera ubi scilicet venit ille gradus latitudinis quia illud signum est locus planete; postea volve speram donec sit gradus solis super orizontem occidentalem signando in oriente gradum equinoctialis; consequenter fac volvi speram donec illud signum quod signasti ad locum planete fuerit, et fac tunc signum super orizontem orientalem in equinoctiali et computa quot gradus infra hec duo signa fuerunt, quos per 15 dividas si vis habere horas equales quibus transactis orietur planeta in nocte illa; *B.* Ad sciendum horam ascensus lune vel cujuscumque stelle volueris de stellis erraticis in quacumque nocte et in quacumque villa volueris. Cum volueris hoc, eleva polum septentrionalem super orizonto [*sic*] in mensura latitudinis ville in qua hoc scire volueris; post hoc, scito gradum lune vel stelle cujus vis scire horam ascensus de gradibus circuli signorum et latitudinem ejus et latus latitudinis ejus, et volve speram donec erit gradus ille de gradibus circuli signorum sub linea meridiei; post hoc computa de gradibus circuli meridiani a loco gradus illius in mensura latitudinis lune vel stelle alterius erratice in latere latitudinis ejus et signabis super speram in loco propinquo illi gradui in quo terminatur numerus; locus enim signatus in spera erit locus lune vel stelle illius erratice in nocte illa; post hoc volve speram donec gradus solis sit in orizonte occidentali et signabis super gradum equinoctialis qui tunc erit in orizonte orientis et volve speram donec sit locus lune vel stelle erratice quem signasti in orizonte orientis et signabis super gradum equinoctialis qui tunc erit in orizonte orientis, et computa quod est inter duas notas et quod est inter eas de gradibus divide per 15 si volueris horas ascensus illius stelle equales, vel per quantitatem unius hore equalis noctis illius si volueris horas ascensus illius inequales, et quod exibit post divisionem utrolibet modo operatus fuerit ostendet tibi horam in qua ascendet luna vel stella illa erratica in nocte illa in villa illa.
7. *Corr. de* 18.

8. Titre du ch. 41: *A.* Ad sciendum occasum planetarum; *B.* In extractione hore occasus lune vel cujuscumque stelle volueris de stellis erraticis in quacumque villa et in quacumque nocte volueris.
9. Nulle part, dans l'une ou dans l'autre version, il n'est question de l'erreur que peut introduire la vitesse du mouvement de la lune.
10. *Corr. de* signum.
11. Ch. 53: *B* In acceptione altitudinis solis in spera quocumque tempore et in quacumque hora volueris. ...et erige stilum unum stantem perpendiculariter super gradum solis ad angulos rectos et adherat spere fortiter cum cera vel cum hujusmodi ...; le texte de *A* ne mentionne pas de stilum, mais il est sans doute fautif, avec, d'une part, une lacune oû aurait dû apparaître le mot *stilum,* et ensuite la même confusion *stilum/signum* que fait le ms lat. 10266.
12. Titre du ch. 58: *A.* Ad sciendum locum lune vel alterius planete; *B.* In extractione loci lune vel cujuscumque stelle volueris de stellis erraticis in nocte in qua poteris accipere maximam altitudinem ejus.
13. *Corr. de* equatio.
14. Ch. 60: *A.* Utrum debeat esse eclipsis lune vel non. Scias latitudinem lune secundum capitulum precedens, et si nullam habeat latitudinem et (complet) altitudinem altiorem cadat in quibusdam zodiaci gradibus tunc eclipsabitur illo mense; et si invenias sibi majorem latitudinem gradu uno et 4 (*corr. de* 54) minutis, scias illam non eclipsari in mense illo; et si minor est latitudo gradu uno et minutis 4, eclipsabitur; *B.* In scientia eclipsis lune an infaillibiliter contingat in mense in quo es. Cum volueris hoc, scito latitudinem lune in nocte 15a sicut didicisti in capitulo precedenti et si inveneris quod nullam habeat latitudinem et quod maxima ejus altitudo (*corr. de* latitudo) in nocte illa cadat super gradum aliquem circuli signorum, scito quod luna eclipsabitur in mense isto; si vero fuerit ejus latitudo plus uno gradu et 4 minutis scito quod luna non eclipsabitur in mense illo; si vero fuerit ei latitudo minus uno gradu et 4 minutis ipsa eclipsabitur absque omni dubitatione.
15. *Corr. de* 13.
16. *Corr. de* 24.
17. Le mot ne figure pas dans la version *B.* Il figure dans les ch. 64 et 65 de la version *A*: ch. 64: Ad sciendum distanciam stellarum fixarum in spera non descriptarum ab equinoctiali. Scias augem altitudinis ejus computando quantum distet a gradu meridionalis in parte in qua invenisti eandem ...; ch. 65: Ad sciendum distanciam stellarum fixarum in spera non descriptarum a zenith capitis. Scias augem illius stelle, postea computa a gradibus meridiani circuli quanta sit latitudo, signum faciendo ibidem...

Jacob Golius and Celestial Cartography

Emilie Savage-Smith and Colin Wakefield

In the Bodleian Library, bound with a Latin essay on the names of the Chinese cycles of years written by the seventeenth-century orientalist Jacob Golius, there are six pages of copperplate engravings which contain an apparently unique set of unmounted paper gores for a celestial globe [Figures 1 to 6].[1] The set of twelve gores and two paper caps for a celestial globe of 340mm diameter was the result of the collaboration of Jacob Golius with his compatriot, the Dutch mapmaker Jacob Aertsz. Colom (1600–1673), who worked in Amsterdam.[2] A terrestrial globe that was intended as a companion piece to this celestial globe was dedicated by Colom to Golius.

The undated gores of the celestial globe were printed in Amsterdam about 1635 and remain unmounted. The only recorded celestial globe to be made from a similar

The authors hope that the honorand of this volume, with his cosmopolitan interests, will appreciate this excursion into the linguistic and scientific concerns of a humanist from an earlier age

[1] Bodleian Library, shelf-mark Vet. B3 b.29. The six copperplate engravings have on them twelve gores and two end-pieces for the celestial globe, four sections for the top of a horizon ring, and two end-pieces for a terrestrial globe. No other set of unmounted gores is recorded. The plates are bound with *De regno Catayo additamentum*, [before 1655]. The authors wish to thank Sylvia Sumira, a specialist in the conservation of globes, and Paul Cook of the National Maritime Museum at Greenwich, for their assistance in examining the globes at Greenwich; Peter van der Krogt of Utrecht University for making some of his studies on Dutch cartographers and their products available prior to publication; Anneka Siertsema for translating certain Dutch passages; and Doris Nicholson of the Bodleian Library. While their assistance has been much appreciated, they are in no way responsible for any errors or omissions.

[2] For the globemaker Colom and his products, including this celestial globe, see Peter van der Krogt, *Globi Neerlandici: De globeproduktie in de Nederlanden*, Utrecht: HES, 1989, pp. 179–183, and the English edition *Globi Needlandici: The Production of Globes in the Low Countries*, Utrecht: HES Publishers, 1993, pp. 196–204 and 539–544. See also Emilie Savage-Smith, 'Celestial Mapping', pp. 12–70 in J. B. Harley and David Woodward, eds, *The History of Cartography, Volume Two, Book One: Cartography in the Traditional Islamic and South Asian Societies*, Chicago: University of Chicago Press, 1992, esp. pp. 62–63. Beneath the head of Ursa Major is the dedication giving the name of the globemaker, Jacob Aertsz. Colom:

From *Learning, Language and Invention: Essays presented to Francis Maddison*, ed. W.D. Hackmann and A.J. Turner. Published by Variorum, Ashgate Publishing Ltd, Gower House, Croft Road, Aldershot, Hampshire, GU11 3HR, Great Britain.

set of gores is now at the National Maritime Museum in Greenwich.[3] In contrast to the unmounted gores, a number of the constellations on the Greenwich globe have been painted a dark green or a translucent pale red, with a few of them highlighted with a dark opaque red. A varnish has also been applied to the globe which over the years has darkened to a deep brown colour. On both the Colom/Golius unmounted gores and the finished globe, the names of all the constellations, major stars, lunar mansions and various circles are given in Arabic as well as in Latin, while the Ptolemaic constellations also have Greek names. This appears to be the earliest European celestial globe to provide the labels in Arabic script as well as in Latin and Greek.[4] The fact that on the unmounted gores the labels for the second brightest star in the skies (α *Carinae*, Canopus) are carefully written by hand on the prow of the ship Argo Navis suggests the possibility that this set of gores was corrected by Golius himself. If so, however, the correction is not reflected in a subsequent emendation of the copperplate from which the gores of the Greenwich globe were printed, for the globe at Greenwich leaves this bright star completely unlabelled.

'Illustrissimis Nobilis- | simis Amplissimis et Potē- | tissimis D.D. Dominis Ordi- | nibus Foederatarum inferio- | ris Germaniae Provinciarū | dignissimis Fidis Patriae | Patribus | Hoc astriferum stellarum arte Coelo | deductarum Caelum Gratus L.M.Q. | D.D.C.Q. | Iacobus Colom [To the most illustrious, noble, splendid and powerful lords, the States General of the United Provinces of the Netherlands, the most worthy of the fathers of the Fatherland, Jacob Colom willingly and rightly gives, dedicates and consecrates with gratitude this star-bearing globe, with the stars which have been skillfully drawn with the burin.]'. The translation is that of Peter van der Krogt, *Globi Needlandici* [English edition], p. 541, slightly emended.

[3] The celestial globe with its companion terrestrial globe are preserved at the National Maritime Museum, Greenwich, inv. nos GLO 0171 and GLO 0170 respectively. See *An Inventory of the Navigation and Astronomy Collections in the National Maritime Museum, Greenwich*, 3 vols, Greenwich, 1971, pp. 16–4 and 16–10, where older inventory numbers are given (G171/36–327c and G170/36–3324c); and van der Krogt, *Globi Needlandici*, pp. 540–542, with an overall illustration of the globe on p. 541 and details of the head of Serpentarius and three cartouches on the globe reproduced on pp. 202 and 542. The Greenwich globe is set in a brass meridian ring graduated by single degrees with every tenth labelled in non-consecutive segments of ninety degrees, two segments beginning the numbering at the pins inserted into the celestial poles and the two opposite segments having the numbering ending at the pins. The meridian ring and globe rest on a wooden horizon ring having the four printed paper sections on top. The horizon ring is supported by four turned legs on a circular wooden base.

[4] The sections of the horizon ring also have the names of the zodiacal houses written in Arabic script. The Italian cartographer Vincenzo Coronelli (1650–1718) provided trilingual labels for constellations on his large celestial globe, which was known to most people through a reduced design for a globe of 1068mm diameter engraved on copperplates and printed in his *Libro dei Globi* of 1693. The Arabic names for the constellations, both Ptolemaic and some of the newer southern constellations, are notably different from those provided by Golius for the Colom/Golius globe. Moreover, Golius supplied Arabic names for major stars and lunar mansions as well as celestial circles, none of which are given Arabic labels by Coronelli. The celestial globe of Golius and Colom appears to have had no influence upon that of Coronelli sixty years later. For a discussion of Coronelli's Arabic terms, see Paul Kunitzsch, 'Zu den dreisprachigen Inschriften einiger Himmelsgloben von V. Coronelli', *Der Globusfreund*, vol. 40/41, 1992, pp. 67–71, English translation pp. 72–76. A facsimile edition of the gores for one of Coronelli's celestial globes having Arabic labels has recently been published: Vincenzo Maria Coronelli, *Libro dei Globi, Venice, 1693 (1701)*, with introduction by Helen Wallis, Amsterdam: Theatrum Orbis Terrarum, 1969.

Engravings for globes by Colom and Golius c. *1635:*
Reproduced from Vet. B3 b.29(2). Copyright Bodleian Library, Oxford
Figure 1. Three gores with Pisces, Aries and Taurus along ecliptic

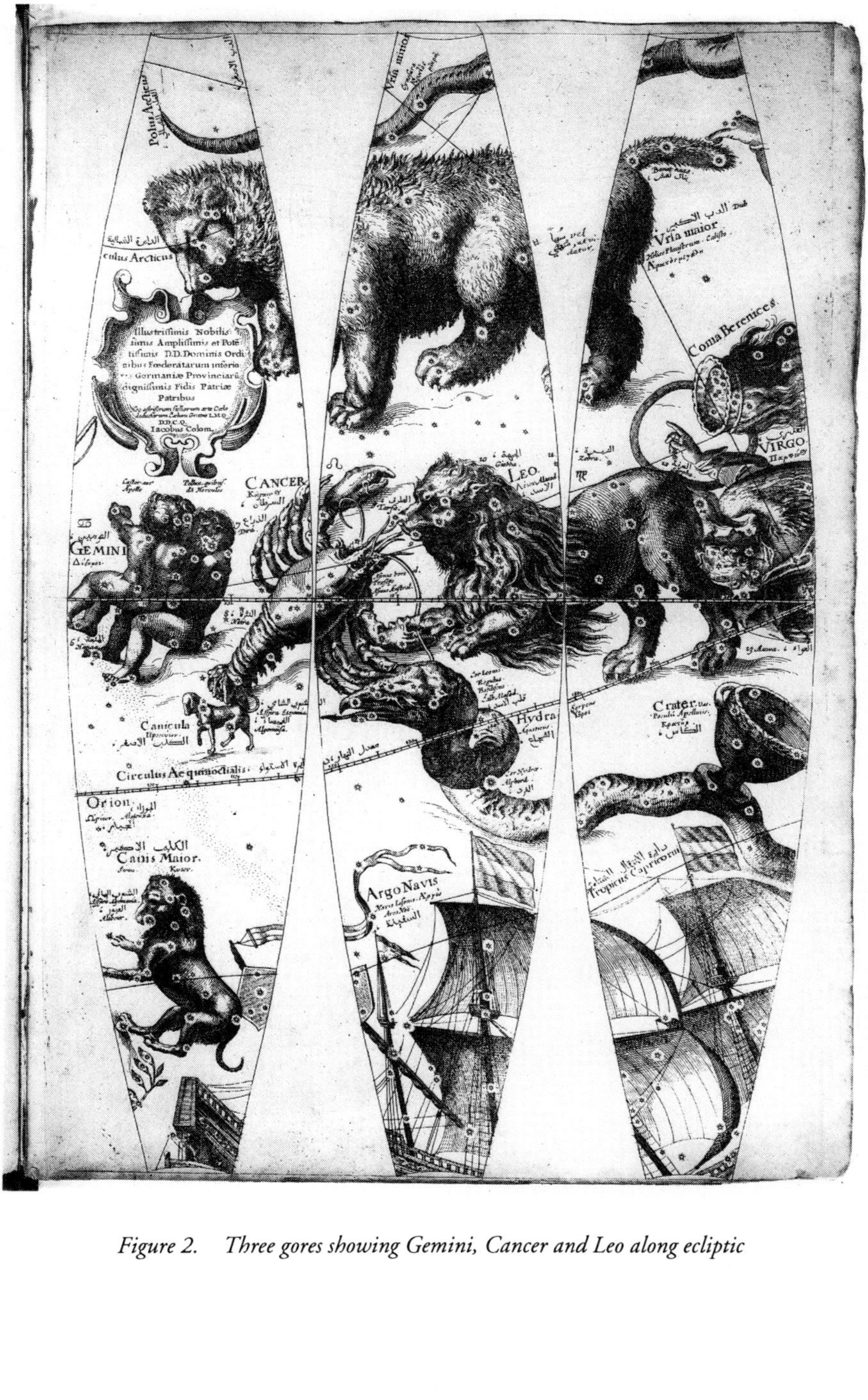

Figure 2. Three gores showing Gemini, Cancer and Leo along ecliptic

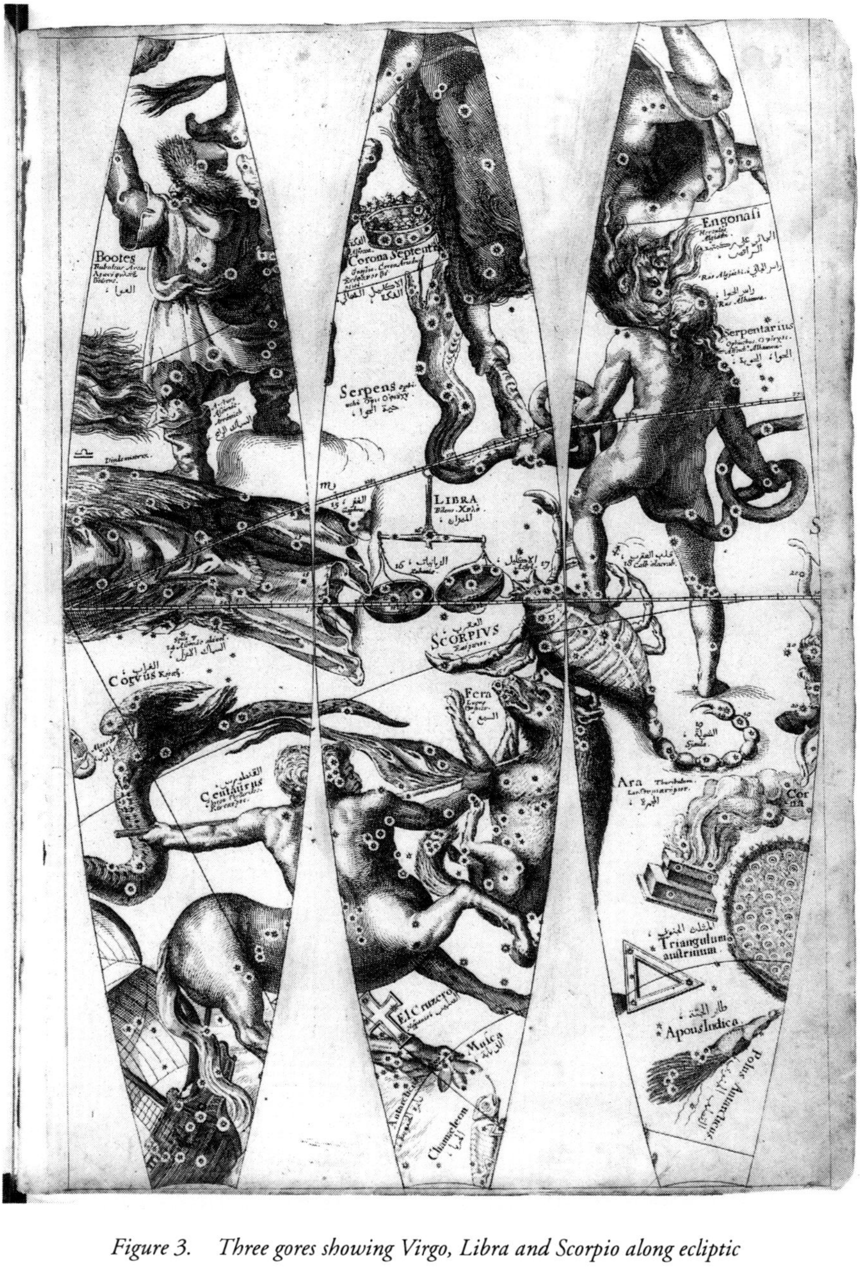

Figure 3. Three gores showing Virgo, Libra and Scorpio along ecliptic

Figure 4. *Three gores showing Sagittarius, Capricorn and Aquarius along ecliptic*

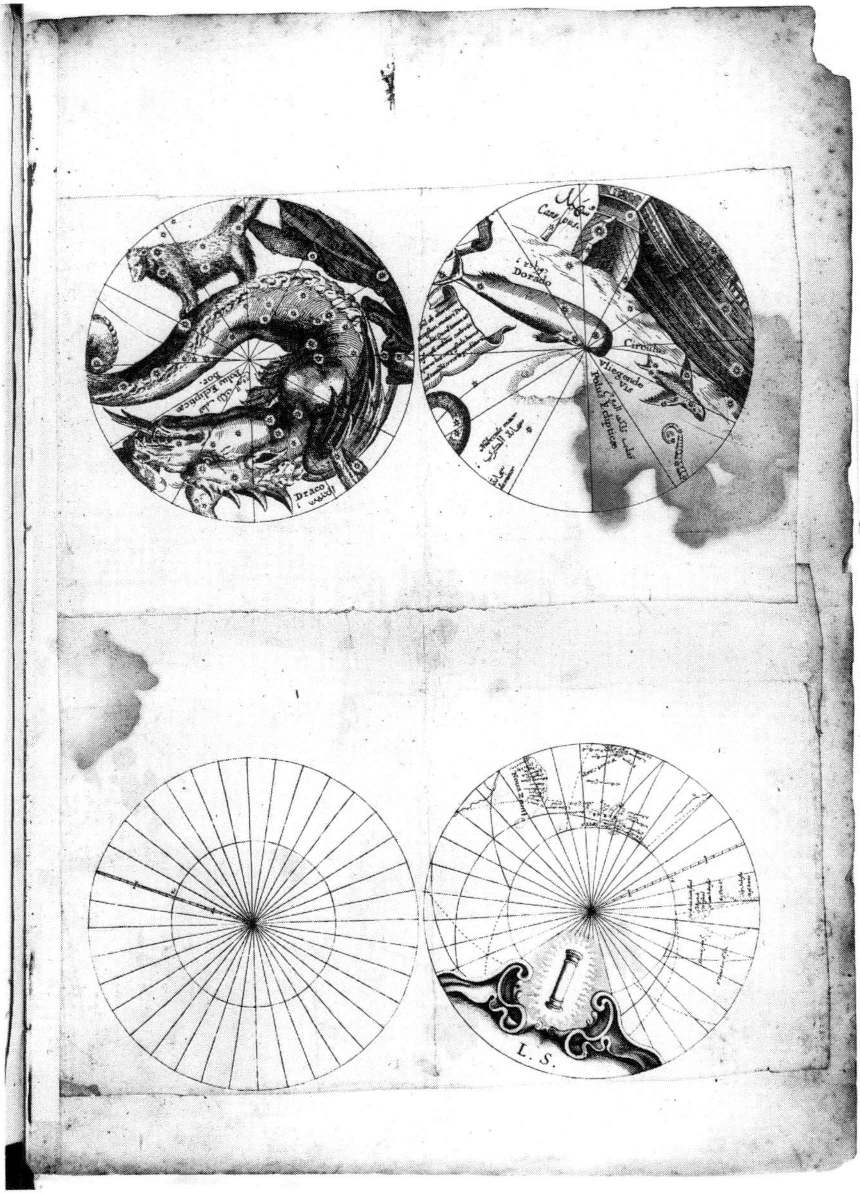

Figure 5. End caps for celestial and terrestrial globes

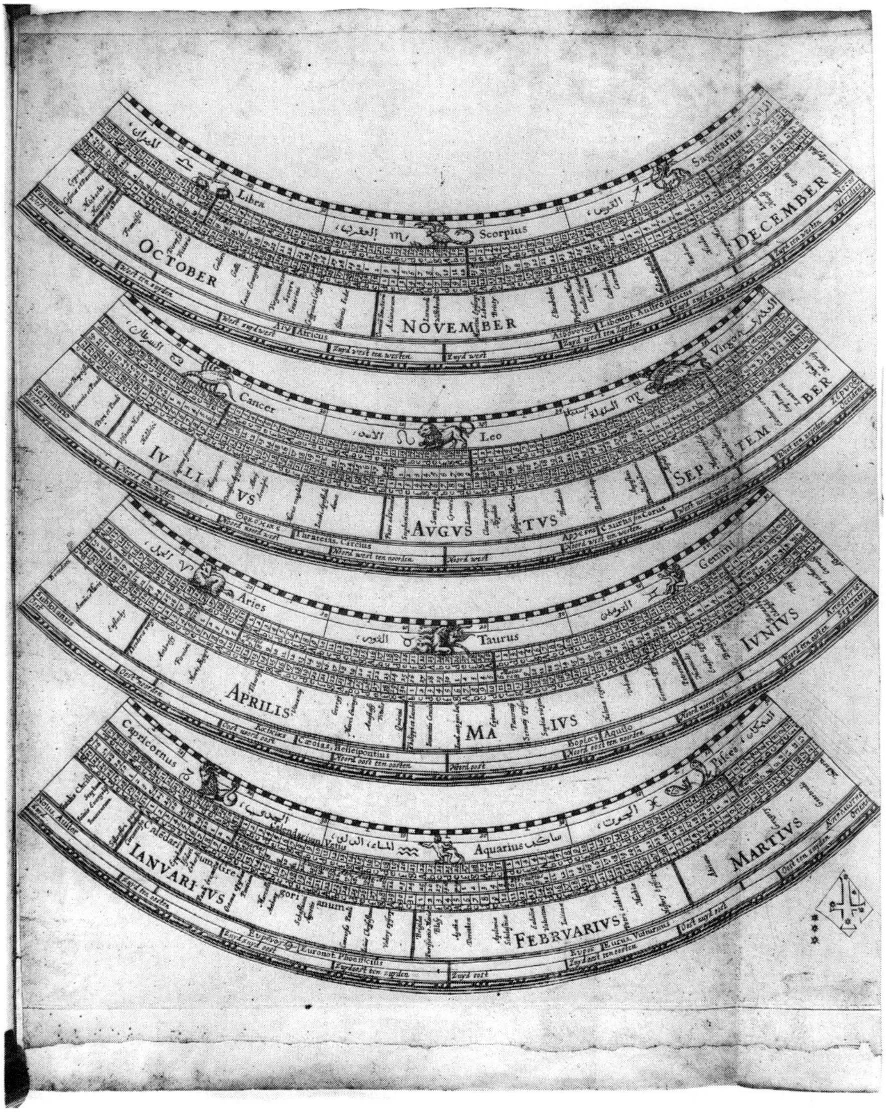

Figure 6. Horizon ring for celestial globe

According to an inscription on the gores, the Arabic terms, which are engraved in Arabic script, were the work of Golius. In an oval cartouche over the constellation Auriga the following title is written:

> GLOBUS | Caelestis. In quo stella- | rum fixarum ex certissi- | mis D. Tychonis Brahe | D. Frederici Houtmanni ali- | orumque Astronomorum discrip- | tionibus accurata dispositio. Pluri- | marum quoque nomina Arabica | operâ Iacobi Golii partim emen | data, partim nunc primum addita | inter quae XXVIII Mansiones | Lunae notis Arithmet juxta | seriem suam expressae | et distinctae | sunt.

> [A celestial globe, on which the fixed stars have been accurately placed according to the precise descriptions of Tycho Brahe and Frederik de Houtman and other astronomers. Through the efforts of Jacob Golius the Arabic names of most of them have been emended, with some now added for the first time, among which are the twenty-eight lunar mansions represented and distinguished by numerical notations in accordance with their sequence.]

The references are to Frederik de Houtman, a sixteenth-century Dutch navigator and astronomer, and to Tycho Brahe (1546–1601), the prominent Danish astronomer whose portrait is included on the globe beneath the southern constellation of Cetus. It is unclear whether Golius is correcting the Arabic terms on an earlier set of globe gores, now lost, or whether the 'nomina Arabica ... emendata' are corrections of the Latinized forms of Arabic names used earlier in Europe.[5] A list of the Arabic labels found on this celestial globe, along with the Latin and Greek labels, is appended to this study. Of particular interest is Golius's rendering of Arabic names for the newly delineated southern constellations, for which there was no medieval Arabic tradition.

The twelve gores of the globe are ecliptical, with the edges of the gores representing the ecliptic latitude-measuring circles. The sphere presents an external projection – that is, the resulting model presents the stars as seen by an observer outside the sphere of fixed stars. The earth was imagined at the centre of the globe, with the stars placed on the surface of the surrounding sphere, and the human constellations facing inward toward the earth. Consequently, the relative positions of the stars on the celestial globe are the reverse, east to west (or right to left), of their appearance when viewed from the surface of the earth. The order of the zodiacal signs is anticlockwise when viewed from above the North Pole.

The globe has the ecliptic indicated by a narrow band graduated by single degrees with every tenth labelled, the numbering repeating every thirty degrees. The celestial equator is also a narrow band graduated by single degrees with every tenth labelled, but the numbering sequence is continuous from the vernal equinox. The globe also has engraved on the gores the equinoctial colure, the north and south tropic circles, and the two celestial polar circles, which are circles having the celestial poles as centres and passing through the ecliptic poles. There are a selection of major stars together with sixty-four constellation outlines.

---

[5] For Latin forms of Arabic star names, see Paul Kunitzsch, *Arabische Sternnamen in Europa*, Wiesbaden: Harrassowitz, 1959, and for Arabic nomenclature see Paul Kunitzsch, *Untersuchungen zur Sternnomenklatur der Araber*, Wiesbaden: Harrassowitz, 1961.

The epoch to which the star positions correspond is about AD 1600. Most of the stars are placed on the globe according to Tycho Brahe's extended star catalogue, with the southern skies augmented by the stars mapped in 1595–1597 by Frederik de Houtman in collaboration with another Dutch navigator Pieter Dircksz. Keyser. Written in a scroll beneath the constellation of Eridanus there is a chart of precessional values for various periods, with the following legend:

> *Stellarū inerrantū centesimo | quoq; anno progressus praeterito | tempori subducendo aut deinceps | elapsuro addendo; accommodandi.* || Latitudinis loca in fixis nonnihil ob alteratum | Eclipticae ad Aequatorem accessum, mutari Pri | mus iam Tycho deprehendit. Quod tamen ut | leuiculum et in immenso vasti ambitus coelo vix | perceptibile, nedū in parvo hoc corpusculo, huma: | no usui destinata vel tantillū erroris praebiturū, | omitti potest.
>
> [*The progress of the fixed stars for every 100 years is to be adjusted by subtracting for past time or adding for subsequent intervals.* || The positions of latitude for the fixed stars [also have] some [change] because of the changing approach of the ecliptic to the equator – an alteration which Tycho first detected. However, because it is insignificant and is scarcely perceptible in the immense circuit of the vast sky, much less on this small miniature model intended for human use, and also [because] it represents such a small error, it can be omitted.]

Nearby is a small scroll stating that seven different symbols are employed on the globe to delineate the six different magnitudes of stars and the nebulae.

In an oval cartouche under the tail of the southern constellation of Cetus there is the following address to the user:

> Habetis hîc, Astrorum studiosi, trecentas et plures | Antarctico mundi vertici viciniores stellas, ex obser- | vationibus secundum iam à Frederico Houtmanno | maiori studio et accommodatioribus instrumentis | ad stellas à Tychone positas factis depromtas, | auctiori numero et accuratiori dispositione vestro | commodo et delectationi depictas.
>
> [Here, students of the stars, you have more than 300 stars near the Antarctic pole of the world, prepared from observations made by Frederik de Houtman with great industriousness and with better instruments next to those stars mapped by Tycho, depicted in greater number and with more accurate placement for your profit and pleasure.]

The constellation designs and the star positions on this set of globe gores by Colom are identical in every respect to those on the revised edition of 1603 of the earliest celestial globe designed by the Dutch cartographer Willem Jansz. Blaeu (1571–1638).[6] On a trip to Hven in the winter of 1595–1596, Blaeu had obtained a copy of Tycho Brahe's extended star catalogue prior to its general circulation about three years later. Blaeu used this star catalogue for his first celestial globe made about 1598, but when the catalogue of one hundred and ninety-six additional southern stars mapped by Frederik de Houtman subsequently became available, Blaeu issued a revised version of his globe

---

[6] This globe was printed again, with only a minor change, in 1621. See Van der Krogt, *Globi Neerlandici* (n.2) [English edition], pp. 492–494 (BLA I state 2 and state 3), and *Globi Neerlandici* [Dutch edition], pp. 181–182. For Willem Jansz. Blaeu, see Van der Krogt, *Globi Neerlandici* [English edition], pp. 140–154 and 488–524 et passim, and Deborah J. Warner, *The Sky Explored: Celestial Cartography 1500–1800*, Amsterdam: Theatrum Orbis Terrarum, 1979, pp. 28–31. The incorporation by Colom of Blaeu's complete design extends to the same identical address to the user that has been given above.

in 1603 incorporating the improved mapping of the southern skies. Of the 'more than three hundred' stars mentioned in the address to the user, one hundred and seven were Ptolemaic stars with revised positions.

The dress and general presentation of many of the constellation figures on Blaeu's globes were decidedly different from those of his predecessors such as Dürer or Mercator. Contemporary Dutch culture was reflected in Blaeu's interpretations, with, for example, Argo Navis rendered as a Dutch ship of the day. Some of the human figures are dressed for a northern European winter. Boötes, for example, wears a large fur hat, a thick cloak and heavy winter boots. In Cygnus there is clearly indicated and labelled the Nova Stella of 1600, which Blaeu had discovered on 18 August 1600.[7]

The same non-Ptolemaic constellations appear on both Blaeu's globe and that of Colom and Golius. In the Northern Hemisphere, there are two constellations introduced in the sixteenth century by the Flemish cartographer Gerard Mercator (1512–1594) but based on Ptolemaic asterisms: Coma Berenices and Antinoüs, the latter being the young friend of the Roman emperor Hadrian, represented as a kneeling figure over the head of Capricorn. In the Southern Hemisphere, there are the new constellations of Columba Noë (Noah's dove) and El Cruzero Hispanis (the Spanish cross), mapped by the Dutch theologian and geographer Petrus Plancius (1552–1622), along with the twelve constellations mapped by de Houtman and Keyser. These latter twelve are Apus (a bird of paradise), Chamaeleon (a chameleon), Piscis Volucris (a flying fish), Dorado (a goldfish), Phoenix, Grus (a crane), Toucan, Hydrus (a small serpent), Indus (a native holding a spear), Pavo (a peacock), Triangulum Australe (the southern triangle), and Musca (a fly).

The novelty of the Colom/Golius version lies in the Arabic stellar nomenclature. Of the non-Ptolemaic constellations, all but Coma Berenices, Antinoüs, and Piscis Volucris were given Arabic names by Golius. It is not known whether Golius himself coined the Arabic names for the newly demarcated constellations of the southern skies, but their inclusion on the engraved gores for the celestial globe appears to be the earliest recorded effort to give Arabic names to the newly outlined asterisms. It is unknown what the precise sources were that Golius employed for the Arabic and Latin forms of the classical constellation and star names or how he determined the Arabic names for the new constellations. By the early sixteenth century many avenues were available by which a European could learn the traditional Arabic stellar nomenclature, at least through Latin translations. The Latin forms of the Arabic terms given by Golius, however, vary slightly from any one European text that has been studied, and it is possible that Golius consulted an Arabic text directly and formed his own Latin transliterations.[8]

---

[7] The labelling of the Nova of 1572 in Cassiopeia, first recorded by Tycho Brahe and included on Blaeu's globe, is omitted on the derivative Colom/Golius globe.

[8] For the European sources available for stellar nomenclature, see the studies by Paul Kunitzsch given above in n. 5 as well as Paul Kunitzsch, *Peter Apian und Azophi: Arabische Sternbilder in Ingolstadt im frühen 16. Jahrhundert* (*Bayerische Akademie der Wissenschaften, philosophisch-historische Klasse, Sitzungsberichte, Bd. 3*), Munich, 1986; Paul Kunitzsch, 'Ṣūfī Latinus', *Zeitschrift der Deutschen Morgenländische Gesellschaft*, vol. 115, 1965, pp. 65–74; and André le Boeuffle, *Les noms latins d'astres et de constellations* (*Collection d'Études Anciennes*), Paris: Société d'édition 'Les Belles Lettres', 1977.

There is no evidence that the new chartings of the southern skies at the end of the sixteenth century were known within the Islamic world until well after this celestial globe was printed. An astrolabe plate engraved in Iran in AH1065/AD1654–55 by Muḥammad Mahdī of Yazd depicts for the first time some of the new southern constellations. This plate reproduces the Northern and Southern Hemispheres from a map printed about 1650 by the Parisian engraver Melchior Tavernier, whose brother Jean-Baptiste Tavernier made six trips to the Near East before his death in 1689.[9] Muḥammad Mahdī, when making his astrolabe plate, rendered the Ptolemaic constellation names into Arabic, but did not attempt to give Arabic names to the non-Ptolemaic constellations,[10] nor does he appear to have been aware of Golius's rendering in Arabic of the new constellation names. Muḥammad Mahdī is known to have made two other astrolabes with similar plates having planispheric star maps, both produced in AH 1070 AD 1659–60 and clearly copies of the earlier plate, though not quite as carefully engraved.[11] His plates appear to have had no subsequent influence in celestial cartography or in celestial nomenclature.

On the companion terrestrial globe by Colom, Golius is referred to as 'Linguae Arabicae & Matheseos Professor', and since he was not granted the chair of mathematics until November of 1629, the terrestrial globe must have been made in 1630 or thereafter. The polar caps of the celestial globe are engraved on the same plate as those containing the polar caps of the terrestrial globe and consequently must have been designed about the same time. It is recorded in the records of the University of Leiden that Golius requested permission to use two large globes owned by the library and that permission for their use was granted on 9 August 1633.[12] It seems likely that their use preceded Golius's coining of Arabic terms for the globe being made by Colom, or certainly preceded the construction of Colom's celestial globe, since, had that globe been available, Golius would not have had to borrow one from the library. It has been suggested by the leading

[9] The current location of Mahdī's astrolabe is unknown. For a comparison of Muḥammad Mahdī's instrument and the map by Melchior Tavernier, see Emilie Savage-Smith, 'Celestial Mapping' (above, n. 2), pp. 65–67 and fig. 2.45–2.48. For the astrolabe-maker Muḥammad Mahdī al-Khādim ibn Muḥammad Amin al-Yazdi, see L. A. Mayer, *Islamic Astrolabists and Their Works*, Geneva: Albert Kundig, 1956, pp. 70–71; A. J. Turner, *The Time Museum, Catalogue of the Collection* vol. 1, *Time Measuring Instruments*, Part 1: 'Astrolabes, Astrolabe-Related Instruments', Rockford, Illinois, 1985, pp. 86–91; Alain Brieux and Francis R. Maddison, *Répertoire des facteurs d'astrolabes et de leurs oeuvres: Première partie, Islam plus Byzance, Arménie, Géorgie et Inde hindoue*, Paris: Centre National des Recherches Scientifiques, in press.

[10] There are two exceptions, the southern triangle (Triangulum Australe) and the peacock (Pavo), which were labelled with common Arabic words such as ṭāwūs, a common word for the easily recognized peacock. There are other differences as well between the Arabic labels on Muḥammad Mahdī's astrolabe plate and those prepared by Golius for the European gores. In addition to some differences in spelling, Muḥammad Mahdī included more star names than Golius did, but omitted the lunar mansions. Muḥammad Mahdī, following Tavernier, also omitted the depiction of the constellation of the southern cross (El Cruzero Hispanis).

[11] Greenwich, National Maritime Museum, inv. no. A64/69–6, and the University of Cambridge, Whipple Museum of the History of Science, inv. no. 1001. Both of these instruments are dateable to the year AH 1070 AD 1659–60 through a Persian chronogram engraved on each *rete* reading: 'It is the mirror of Alexander and the mirror representing the entire universe'. By adding the numerical values of the Persian letters forming the statement, the date 1070 is obtained

[12] Van der Krogt, *Globi Neerlandici* (n.2), English edition p. 202 and Dutch edition p. 182.

authority on Dutch cartography, Peter van der Krogt, that Golius asked Colom to make the globe for him and that Golius would not have approached Colom to do so until after Colom had established a name as a publisher.[13] Since Colom's reputation was not well-established until 1635, following this line of reasoning it would seem likely that the pair of globes were produced in 1635 or shortly thereafter. If the celestial globe was engraved about 1635, it indicates that Golius was interested in Arabic star names well before his edition and Latin translation of an Arabic astronomical compendium by the ninth-century astronomer al-Farghānī, published two years after his death and more than thirty years after the publication of the globe gores.[14]

Jacob Golius (Jacob Gool) was without doubt the most outstanding Orientalist in seventeenth-century Europe.[15] He was born in 1596 at The Hague and from 1612–16 attended the University of Leiden, where he studied classics, philosophy, medicine, and mathematics. On leaving the University he spent a further two years at his father's house privately pursuing his study of Greek philosophy and mathematics. In 1618 he enrolled again at the University to read Arabic with Thomas Erpenius (1584–1624), the eminent professor of Oriental languages at Leiden. Golius's interest in Arabic sprang from his awareness that much of the Greek scientific and mathematical heritage in which he was interested had been translated and commented upon by the Arabs.

After a short spell in France teaching Greek at the Huguenot University of La Rochelle, in 1622 the States General directed that Golius should accompany a Dutch trade mission to Morocco. Although his official capacity was that of engineer, it was now that he took the first steps in assembling a collection of Arabic and Oriental manuscripts which was to become the greatest of its time in Europe. It was partly at the instance of Erpenius that Golius had begun collecting, but unfortunately Erpenius himself was unable to make use of the manuscripts. Shortly after Golius's return from Morocco in July 1624, Erpenius fell ill and he died in the following November. In May 1625 Golius was named successor to Erpenius at Leiden, and not long afterwards he obtained permission from the University to spend time in Syria to improve his knowledge of Arabic and acquire manuscripts. During his residence there, Golius was an official at the Dutch Consulate in Aleppo. He left Aleppo in May 1627 and travelled overland to Con-

[13] Van der Krogt, *Globi Neerlandici* (n.2), English edition pp. 202–203 and Dutch edition pp. 182–183.

[14] J. Golius, editor, *Muhammedis Fil. Ketiri Ferganensis, Qui vulgo Alfraganus dicitur, Elementa Astronomica, Arabicè & Latinè. Cum Notis ad res exoticas sive Orientales, quae in iis occurrunt*, Amsterdam, 1669; reprinted, al-Farghānī, *Jawāmiʿ ʿilm al-nujūm wa-uṣūl al-ḥarakāt al-samāwīyah, herausgegeben als Elementa Astronomica mit lateinischen Übersetzung von Jacob Golius* [Veröffentlichungen des Institutes für Geschichte der Arabisch-Islamischen Wissenschaften, Reihe B, Nachdrucke. Abteilung 2], Frankfurt, 1986. The Latin transliterations of the lunar mansions given by Golius in his translation of chapter 20 of al-Farghānī's treatise are not the same as the Latin names for the lunar mansions given by Colom and Golius in the trilingual labels on the celestial globe. The non-Ptolemaic constellations are not named or discussed by al-Farghānī or by Golius in his translation and commentary. Abū al-ʿAbbās Aḥmad ibn Muḥammad al-Farghānī is known to have been active between AH 218/AD 833 and AH 247/AD 861 and is a major source of information regarding the pre-Islamic concepts of lunar mansions; see Fuat Sezgin, *Geschichte des arabischen Schrifttums, Band 6: Astronomie bis ca. 430 H.*, Leiden: E. J. Brill, 1978, pp. 149–151.

[15] This account of Golius's life and works is drawn in large measure from J. J. Witkam, *Jacobus Golius (1596–1667) en zijn handschriften* (*Oosters Genootschap in Nederland*, 10), Leiden: E. J. Brill, 1980.

stantinople where, through the good offices of the Dutch representative at the Porte, Cornelis Haga, he was introduced to local scholars and notables who assisted him in the acquisition of further manuscripts. Golius returned to Holland in 1629 to settle down to academic life at Leiden where, in the same year, the University honoured him with the chair of mathematics. He held both chairs at Leiden until his death on 28 September 1667.

The list of Golius's publications is not long, but the influence of his work was far-reaching. He had the the distinction of being the first person to publish, in 1629, an Arabic poem, the *Lāmīyat al-'Ajam* of al-Ṭughrā'ī (1063–1120), which appeared in *Proverbia quaedam Alis, imperatoris Muslimici, et Carmen Tograï, poetae doctiss. nec non dissertatio quaedam Aben Sinae.* In 1636 his edition of Ibn 'Arabshāh's life of Tīmūr (Tamerlane) was published, *Ahmedis Arabsiadae vitae & rerum gestarum Timuri, qui vulgo Tamerlanes dicitur, historia*. Golius's abiding fame rests on his *Lexicon Arabico–Latinum*, which appeared in 1653. It superseded previous European attempts at Arabic lexicography and remained the standard work until the nineteenth century. In addition, Golius edited for publication Erpenius's *Historia Saracenica*. It came out in 1625, a year after Erpenius's death. And Golius was responsible for the re-editing, in 1656, of Erpenius's famous Arabic grammar, published as *Arabicae linguae tyrocinium.* Two works of Golius were published posthumously, both in 1669. The first was the edition and Latin translation of several chapters from al-Farghānī, in which Golius also demonstrated an interest in Arabic star nomenclature. The second was Golius's Persian–Latin dictionary, which was incorporated into the *Lexicon Heptaglotton* of Edmund Castell (1606–86), the Sir Thomas Adams Professor in Arabic at the University of Cambridge. Finally, there is the undated ten-page essay entitled *De regno Catayo additamentum*, which in the Bodleian copy is bound with the plates for the celestial globe. In 1655 this contribution was published in Amsterdam in Martinus Martinius's *Novus atlas Sinensis.* Martinius was a Jesuit who lived and worked in China and taught Golius Chinese whilst he was in Holland awaiting an East India Company ship to take him back to China. It was probably from Martinius that Golius obtained the Chinese manuscripts and printed books that were to form part of his large private library.

One of Golius's major projects remained unpublished – an edition of the Arabic translation of the *Conica* of Apollonius of Perga. Golius had acquired a magnificent manuscript of the work (dated AH462/AD1070) in Aleppo in 1627 and had also borrowed Christianus Ravius's copy of 'Abd al-Malik al-Shīrāzī's epitome of the *Conica.*[16] This he had transcribed for his own use by a certain Nicolaus Petri, a Christian from Aleppo whom Ravius had originally brought to Europe. It may be that Leiden University had given Golius the chair of mathematics as an encouragement to the publication of this important work, of which Books 5–7 are preserved only in Arabic translation. Europe had to wait until 1710 when Edmund Halley finally published

[16] Now also in the Bodleian Library: shelfmark MS Thurston 3. See K.M.P., 'The Wanderings of Apollonius', *The Bodleian Quaterly Record*, vol. 3, 1920–22, pp. 152–153.

the *Conica*, making use of the manuscript which Golius had obtained in Aleppo.[17] But by then this manuscript, and the greater part of Golius's private collection, had passed into the hands of Narcissus Marsh, Archbishop of Armagh (1638–1713).

The existence in Leiden of two 'Golius collections' during the lifetime of the scholar has been discussed by J.J. Witkam.[18] The 'public' collection comprises the two hundred or so manuscripts he assembled for the Library of Leiden University, where they remain to this day numbered Codd. Or.1–211. Golius also amassed a larger 'private' library of manuscripts which he kept at his house and reserved for his own use. The titles in the catalogue printed by Antonius Vitray in Paris in 1630 correspond in the main with the manuscripts assigned to the University Library.[19] This catalogue in a revised form, with titles in Arabic type, was published ten years later as part of the general catalogue of the whole of Leiden University Library.[20] When Golius died in 1667 Leiden University expected that his private library of manuscripts would pass to the University to join the others that were already in the University Library. Since his will contained nothing to this effect, however, his private collection passed to his heirs. A catalogue of the collection was printed by the Elzeviers in Leiden in 1668 in an attempt to stimulate interest but, although there was some enthusiasm in Leiden and Cambridge, no buyer was forthcoming, and it was not until almost thirty years after Golius's death that his heirs put the oriental manuscripts (and Chinese printed books) up for auction.[21] This delay contrasts with the disposal of Golius's western printed books, which were sold at auction the year after his death.

At the auction in Leiden on 16 October 1696, Narcissus Marsh became the proud owner of some 70 per cent of Golius's private library of manuscripts and Chinese printed books. Acting on behalf of Marsh at the auction was Dr Edward Bernard, Savilian professor of astronomy at Oxford (1673–91). Bernard was not unfamiliar with the Golius collection for in 1668 he had been allowed access to the Apollonius manuscript to copy out the text of the 'missing' books. Narcissus Marsh died in 1713 and his manuscripts were bequeathed to the Bodleian Library where they were received in

[17] *Apollonii Pergaei Conicorum libri octo. (Libri iv priores... ex codd. MSS. Graecis edidit E. Halleius. Libri tres posteriores, Arabico sermone in Latinum Conversi. Opera et studio E. Halleii)*, 2 parts, Oxford, 1710.

[18] See n.15 above.

[19] *Catalogus rarorum librorum, quos ex Oriente nuper advexit, et in publica Bibliotheca inclytae Leydensis Academiae deposuit Clariss. et de bonis Artibus meritiss. Iacobus Golius in illa eadem Academia, et linguarum Orientalium, et Matheseos Professor insignis*, Paris, 1630. The catalogue was based on details supplied by Golius to the French scholar Pierre Gassendi (1592–1655), who appears to have arranged its publication in Paris. See Anthony Turner with Nadine Gomez, *Pierre Gassendi, explorateur des sciences. Catalogue de l'exposition quatrième centenaire de la naissance de Pierre Gassendi, Musée de Digne 19 mai au 18 octobre 1992*, Digne-les-Bains, 1992, p. 66.

[20] *Catalogus Bibliothecae Publicae Lugduno-Batavae*, 2 parts, Leiden, 1640.

[21] The 1696 sale catalogue, printed in Leiden, exists in two forms. The earlier printing gave the book titles in both Latin and in Arabic type and had the following title-page: *Catalogus Insignium in omni facultate, linguisque, Arabica, Persica, Turcica, Chinensi &c. Librorum M.SS. quos Doctissimus Clarissimusque Vir D. Jacobus Golius, dum viveret: Mathesios & Arabicae Linguae in Acad. Lugd. Batav. Professor Ordinarius, Ex variis Regionibus magno studio, labore & sumptu, collegit. Quorum auctio habebitur in Aedibus Johannis du Vivie, Bibliopolae. Ad diem XVI. Octobris St. Novo, ad punctum horae nonae.* It was then reset in a smaller format with essentially the same wording on the title-page but without the Arabic book titles.

the following year. So it was that the Bodleian Library in Oxford became the final resting-place of the greater part of the Dutch scholar's private collection of oriental manuscripts.

Mention has already been made of Golius's manuscript of Apollonius, now one of the treasures of the Bodleian Library.[22] Another outstanding item is the early copy (dated AH 400/AD 1009–10) of the major guide to constellation diagrams in the Islamic world, *Kitāb Ṣuwar al-kawākib al-thābitah,* written in the tenth century by 'Abd al-Raḥmān al-Ṣūfī, a court astronomer in Isfahan.[23] Further notable Arabic manuscripts deriving from Golius are Galen's *Anatomica*, in fifteen books,[24] and a calligraphic copy of Ibn Miskawayh's *Jāwīdān Khirad* dated AH 439/AD 1047.[25] And there are many other rare and interesting works, far too numerous for individual mention, representing the traditional fields of Arabic study: mathematics, astronomy, zoology, the occult, lexicography, grammar, poetry, history, and philosophy. Nor should we forget a number of Chinese printed books and manuscripts, a Malay–Dutch dictionary[26] and an outstanding illuminated and illustrated Persian manuscript of the *Būstān* of Sa'dī of Shiraz (thirteenth century). It is undated but belongs to the period 1515–20 and by the middle of the sixteenth century had passed into the possession of the Ottoman Sultans.[27] Among the manuscripts in the Bodleian Library are also several in Golius's own hand, including the autograph of his Persian–Latin dictionary which was published in Edmund Castell's *Lexicon.*[28] There is also a Turkish–Latin dictionary compiled and written by Golius,[29] and an incomplete work entitled *Cornucopiae naturae et artis* comprising vocabulary relating to medicine and biology, mainly Arabic but with some Persian and Turkish.[30]

---

22 MS Marsh 667. For a description of the manuscript, see A.F.L. Beeston, 'The Marsh Manuscript of Apollonius's Conica', *The Bodleian Library Record,* vol. 4, 1952, pp. 76–77; note, however, that the *abjad* date should be rendered AH 462 (not 472).

23 MS Marsh 144.

24 MS Marsh 158, fols 406^{v}–650^{r}.

25 MS Marsh 662.

26 MS Marsh 712.

27 MS Marsh 517.

28 MS Marsh 213.

29 MS Marsh 193.

30 MS Marsh 451. This title is now nowhere to be found in the manuscript. The Latin title in the sale catalogue of 1696 (p. 23) is: *Cornucopiae naturae & artis Arab. cum partiali tantum versione Lat. notisque J. Golii.* In A. Nicoll and E.B. Pusey's catalogue, the manuscript is described as follows: *Codex chartaceus, spissus, forma in folio, complectens Jacobi Golii Adversaria Alphabetica, quibus ipse in tegumento adscripsit titulum: 'Cornucopiae Naturae et Artis, inchoatae Kalendis Novembris 1644,'* item '*Lexicon Naturae et Artis* ....' The 'tegumentum' on which Golius had written the title was evidently some sort of loose wrapper; in the sale catalogue the manuscript is entered among the unbound items (no. 17 of the section 'Libri M.S. Incompacti in Folio'). The wrapper with its inscription must have been discarded when the volume was rebound in its present library binding in the second half of the nineteenth century. In an interesting note on folio 1 verso of the manuscript Golius mentions his brother Petrus as one of his sources for the dictionary. Petrus was a Carmelite monk who had lived in Syria for many years. Golius' note is reproduced in A. Nicoll and E.B. Pusey's catalogue: '*Frater germanus meus et unicus Petrus Golius cui discalceatorum ordinem ingredienti nomen Coelestini de S. Lidwina inditum fuit. Plusquam duodecennium in Syria*

A small number of Golius manuscripts in the Bodleian Library came by routes other than the bequest of Narcissus Marsh. One such is his copy of the *Câmasbnâme,*[31] a Turkish romance, which subsequently fell into the possession of the nineteenth-century English orientalist Sir William Ouseley (1767–1842). One year after his death, the whole of Sir William's collection, including the *Câmasbnâme,* was purchased for the sum of £2 000 by the Bodleian Library.

The route by which the set of unmounted celestial globe gores, and the undated printing of the essay on the names of the Chinese cycles of years with which it is bound, reached the Bodleian is unknown, but it is recorded as being in the Bodleian Library in 1738.[32] An unknown small hand of possibly the eighteenth century, using a brown ink, has annotated one star in the constellation Ursa Major with the legend '*suhā* vel *suhá,* ut videtur', referring to a small star overlooked by Ptolemy (g *Ursa Majoris,* Alc or) by which Arab astronomers said that men could test their vision. On one of the engraved pieces for the horizon ring, the same hand has rewritten one of the Arabic names for Virgo, *al-sunbulah,* so as to make the spelling clearer.[33] It seems certain that the hand which produced these notes is a different and later hand than the one that labelled Canopus. It is possible that the Latin and Arabic handwritten labels of Canopus are in Golius's own hand.

This unmounted set of gores, possibly annotated by Golius himself, must be considered as highly significant, for along with the single recorded example of the completed globe now at Greenwich, it is important evidence of Golius's collaborative effort with the mapmaker Colom and his early interest in Arabic stellar nomenclature. The serious modern European study of Islamic astronomy, by scholars who could read the original texts, began with the work of Jacob Golius in Holland.[34] This celestial globe exemplifies the scientific and linguistic pursuits of early modern scholarship, as well as the curiosity and learning of one of the leading seventeenth-century orientalists, whose manuscript collection now forms an important part of the holdings of the Bodleian Library.

---

vixit, et adhuc vivit, linguarum rerumque Orientalium perquam gnarus, Coenobitarum, qui in monte Libano et Tripolitana dioecesi degunt, antistes.' See *Catalogi Codicum Manuscriptorum Orientalium Bibliothecae Bodleianae Pars Secunda Arabicos Complectens; confecit Alexander Nicoll; editionem absolvit et Catalogus Urianum aliquatenus emendavit E.B. Pusey,* Oxford, 1835, p. 350, 'Codices Arabici Mohammedani 356.'

[31] MS Ouseley 103. It is 'Libri Misc. MS In Quarto, &c 29' on p. 18 of the 1696 sale catalogue. For five Golius manuscripts (four Arabic and one Turkish) presented to the Library in 1700 by Moses Amyrault, see *A Summary Catalogue of Western Manuscripts in the Bodleian Library at Oxford,* 7 vols, Oxford 1895–1953 (reprinted Munich, 1980), vol. 5, pp. 295–296 nos. 27597–27601.

[32] *Catalogus impressorum librorum Bibliothecae Bodleianae in Academia Oxoniensi,* 2 vols, Oxford, 1738. It is entered on p. 506 of vol. 1 with its earlier shelf-mark, Arch. Bodl. C. 66.

[33] On the horizon ring, in contrast to the sphere itself, the zodiacal house of Virgo is given two Arabic names: *al-'adhrá* and *al-sunbulah.* The latter name, a more common term, is omitted on the sphere.

[34] See David A. King, 'Some Remarks on Islamic Scientific Manuscripts and Instruments, and Past, Present, and Future Research', pp. 115–143, in John Cooper, ed., *The Significance of Islamic Manuscripts: Proceedings of the Inaugural Conference of Al-Furqān Islamic Heritage Foundation (30th November-1st December 1991),* London: Al-Furqān Islamic Heritage Foundation 1992, esp. p. 118.

APPENDIX: Arabic, Latin and Greek Labels on the Celestial Globe by Golius and Colom

Arabic labels	Latin and Greek labels	Modern identifications
	Northern constellations	
al-dubb al aṣghar	Vrsa minor. Cynosura. 'Αρκτὸς μικρά.	Ursa Minor
al-dubb al-akbar	Vrsa maior. Helice Plaustrum. Calisto. 'Αρκτὸς μεγάλη.	Ursa Major
al-tinnīn	Draco. [no Greek]	Draco
al-multahib, Qīqā'ūs	Cepheus. Κηφεῦς.	Cepheus
al-'awwā'	Bootes. Bubuleus Areas. 'Αρκτόφυλαξ. Βοώτης.	Boötes
al-iklīl al-shamālī, al-fakkah	Corona Septentri. Gnossin. Coron. Ariadn. Στέφανος βάρειος.	Corona Borealis
al-jāthī 'alá rukbatayhi, al-rāqiṣ	Engonasi. Hercules. Algiáthi. [no Greek]	Hercules
al-shulyāq, al-nasr al-wāqi'	Lyra Arionis. Vultur cadens. Λύρος.	Lyra
al-dajājah	Cygnus. Olor. Gallina. Ὄρνις.	Cygnus
dhāt al-kursī	Cassiopea mater Andromedae. Κασσιοπεῖα.	Cassiopeia
ḥāmil ra's al-ghūl, Barsā'ūsh	Perseus. Abantiades. Περσεύς. Celeub.	Perseus
mumsik, al-a'innah	Auriga. Erichtonius. Heniochus. Ἡνίοχος.	Auriga
al-ḥawwā, al-ḥawīyah	Serpentarius. Ophiuchus. 'Οφίουχος. Alfeichus. Alhauwa.	Serpentarius; Ophiuchus
ḥayyat al-ḥawwā	Serpens Ophiuchi. Ὄφις 'Οφιούχου.	Serpens
al-sahm	Sagitta. Telum. 'Οισός [*sic*].	Sagitta
al-'uqāb, al-nasr al-ṭā'ir	Aquila. Vultur volans Iovis ales. 'Αετός.	Aquila

APPENDIX: cont.

Arabic labels	Latin and Greek labels	Modern identifications
al-dalfīn	Delphinus Arionis. Δελφιν.	Delphinus
qiṭ'at al-faras	Equulus. Equi sectio. Ἵππου προτομή.	Equuleus
al-faras al-akbar	Pegasus. Equus alatus. Ἵππος. Alpheras.	Pegasus
al-mar'ah al-musalsalah	Andromeda. Ἀνδρόμεδα.	Andromeda
al-muthallath	Triangulum. Triquetrum. Deltroton. Τρίγονος aut Δελτῶτον.	Triangulum
[no Arabic given]	Antinous. [no Greek]	Antinous
[no Arabic given]	Coma Berenices. [no Greek]	Coma Berenices
	Zodiacal Signs	
al-ḥamal	Aries. Κριός.	Aries
al-thawr	Taurus. Ταῦρος.	Taurus
al-taw'amayn	Gemini. Δίδυμοι. Castor aut Apollo I Pollus quibusdā Hercules.	Gemini
al-saraṭān	Cancer. Κάρκινος.	Cancer
al-asad	Leo. Λέων. Alased.	Leo
al-'adhrá	Virgo. Παρθένος. Vindemiatrix.	Virgo
al-mīzān	Libra. Bilans. Χηλή.	Libra
al-'aqrab	Scorpius. Σκόρπιος.	Scorpio
al-qaws, al-rāmī	Sagittarius. Τοξεύτης.	Sagittarius
al-jady	Capricornus. Ἀιγοκερεύς.	Capricorn
sākib al-mā', al-dalw	Aquarius. Ganimedes, puer Idueus pincerna Iovis. Ὑδκόχοος.	Aquarius
al-samakatān, al-ḥūt	Pisces. Ἰχθύες.	Pisces

APPENDIX: cont.

Arabic labels	**Latin and Greek labels**	**Modern identifications**
	Southern Constellations	
Qīṭus	Cetus. Balena Pristis. Leo Marinis. Κῆτος.	Cetus
al-jawzā', al-jabbār	Orion. 'Ορίων. Algjeuza.	Orion
al-nahr	Fluvius. Eridanus. Nilus. Ποταμός.	Eridanus
al-arnab	Lepus. Λαγωός.	Lepus
al-kalb al-akbar	Canis Maior. Sirius. Κύων.	Canis Major
al-kalb al-aṣghar	Canicula. Προκύν.	Canis Minor
al-safīnah	Argo Navis. Navis Iasonis. 'Αργώ. Arca Noë.	Arvo Navis
al-shujā'	Hydra. Serpens Aquaticus. 'Ύδρος.	Hydra
al-kās	Crater. Vas. Poculū Apollinis. Κρατήρ.	Crater
al-ghurāb	Corvus. Κόραξ.	Corvus
al-qinṭūrus	Centaurus. Chiron. Phyllirides. Κένταυρος.	Centaurus
al-sab'	Fera. Lupus. Θηρίον.	Lupus
al-mijmarah	Ara. Thuribulum. Lar. Θυμιατήριον.	Ara
al-iklīl al-janūbī	Corona Austrina. Uramseus. [Στέ]φανος νότιος.	Corona Australis
al-ḥūt al-janūbī	Piscis Austrinus. 'Ιχθὺς νότιος.	Piscis Austrinus
	Non-Ptolemaic Southern Constellations	
Dūrādah	Dorado	Dorado
ḥamāmat [al-]Nūḥ	Columba Noë	Columba Noë
al-qaqnus	Phoenix	Phoenix
al-kurkī	Grus	Grus
al-'aq'aq al-hindī	Pica Indica ab Indis Lang.	Pica Indica

APPENDIX: cont.

Arabic labels	Latin and Greek labels	Modern identifications
ḥayyat al-baḥr	Hydrus	Hydrus
al-hindī	Indus	Indus
al-ṭāwūs	Pauo	Pavo
ṭā'ir al-jannah	Apous Indica	Apus
al-muthallath al-janūbī	Triangulum astrinum	Triangulum Australe
al-ḥirbā'	Chamaeleon	Chamaeleon
al-dhubābah	Musca	Musca
al-ṣalīb	El Cruzero Hispanis	El Cruzero Hispanis
[no Arabic given]	Vligende Vis	Piscis Volucris
	Stars and Nebulae	
al-mīzar	Mizar	β *Andromedae*
al-'anāq	Alanâc	space between εδρ *Arietis* and α *Persei*
ra's al-ghūl	Caput Medusae. Ras Algoul.	β *Persei* (Algol)
al-janb	Algjenb	α or γ *Persei*
al-'ayyūq	Capella. Hirens. Alajouc.	α *Aurigae* (Capella)
al-jadyān	Hoedi	ζη *Aurigae*
rijl al-jabbār	Rigel Algiebbar	β *Orionis* (Rigel)
al-shi'rá al-yamānī	Assjira Aljemanía	α *Canis Majoris* (Sirius)
al-'abūr	Alabour	β *Canis Majoris*
al-shi'rá al-shāmī	Assjira Essjamía	α *Canis Minoris* (Procyon)
al-ghumayṣā'	Algomeisa	β *Canis Minoris*
[no Arabic]	Asinus boreal. Praesepe. Asinus Austral.	γδ *Cancri* and M44 (Praesepe)
qalb al-asad	Cor Leonis. Regulus. Basiliseus. Calb Alased.	α *Leonis* (Regulus)
al-fard	Cor Hydrae. Alpherd	α *Hydrae* (Alphard)
banāt [al-]na'sh	Benat nas.	ηζε *Ursae Majoris*
al-simāk al-rāmiḥ	Arcturus. Assimāe-Arrámichi	α *Boötis* (Arcturus)
al-ghurāb	Algorab	δ *Corvi* (Algorab)
al-fakkah	Alfecca	α *Coronae Borealis* (Alphecca)
ra's al-jāthī	Ras Algjáthi	α *Herculis* (Rasalgethi)
ra's al-ḥawwā	Ràs Alhouwa	α *Ophiuchi* (Ras Alhague)

APPENDIX: cont.

Arabic labels	**Latin and Greek labels**	**Modern identifications**
al-ridf	Arridf.	α *Cygni* (Deneb)
sa'd al-zarrā' [sa'd al-bāri']	Saad-azzerá	λμ *Pegasi*
al-mankib	Menceb	β *Pegasi*
sa'd	Saad	η *Aquarii*
fam al-ḥūt	Fumalhout	α *Piscis Austrini* (Fomalhaut)
al-majarrah	Via Lactea	The Milky Way
saḥābat al-kubrá	Nubecula maior	[beneath Hydrus]
saḥābat al-ṣughrá	Nubecula minor	[beneath Hydrus]
[no Arabic given]	Nova stella anni 1600	
suhayl*	Canopus*	α *Carinae* (Canopus)

* Added by hand in ink

Lunar Mansions

1. al-sharaṭān	Sjertán.	βγ or αβγ *Arietis*
2. al-buṭayn	Butain.	εδρ *Arietis*
3. al-thurayyā	Tureija. Pleiades.	The Pleiades
4. al-dabarān	Aldeberân. Hyades.	α *Tauri* (Aldebaran)
5. al-haq'ah	Hacaa.	$\lambda\phi^{1,2}$ *Orionis*
6. al-han'ah	Hanaa.	γξ *Geminorum*
7. al-dhirā'	Dirá.	αβ *Canis Minoris* or αβ *Geminorum*
8. al-nathrah	Netra.	γδ *Cancri* and M44 (Praesepe)
9. al-ṭarf	Tarfa	λ *Leonis* and κ *Cancri*
10. al-jabhah	Giebha.	ζγηα *Leonis*
11. al-zubrah	Zebra.	δθ *Leonis*
12. al-ṣarfah	Sarpha.	β *Leonis*
13. al-'awwā'	Auwa.	βηγδε *Virginis*
14. al-simāk al-a'zal	Spica. Assimâc aláxel.	α *Virginis* (Spica)
15. al-ghafr	Gaphra.	ικλ *Virginis*
16. al-zubāniyāt	Zubaniae.	$\alpha^{1,2}\beta$ *Librae*
17. al-iklīl	Aclyl.	βδπ *Scorpionis* [the tradition is not consistent]
18. qalb al-'aqrab	Calb 'alacrab.	α *Scorpionis* (Antares)
19. al-shawlah	Sjaula.	λυ *Scorpionis*

APPENDIX: cont.

Arabic labels	Latin and Greek labels	Modern identifications
20. al-na'ā'im	Naaümae.	γδεησφτζ *Sagittarii*
21. al-baldah	Belda.	[open space]
22. sa'd al-dhābiḥ	Sado-ddabig.	$\alpha^{1,2}\beta$ *Capricorni*
23. sa'd bula'	Sado-bula.	νε *Aquarii*
24. sa'd al-su'ūd	Sado soûd.	βξ *Aquarii*
25. sa'd al-akhbīyah	Sado-lachbía.	γπζη *Aquarii*
26. al-fargh al-muqaddam	Farga anterior.	αβ *Pegasi*
27. al-fargh al-mu'akhkhar	Farga posterior	γ *Pegasi* and α *Andromedae* [or δ *Pegasi*]
28. baṭn al-ḥūt	Batno 'lhūt.	β *Andromedae*

Other Arabic Labels on the Celestial Map

dā'irat al-janūbīyah	Circulus Antartica	South celestial polar circle
dā'irat al-shamālīyah	Circulus Arcticus	North celestial polar circle
dā'irat al-intiqāl al-shatawī	Tropicus Capricorni	Tropic of Capricorn
dā'irat al-intiqāl al-ṣayfī	Tropicus Cancri	Tropic of Cancer
dā'irat al-burūj	Zodiacus	Ecliptic; zodiac
mu'addil al-nahār, dā'irat al-istiwā'	Circulus Aequinoctialis	Celestial equator
quṭb falak al burūj	Polus Eclipticae Bor.	North ecliptic pole
quṭb falak al burūj	Polus Eclipticae	[south] ecliptic pole
al-quṭb al-shamālī	Polus Arcticus	North celestial pole
al-quṭb al-janūbī	Polus Antarticus	South celestial pole

A propos d'un emblème de Tycho Brahe dans les *Mechanica*

A.Ph. Segonds

Dans les dernières années on semble enfin commencer de réaliser l'importance qu'a eue l'œuvre de Tycho Brahe non seulement pour l'astronomie du XVIè siècle, mais aussi et surtout pour le développement institutionnel de la discipline. Témoin de cet intérêt renouvelé la publication d'une nouvelle biographie de Tycho,[1] et des articles où est examinée la principale œuvre de Tycho:[2] la création (entre 1576 et 1584), l'installation et le développement du premier véritable institut de recherche scientifique de l'époque moderne, Uraniborg, dans l'île de Hven, avec son observatoire annexe de Stellæborg.

Depuis le début, Tycho a désiré faire connaître au reste du monde cette entreprise; il le devait non seulement pour honorer son souverain qui le subventionnait généreusement mais aussi pour affirmer sa place parmi les savants européens et, ainsi, faire taire les critiques domestiques, indignés que tant d'argent pût être dépensé sur un projet socialement aussi peu intéressant.

Pour se défendre et faire connaître son entreprise, dès 1584 Tycho avait réalisé une gravure sommaire représentant ses bâtiments; en 1591, il avait réalisé sur ses propres presses un ouvrage en onze feuillets, ancêtre des *Mechanica* de 1598, oû les bâtiments d'Uraniborg et les instruments étaient représentés et accompagnés d'une description sommaire:[3] nous possédons encore un exemplaire de cette publication, conservée

[1] Voir V.E. Thoren, *The Lord of Uraniborg. A Biography of Tycho Brahe*, Cambridge, 1990. Bien informée et agréablement écrite, cette biographie ne fait cependant pas définitivement oublier celle de J.L.E. Dreyer, *Tycho Brahe. A Picture of Scientific Life and Work in the Sixteenth Century*, Edinburgh, 1890 (réimpr. Gloucester, Mass., 1977), qui repose sur une connaissance incomparable de l'œuvre de Tycho.

[2] Cf. par exemple O. Hanaway, 'Laboratory Design and the Aim of Science: Andreas Libavius versus Tycho Brahe', *Isis*, lxxvii, 1986, pp. 585–610; J. Shackelford, 'Tycho Brahe, Laboratory Design, and the Aim of Science: Reading Plans in Context', *ibid.*, lxxxiv, 1993, pp. 211–230.

[3] Cet exemplaire est décrit et reproduit par J.L.E. Dreyer dans son édition critique de Tycho (citée la note suivante), vol. 5, p. 367.

From *Learning, Language and Invention: Essays presented to Francis Maddison*, ed. W.D. Hackmann and A.J. Turner. Published by Variorum, Ashgate Publishing Ltd, Gower House, Croft Road, Aldershot, Hampshire, GU11 3HR, Great Britain.

aujourd'hui à Bratislava. Un autre exemplaire avait très certainement été adressé au Landgrave de Hesse, un des correspondants les plus assidus de Tycho, puisque leur correspondance (et celle de Rothmann son mathématicien) emplit tout le volume des *Epistolæ astronomicæ*, publié par les soins de Tycho en 1596. Comme apparemment Rothmann n'avait pas remis cette description à son destinataire, Tycho décida d'en adresser un second exemplaire avec une lettre en allemand, oû il décrivait Uraniborg. En 1596 lors de la publication des *Lettres astronomiques*, le document est reproduit et accompagné d'un double texte, allemand et latin, qui ne se recouvrent pas toujours.[4] Ainsi dès 1596, le projet de Tycho avait reçu une assez vaste diffusion dans le milieu savant.

L'année suivante, après son départ précipité du Danemark et alors qu'il ne disposait plus d'un mécène, Tycho décida de publier, en guise de cadeau de nouvel an, une description complète de ses installations, maintenant abandonnées, de ses instruments, de ses principaux résultats et de ses buts: ainsi espérait-il susciter un nouvel intérêt pour ses travaux et retrouver un mécène. On le sait, l'entreprise aboutit à la publication d'un somptueux ouvrage, tiré à tout petit nombre d'exemplaires, et adressé seulement aux amis les plus choisis ou aux mécènes potentiels:[5] les *Astronomiæ instauratæ mechanica* (*Description des instruments employés dans la restauration de l'astronomie*) publiés à Wandesburg, à côté de Hambourg, au début de l'année 1598. Somptueusement reliés, augmentés d'un catalogue de mille étoiles manuscrit et souvent enrichis d'un portrait peint à l'aquarelle de Tycho et de rehauts de couleur faits à la main dans les figures des instruments, ces ouvrages font la gloire des bibliothèques et le bonheur des iconographes.

Traduisant et annotant cet ouvrage en vue d'une nouvelle édition, mon attention a été retenue par un passage de la description du premier instrument décrit dans la collection:[6] un petit cadran de cuivre doré construit, note-t-il dès le début, «il y a bon nombre d'années» [Figure 1]. D'autres détails dans la description montrent qu'il s'agit, aux yeux de Tycho, d'un instrument dépassé: en particulier, il était notablement

[4] Nous citerons toujours nos textes de Tycho Brahe d'après l'édition de J.L.E. Dreyer, *Tychonis Brahe Dani. Opera Omnia*, Copenhague, 1913–1929 (réimpression Amsterdam, 1972): nous emploierons l'abréviation TBOO suivi du volume et de la page. – Il existe au moins deux fac-similés des *Mechanica* de Tycho: l'un publié en 1900 par l'Académie Danoise, et l'autre en 1967 à Bruxelles. Mentionnons aussi la traduction anglaise (incomplète, malheureusement) de cet ouvrage publiée sous le titre de *Tycho Brahe's Description of his Instruments and Scientific Work*, København, 1946, par H. Ræder, E. Strömgren et B. Strömgren. – Ce qui s'est publié à Paris sous le nom de «traduction française» des *Mechanica* ne mérite absolument pas ce nom et ne sera donc cité nulle part.

[5] Liste des exemplaires conservés et connus par exemple dans W. Norlind, *Tycho Brahe: En Levnadsteckning. Med nya bidrag belysande hans liv och verk*, Lund, 1970, pp. 286–293.

[6] On trouvera le texte des *Mechanica* au vol. 5 de l'édition Dreyer. – Notre texte figure aux pp.13–15 (le passage visé dans le texte: p. 13.3). Le texte de la description adressée au Landgrave de Hesse (lettre daté du 26 septembre 1591) figure au vol. 6, pp. 243–246 (lettre allemande), 246–249 (traduction latine: *sequuntur illæ ipsæ literæ in latium sermonem conuersæ propter exteros*, p. 246.24–25); p. 250–262.13 (description des instruments et des bâtiments en allemand); p. 262.14–295 (description des instruments et des bâtiments en latin).

QVADRANS MINOR
ORICHALCICUS INAURATUS.

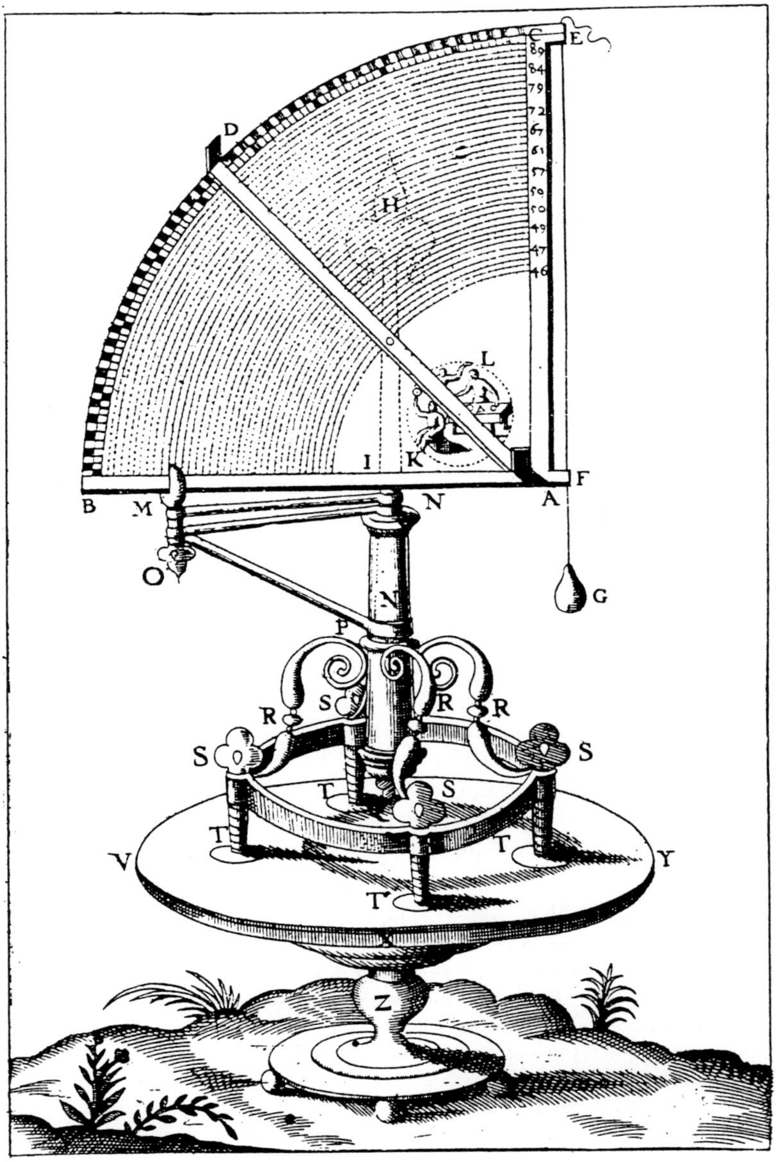

Figure 1. Petit quadrant de cuivre doré, figure tirée de Tychonis Brahe, Astronomie Instauratae Mechanica, *Wandesburgi, MDIIC, 4ᵛ. L'emblème est sommairement représenté en KL.*

plus petit que les autres instruments (puisque son rayon ne dépasse pas 40 centimètres), et était pourvu d'un système de division inspiré du mathématicien portugais, P. Nuñez, qui s'était avéré incommode et pratiquement inutilisable; enfin, il n'avait pas été équipé des pinnules nouvelles de Tycho. La lettre au Landgrave de Hesse nous apprend au surplus qu'il était conservé dans une des salles souterraines de Stellæborg, où il était utilisé de temps en temps sur une petite table de pierre, située à l'extérieur (marquée M dans la planche de la vue droite de Stellæborg[7]). Après la description de l'instrument et avant d'en venir à son usage, Tycho insère ce long texte que nous allons citer complètement:[8]

> Quant à l'image LK que l'on peut voir, en quelque façon, comprise dans une forme circulaire entre le centre et le plus petit arc de cercle, elle a été gravée à titre d'ornement et aussi pour que, par sa signification hiéroglyphique, elle enseigne quelque chose, parce que, autrement, cet espace serait demeuré vide. Sur cette image (pour l'indiquer aussi en passant) sous un arbre d'un côté vert et feuillu, un jeune homme avec une couronne de laurier, est assis sur une pierre carrée, et d'une main, il tient un globe céleste, de l'autre, un livre sur un gazon aussi verdoyant; il étend ses pieds sur des herbes qui cachent la racine de l'arbre. De l'autre côté, c'est-à-dire du côté gauche de l'arbre, la racine est flétrie, les branches stériles ont perdu leurs feuilles; au milieu il y a une table qui présent quelque chose de presque tout ce que les mortels tiennent en estime sur la scène du monde: par exemple, une corbeille pleine de pièces d'or, des sceptres, des couronnes, des blasons, des chaînes d'or, des pierres précieuses, une parure féminine, des gobelets, des cartes, des dés et autres choses de cette sorte, et tout cela un squelette, qui représente la mort, l'entoure de ses mains et de ses pieds, comme s'il voulait le prendre pour lui. Au dessus de toute la figure, un vers en explique le sens allégorique: NOUS VIVONS PAR L'ESPRIT, TOUT LE RESTE APPARTIENDRA A LA MORT, [placé] de telle sorte que la première partie du vers, sur le côté droit et vivant de l'arbre, où siège le jeune philosophe, pende des rameaux comme s'il était dans un cartouche, tandis que de l'autre côté, vient l'hemistiche restant.
>
> Par là, nous avons voulu signifier qu'une solide doctrine (particulièrement la sublime connaissance des choses célestes) donne sur cette terre perpétuité de vie et de mémoire, tandis que tout le reste est frivole et destiné à périr, et périt avec le corps de l'homme. Mais lorsque nous avons regard à l'éternité du royaume des Cieux, alors cela même qui était du côté de l'arbre florissant passe à la seconde [place], puisqu'il n'y a nul rapport entre le fini et l'infini. Et alors le vers inscrit change et devient: NOUS VIVONS EN CHRIST, TOUT LE RESTE APPARTIENDRA A LA MORT, de telle sorte que la première partie est adaptée à l'arbre florissant, la seconde à la partie stérile. A l'intérieur même de la frondaison de l'arbre, il y a quelque indication hiéroglyphique de la vie et de la doctrine du Christ. Tout le reste demeure comme auparavant, si ce n'est que les

[7] Planche intitulée: *Orthographia Stellæburgi* dans *Mechanica*, éd. cit., p. 146.

[8] Voir le texte dans les *Mechanica*, éd. cit., p. 14.3–34; on trouve les textes parallèles dans la lettre au Landgrave respectivement p. 261. 1–4 (texte allemand: la description de l'instrument se réduit à ceci: «XXVII. Il y a aussi un petit *quadrant* de cuivre entièrement doré qui depuis *le centre jusqu'à la circonférence* mesure une *coudée*; il se *meut* sur un piedestal de fer. On notera que le numéro ne correspond pas à celui du latin, qui est XXV,. – Dans les traductions de textes allemands, on imprime en italiques les très nombreux mots de latin); pp. 284.11–285.28 (texte latin de la lettre au Landgrave, souvent plus détaillé que celui des *Mechanica*; nous donnerons en note les principales divergences).

doctrines et les inventions humaines passent vers la partie gauche (ou postérieure) de l'arbre, c'est-à-dire la partie stérile. Par quoi nous avons voulu signifier que rien ne peut faire le bonheur de l'homme et lui donner l'immortalité pour toujours, sinon le mérite du Fils de Dieu, le Christ Rédempteur, et l'observance de sa vie et de sa doctrine. Il nous a plu d'exposer tout cela ici, peut-être en nous écartant de notre sujet, à cause de l'image jointe.

Ce long texte, qui a bien aux yeux de Tycho valeur de digression, est destiné à faire connaître une petite image circulaire, gravée sur un ancien instrument de Tycho, plus guère utilisé. Il s'agit selon les propres termes de Tycho d'une *hieroglyphica..incisio*, c'est-à-dire d'une gravure à sujet emblématique puisque, on le sait, les emblèmes, au XVI et au XVIIè siècle, ont partie liée avec les hiéroglyphes: dans les deux cas il s'agit, croyait-on à l'époque, d'un discours condensé, où l'image transmet immédiatement un contenu, que le *motto* ne transmet que médiatement[9]. Le fait est, donc, que Tycho fait graver sur un de ses instruments un emblème, et qu'il le pourvoit d'un motto: *vivimus ingenio, cetera mortis erunt.* L'illustration elle-même oppose clairement les deux mondes: celui de l'esprit, symbolisé par un jeune homme les yeux tournés vers le ciel, montrant d'une main une sphère céleste (l'observation) et de l'autre, un livre (la transmission), et celui des biens de ce monde: argent, pouvoir, sexe, honneurs, boissons et jeux, qui sont la proie de la mort.

Le *motto* lui-même est assez facile à identifier: il s'agit d'un vers tiré d'une pièce attribuée à un auteur anonyme (quelquefois identifié, à tort semble-t-il, avec Virgile) et intitulée: *Elegia in Maecenatis obitum*, où l'on lit le vers suivant:[10]

Viuitur ingenio, cetera mortis erunt.
(On vit par l'esprit, tout le reste appartiendra à la mort.)

La portée de ce texte est parfaitement claire: seul échappe à la mort l'esprit, tout le reste, c'est-à-dire le corps et les biens du corps sont voués à la disparition, à l'anéantissement. Et encore la survie dont il est question est-elle celle que procurent, dans le souvenir des hommes, la gloire ou des exploits, au total une pensée parfaitement païenne, sans vraie considération d'une survie personnelle.

Un pareil texte devait évidemment attirer l'attention des humanistes du XVIè siècle, puisqu'il semble leur garantir l'immortalité par leurs travaux, tandis qu'elle est refusée aux grandeurs de ce monde, aux puissants de ce monde. C'est ainsi que nous le retrouvons utilisé dans un des plus célèbres portraits de Dürer, sous sa forme canonique (*uiutur ingenio…*), celui de Willibald Pirckheimer,[11], réalisé vers 1524 et

[9] Sur les rapports entre emblème et hiéroglyphes au XVI-XVIIè siècle, voir par exemple A. Saunders, *The Sixteenth-Century French Emblem Book. A Decorative and Useful Genre* (Travaux d'Humanisme et Renaissance, vol. 224), Genève, 1988, pp. 71–97.

[10] *Elegia in Maecenatis obitum*, I 38 éd. E.J. Kenney dans W.V. Clausen et F.R.D. Goodyear et *alii*, *Appendix Vergiliana*, Oxford, 1966, p. 89. – Sur ce texte et son attribution à Virgile, cf. sv. *Vergilius Maro* dans PW VIII A1, col. 1163–1169 (par K. Büchner). L'attribution à Virgile de ces élégies est très répandue à la Renaissance. – Nous expliquerons *infra* p. 275–276 la divergence de leçon que nous nous bornons à relever pour l'instant.

[11] Sur Pirckheimer, cf. W.P. Eckert-C. von Imhoff, *Willibald Pirckheimer. Dürers Freund im Spiegel seines Lebens, seiner Welt und seiner Umwelt*, Köln, 1970, pp. 80–84 (sur le portrait); voir aussi le très riche travail de N. Holzberg, *Willibald Pirckheimer. Griechischer Humanismus in Deutschland*, München, 1981.

très largement diffusé au XVIè siècle, parce qu'il était censé représenté une figure type d'humaniste chrétien. Cette légende n'a pas échappé à E. Panofsky[12] dans sa grande monographie sur Dürer et, en note, il proposait deux autres exemples remarquables de l'emploi de ce motto: Vésale et Tycho Brahe.

Dans le cas de Vésale,[13], il s'agit d'une étonnante planche tirée du *De humani Corporis Fabrica libri septem*, dont la première édition parut à Bâle en 1543, la même année que le *De revolutionibus orbium cœlestium* de Nicolas Copernic. On trouve au livre premier, à la planche suivant la page 124, une représentation d'un squelette tenant d'une main un crâne, tandis qu'il s'appuie de l'autre pensivement sur une base architecturale, portant la légende: 'VIVITVR /INGENIO/CÆTERA MORTIS /ERVNT'. Un des grands exégètes de Vésale, C.D. O'Malley, caractérise ainsi la scène : 'The Skeletal Hamlet soliloquizing beside the tomb upon some poor Yorick is perhaps the most greatly admired figure of the osteologic series'. On sait maintenant qu'une bonne partie des bois gravés de l'ouvrage de Vésale, et en tout cas ceux de la partie ostéologique, proviennent de l'atelier du Titien.[14]

En dehors de l'emploi en commun du même motto ces trois documents n'ont rien d'autre en commun qu'une exaltation, somme toute, banale de l'esprit par rapport à la mort.

Nous allons maintenant présenter un autre document qui a des liens incontestables avec Tycho Brahe, mais diffficiles à préciser. Il s'agit d'un emblème figurant dans un des recueils les plus admirés du XVIIè siècle: le *Nucleus Emblematum*[15] de Gabriel Rollenhagen,[16] publié à Arnheim en 1611, avec des gravures justement admirées dues

---

[12] Cf. E. Panofsky, *The Life and Art of Albrecht Dürer*, 2 vols, Princeton, 1945, p. 29 et n. 213 (la tr. fr., *La vie et l'art d'Albrecht Durer*, Paris, 1987, ne contient pas les notes!).

[13] Sur Vésale, cf. C.D. O'Malley, *Andreas Vesalius of Brussels (1514–1564)*, Berkeley-Los Angeles, 1965, voir pp. 124–129 et 130. Sur les illustrations du *De Humani Corporis Fabrica*, cf. C.D. O'Malley-J.B de C.M. Saunders, *The Illustrations from the Works of Andreas Vesalius*, Cleveland-New York, 1950.

[14] Cf. C.D. O'Malley *Op. cit.* (n. 13), p p.25–29; voir aussi D. Rosand et M. Muraro, *Titian and the Venetian Woodcut, A Loan Exhibition Organized and Circulated* [...]1976–1977, s.l., pp. 211–233. – La même planche est reemployée dans l'*Epitome de humani corporis fabrica*, paru à Bâle en 1543, mais avec une légende différente: deux vers saisissants, tirés de Silius Italicus, *Punica*, XXI 243–244 (Solvitur omne decus leto niveosque per artus /It stygius color, et formæ populatur honores ['S'efface toute grâce sous l'effet de la mort, sur les membres de neige court la couleur du Styx, elle va ravageant la beauté de son corps': trad. M. Martin, coll. Budé, Paris, 1984]).

[15] L'ouvrage, publié en deux parties, respectivement en 1611 et en 1613 porte le titre suivant: *Nucleus/ Emblematum se-/ lectissimorum, quæ/ Itali vulgo impresas/ vocant priuata in-/ dustria studio sin-/ gulari, undique/ conquisitus, / non paucis venustis in-/ uentionibus auctus, / additis carminibus, / illustratus./ A Gabriele Rollenhagio / Magdeburgense. / Coloniæ/ E Musæo cœlatorio / Crispiani Passæi / Prostant/ Apud Ioannem Iansonium/ Bibliopolam Arnhemiensem.* Nous utilisons la réimpression parue à Paris en 1989 (Les recueils d'emblèmes et les traités de Physiognomonie de la Bibliothèque Interuniversitaire de Lille n°3).

[16] Sur Gabriel Rollenhagen, cf. G. Dünnhaupt, *Personalbibliographien zu den Drucken des Barocks*, 6 vols, Stuttgart, 1990–1993, vol. 5 [1991], pp. 3466–3475, où l'on trouvera une bibliographie de toutes les publications de Gabriel Rollenhagen ainsi qu'une liste de travaux modernes. Parmi ceux-ci retenons, W.Harms, 'Der Fragmentcharakter emblematischer Auslegungen und die Rolle des Lesers. Gabriel Rollehagen's Epigramme', dans *Deutsche Barocklyrik*, hrsg. M. Bircher et A.M. Haas, Bern, 1973, pp. 49–64; l'article dans ADB (29) [1889], pp. 84–87 (W. Seelmann) mérite encore d'être consulté.– Voir aussi W. Harms,

à Crispin (I) de Passe.[17]. L'ouvrage s'ouvre donc par un portrait gravé de Gabriel Rollenhagen, alors âgé de 27 ans (il est, en effet, né le 22 mars 1583), suivi d'une préface, et présente comme premier emblème celui-ci [Figure 2]:

VIVITVR INGENIO CÆTERA MORTIS ERVNT.

avec comme *subscriptio* (ce que Rollenhagen a présenté dans sa page de titre comme *additis carminibus*):

Disce bonas artes, et opes contemne caducas
VIVITVR INGENIO CÆTERA MORTIS ERVNT.

On aura remarqué que Rollenhagen reprend le vers pseudo-virgilien (dans sa forme canonique) et comme motto et comme partie de sa *subscriptio*, selon une technique qui lui est familière. C'est qu'il y a, chez lui, un lien particulièrement fort entre l'image et le motto. Or si l'on considère de près l'image, on retrouve absolument tous les éléments énoncés par Tycho Brahe dans le texte cité plus haut: l'arbre au centre de l'image, avec un côté verdoyant et un côté stérile; les deux personnages (le savant, le squelette); les objets (livre et globe d'un côté; sceptre, couronne, collier, gobelets, armes nobiliaires, gemmes); même le siège sur lequel est assis le jeune homme (un édicule en pierre) la présence d'herbes à ses pieds tandis que du côté de la mort on distingue des racines flétries – tout cela reprend les éléments de la description de Tycho Brahe. Sur un point pourtant (important aux yeux de Tycho), l'emblème de Rollenhagen ne concorde pas avec la description de Tycho: la légende est mal placée, puisque la partie *vivitur ingenio* est du côté de la mort, tandis que *cætera mortis erunt* est du côté de l'immortalité. Or l'insistance de Tycho sur la disposition des légendes est considérable et nous allons la retrouver un peu plus loin. Allons même plus loin: il y a très certainement une erreur dans la disposition de l'emblème de Rollenhagen, probablement due au graveur (ou même à l'imprimeur). L'image est, en effet, présentée à l'envers. Dans son état actuel, le philosophe est à gauche, la mort, à droite; or Tycho dit très clairement dans sa description que: *A læua parte Arbor habet & radicem emarcidam* (p. 284.31–32; *Mechanica*, p. 14.9–10: '*ab alia* [*parte*], *læva videlicet arboris*'), c'est-à-dire que la partie morte de l'arbre est à gauche et sa partie vive, à droite, selon la symbolique traditionnelle du droit et du gauche.

Pour l'instant, la question qui se pose est la suivante: qui est le modèle de qui?, puisqu'il est assuré qu'il existe un rapport entre le texte de Tycho et l'emblème de Rollenhagen? Doit-on croire qu'il existe un modèle commun aux deux auteurs, un recueil d'emblèmes antérieur à Tycho et à Rollenhagen où l'un et l'autre, indépendamment l'un de

Homo viator in bivio. Studien zur Bildlichkeit des Weges (Medium ævum 21), München, 1970, pp. 52 et 93, qui a reconnu dans cet emblème un aboutissement du célèbre *Hercules ad bivium* [cf. E. Panofsky, *Hercules am Scheidewege u. andere antike Bildstoffe in der neueren Kunst* (Studien der Bibl. Warburg, 18), Leipzig-Berlin, 1930] et a reconnu le caractère potentiellement chrétien de cet arbre, qui évoque le bois de la Croix du Christ (cf. pp. 90–97), alors qu'il ne connaissait pas le second emblème de Tycho.

[17] Sur Crispin de Passe (Crispin l'ancien) [1564–1637], voir Thieme-Becker, *Allgemeines Lexikon der Bildenden Künstler*, vol. 26 [1932], pp. 280–281.

Figure 2. Emblème tiré de G. Rollenhagen, Nvclers Emblematvm Selectissimorvm [...], *Arnheim, 1611, p. 1. On notera que le chiffre du foliotage, en haut à gauch, est inversé, ce qui confirme que la planche est montée à l'envers.*

l'autre, auraient puisé leur inspiration? ou bien, puisque Tycho est antérieur à Rollenhagen, faut-il admettre que Rollenhagen s'est inspiré de Tycho?

Qu'il ait existé un recueil d'emblèmes antérieur, et source de nos deux auteurs, peut, je crois, être rejeté: un examen de tous les travaux bibliographiques sur les livres d'emblèmes[18] m'a montré que ce motto (*vivitur ingenio, cætera mortis erunt*) est bien connu de toute la littérature des proverbes,[19] mais qu'il n'existe pas d'autre emblème que celui de Rollenhagen à l'avoir repris pour motto.

Est-il possible que Rollenhagen se soit inspiré de Tycho dans son emblème? C'est chose bien possible, puisque la communauté d'inspiration ne fait aucun doute; puisque l'ouvrage de Tycho où figure cet emblème était bien connu au début du XVIIè siècle.[20] Mais il y a encore plus: on sait que Tycho et Georg Rollenhagen,[21] le propre père de Gabriel Rollenhagen, étaient étroitement liés et qu'ils correspondaient régulièrement.[22] On constate que celui-ci admirait énormément Tycho: il est donc bien possible que cette admiration ait été transmise à Gabriel, son fils.

Dans la description de 1591,[23] l'emblème est présenté d'une manière plus complète, qui nous permettra de comprendre d'où vient la leçon 'vivimus' dans la citation du pseudo-Virgile, au lieu de 'vivitur'. Le texte se présente ainsi:

> Tycho emploie encore en d'autres occasions ce symbole hiéroglyphique, non plus pour les choses qui passent, comme ici, mais en l'adaptant pour les choses éternelles et plus importantes. Dans ce cas, en effet, dans la partie vivante et pleine de feuilles de l'arbre, on trouve une représentation hiéroglyphique de la vie et de la doctrine du CHRIST, au

---

[18] Voir par exemple A. Henkel-A. Schöne, *Emblemata. Handbuch zur Sinnbildkunst des 16. und 17. Jahrhunderts*, 2 éd., Stuttgart (Supplément, 1976): voir aussi J.M. Diaz de Bustamente, *Instrumentum Emblematicum curavit* (coll. Alpha-Omega. Reihe B, vol. 5), 2 vols, Hildesheim, 1992: ni l'un ni l'autre de ces ouvrages considérables ne connaît d'autre emblème avec le motto *'vivitur ingenio, cætera mortis erunt'*, hors celui de Rollenhagen.

[19] Voir Hans Walther, *Lateinische Sprichwörter und Sentenzen des Mittelalters in Alphabetischen Anordnung*, 5 vols, Göttingen, 1963–1967, voir vol. 5, n° 34038, p. 882, où l'on trouvera une liste d'attestations de cette sentence, mais aucune qui provienne d'un recueil d'emblèmes: toutes les attestations invoquées sont tardives (fin XVIIè siècle-XVIIIè siècle) et s'avèrent souvent n'être que des citations directes du pseudo-Virgile.

[20] La description de l'emblème de Tycho figurait non seulement dans les *Epistolæ astronomicæ* de 1596 et dans les *Mechanica* de 1598, mais aussi dans une seconde édition de cet ouvrage, publiée en 1602 et tirée à bien plus grand nombre.

[21] Sur Georg Rollenhagen [1542–1609] (souvent confondu avec son fils, à cause de l'identité des initiales), voir G. Dünnhaupt, *Personalbibliographien* (n.16), vol. 5, pp. 3476–3491; voir aussi l'article dans l'ADB (29) [1889], pp. 87–93 [W. Seelmann].

[22] Sur les rapports G. Rollenhagen–Tycho Brahe, voir E. Rosen, *Three Imperial Mathematicians*, New York, 1986, *passim*. – Leur correspondance est édité aux volumes 7 et 8 des TBOO (par exemple: vol. 8, pp. 50–52 [avril 1598]; p. 72[18 juin 1598]; p. 327 [9 juin 1600]; pp. 416–417 [18 avril 1601]; pp. 420–421 [15 mai 1601]; les lettres de Tycho sont en général plus longues: cf. pp. 293–295 [5 avril 1600]; pp. 369–375 [26 septembre 1600]. Une bonne partie de ces lettres traitent de l'affaire Ursus: en effet, l'ennemi mortel de Tycho, au cours de son voyage en Allemagne, avait séjourné au Gymnase de Magdebourg (dont Rollenhagen était le directeur) et y avait même enseigné. Les autres traitent du sort des effets de Tycho laisssés à Magdebourg au cours de son voyage vers Prague. Les relations Tycho–Rollenhagen était donc assez étroites.

[23] Sur ce texte, voir plus haut, p. 268 et n. 6.

> dessus de laquelle est inscrit: NOUS VIVONS EN CHRIST. Au contraire, dans le présent exemple, les études de la philosophie et leur représentation hiéroglyphique passent de l'autre côté de l'arbre, là où il est sec et stérile, sous l'empire de la mort, tandis que l'on retrouve la même inscription qu'auparavant dans l'emblème des choses qui passent, de telle sorte que le vers entier, dans cette figuration hiéroglyphique qui a regard à l'éternité, est finalement le suivant: NOUS VIVONS EN CHRIST, TOUT LE RESTE APPARTIENDRA A LA MORT.
>
> De la sorte, au moyen de ces symboles hiéroglyphiques, Tycho veut faire entendre que la doctrine solide et la connaissance des choses tant célestes que terrestres, et même la philosophie toute entière, dans cette vie mortelle et dans ce monde visible et qui passe, arrachent les hommes à leur mortalité bien mieux que toutes les activités auxquelles les hommes de ce monde se consacrent le plus souvent; néanmoins, lorsqu'il s'agit d'obtenir l'éternité dans une suprême béatitude, tout cela n'aide en rien, mais toutes ces [connaissances], eu égard à cette [éternité], sont exposées à la mort et à la disparition, alors que c'est la bonté de Dieu et sa miséricorde, obtenues par le mérite de son fils, notre Rédempteur, le CHRIST, l'observance et l'imitation de sa vie et de sa doctrine qui seules comptent et perdurent.

Dans cette nouvelle description d'emblème qui, rappelons-le, ne figure pas dans les *Mechanica*, on voit que, dans l'esprit de Tycho, il s'agissait d'un ensemble de deux emblèmes qui, à eux deux, devait transmettre la sagesse totale de Tycho et lui donner son sens ultime. Dans le premier emblème, comme on l'a vu, Tycho entendait montrer la supériorité de l'idéal contemplatif sur les valeurs du monde (pouvoirs et argent): c'est une sorte de revendication de sa vocation contre les aspirations de sa classe sociale d'origine, dont il y a souvent trace dans ses écrits (voir *infra*, p. 278). Dans le second emblème, cette fois, l'idéal de contemplation est confronté au christianisme et est trouvé aussi déficient que, dans le premier cas, les biens du monde, puisque, comme le dit Tycho, «il n'y a nul rapport entre l'infini et le fini».[24] Malheureusement, la description ne précise ni où ni quand ce second emblème était employé par Tycho, ni même s'ils étaient côte à côte quelque part dans la demeure de Tycho.

Un point mérite de retenir notre attention: c'est que la devise fait allusion à une «Vie en Christ» (*Vivimus in Christo*), c'est-à-dire selon un idiotisme fréquent dans la langue du Nouveau Testament: «nous vivons par le Christ», et c'est manifestement ce choix qui a amené Tycho à 'corriger' la citation du texte du pseudo-Virgile, de manière que les deux mottos se correspondent rigoureusement.

Cette même description de 1591 précise encore, pour finir, que dans Uraniborg et Stellæborg les devises et inscriptions abondent:

> En outre, [Tycho] a pris soin de faire graver, tant plus haut dans le chateau lui-même, en divers endroits, que même sur certains de ses instruments, d'autres expressions sententieuses en forme d'emblèmes, qui laissent entendre que la vérité se trouve profondément enfouie dans la moelle des choses, et non pas dans leur écorce. Je ne juge pas utile de les citer toutes ici.

---

[24] Sur ce célèbre axiome de la philosophie scolastique, voir Aristote, *De cælo*, A 6, 274a7–8.

Nous sommes, en effet, en mesure de rassembler, grâce à plusieurs documents écrits, un très grand nombre d'inscriptions et de symboles ayant figuré dans le chateau d'Uraniborg[25] et grâce auxquelles Tycho sacrifiait à son didactisme (bon nombre de ces textes sont des épigrammes, où Tycho fait connaître son sentiment sur tel ou tel personnage, tel ou tel instrument etc.), ou au contraire, ne s'adressait qu'aux «Happy few» capables de comprendre. Ainsi, toujours dans la description de 1591, Tycho (ou son copiste allemand) note-t-il à propos d'un petit ensemble énigmatique de Stellæborg:[26]

> L'Hypocauste dont je parle actuellement [*scil.* la pièce principale de Stellæborg qui était entièrement décorée de peintures 'signifiantes'] est entièrement souterrain, mais il est couronné d'une voûte de pierre qui s'élève, à l'extérieur, au niveau du sol, de telle sorte qu'elle a l'apparence d'une petite colline, recouverte de terre et de gazon verdoyant, par quoi [Tycho] représente en quelque manière un petit Parnasse pour les Muses, parmi lesquelles Uranie est la plus importante, et ce seulement pour ceux qui sont capables de comprendre et qui sont les élèves d'une philosophie plus sublime, car c'est pour la même raison que les poètes ont recouvert du voile de la fable leurs [inventions], pour que le vulgaire sans instruction ne puisse pas comprendre.

On voit donc que toute la réalisation de Tycho est marquée du sceau de sa personnalité, aussi bien les instruments que la disposition des lieux, le choix des allégories réalisées au moyen de statues ou de décors peints, ou encore les devises gravées aux points névralgiques de l'ensemble et destinées à instruire le visiteur sur l'intention réelle du réalisateur.

Pour terminer, je voudrais transcrire une de ses devises, dans laquelle Tycho donnait une expression parfaitement appropriée de ce qu'il avait voulu réaliser, et dont on verra qu'elle est bien en accord avec l'emblème qu'il avait fait graver sur son quadrant. Au dessus donc du portail d'entrée à l'escalier menant aux installations souterraines de Stellæborg on trouvait les statues de trois lions couronnés, et, en dessous d'eux, la devise suivante:

NEC FASCES NEC OPES
SOLA ARTIS
SCEPTRA PERENNANT.

Les trois lions sont de taille différentes et celui du milieu, surélevé par rapport aux autres, tient un sceptre: ils sont une représentation allégorique des royautés du pouvoir

[25] Une partie seulement de ces inscriptions a été reproduite dans les *Mechanica*; on en trouve davantage dans la description de 1591 (dans la version latine et non pas dans l'allemand; voir TBOO, vol. 6, pp. 269–271 par ex., ou pp. 272, 273, 274–275, 276 etc.) et enfin dans un ouvrage du XVIIè siècle de J. Resenius, *Inscriptiones Haffnienses*, Hauniæ, 1668, qui donne une description complète de l'état de Uraniborg au milieu du XVIIè siècle et publie un grand nombre d'inscriptions relevées *in situ*.

[26] Voir TBOO, vol. 6, pp. 273.33–274.2; je donne ici en latin la formule la plus saisissante: *idque intelligentibus saltem & Philosophiæ sublimioris alumnis; ex quo etiam Poetæ hæc fabularum involucris texerint, ne vulgus imperitum capere posset.* On trouverait en maints endroits de l'œuvre de Tycho l'idée qu'il ne convient pas de tout révéler de la pratique scientifique, d'en réserver la transmission à une petite élite. – Il désigne souvent par *philosophia sublimior* ce type de spéculations réservées à l'élite: par exemple l'Alchimie, ou la philosophie paracelsiste, ou encore la partie prédictive de l'astrologie etc.

(*fasces*) et de la fortune (*opes*), dominées par celle de l'art, c'est-à-dire de la science. Autrement dit, Tycho signifiait à ses visiteurs (nobles généralement) que 'Ni les faisceaux [des charges] ni l'argent ne donnent l'immortalité, mais seul le sceptre de l'art', manifestant ainsi d'une manière éclatante son refus de choisir la vie que semblait devoir lui dicter son origine sociale. Et dans la description de 1591, qui nous fait connaître cette devise, on trouve le bref commentaire suivant:[27]

> Sous ces Lions on trouve l'inscription suivante, réalisée sur une pierre spéciale, dans laquelle on répond brièvement et véritablement, au moins eu égard au monde qui passe, aux objections silencieuses de ceux qui, par ignorance, s'imaginent peut-être que tous ces appareils si nombreux et si somptueux sont inutiles.

Dans les *tacitæ obiectiones* il faut entendre les commentaires faits par ses illustres visiteurs – ou simplement devinés par Tycho – : à quoi tout cela peut-il bien servir? pourquoi tant d'argent dépensé en vain/ Et la réponse de Tycho est très claire: tout cela concourt à la recherche de l'immortalité, cela même qu'annonçait notre emblème (*Ingenio vivimus*). Mais on aura très certainement remarqué la subtile nuance qu'introduit Tycho dans son commentaire: cette recherche ne vaut que *quoad Mundum temporaneum* (eu égard au monde qui passe), car ces valeurs ne valent que eu égard à notre monde qui est destiné à passer, mais eu égard à la véritable immortalité, seul compte l'enseignement du Christ, ainsi que nous l'avions vu plus haut dans le second emblème de Tycho. On a donc la preuve que cet emblème qui a été notre point de départ gît bien au centre même de la vision du monde de Tycho, cette vision qu'il a voulu concrétiser par toute sa réalisation d'Uraniborg.

[27] Voir TBOO, vol. 6, p. 272.4–10.

Elucidatio fabricae ususque: Rambling Among the Beginnings of the Scientific Instrument Bookshelf

A.V. Simcock

Astrorum quicunque vias inquirere gaudet,
Hoc opus exiguo vendicet aere sibi.[1]

Imagine a library containing every printed book about scientific instruments. And imagine, instead of beginning at Abbatt, Richard,[2] that the librarian has gone mad and arranged it all in chronological order. In the brief interregnum between her being dragged away, wild-eyed and muttering curses against the alphabet, and the arrival of a flying squad of chartered librarians fully conversant with OLIS and Dewey-decimal,[3] we must try to learn something of the history of the scientific instrument book. I know that studying the history of books according to the initial letter of the author's surname has endless insights to offer;[4] but just in these few moments of imaginary release, let us pretend, *pace* Einstein, that time is the most important framework within which to understand the order of things. And in the small amount of it allowed on this occasion, I doubt whether we can do more than ramble among the beginnings, among the first few inches of books on the top shelf.[5]

[1] Epigraph to Johann Stöffler, *Elucidatio fabricae ususque astrolabii*, Oppenheim, 1513 and 1524 (hereafter cited as Stöffler). The second line has a double meaning, as all self-respecting epigraphs should: the *opus* is the book, which you can obtain for a small amount of brass, and the science of astronomy, which you can possess in a small piece of brass, i.e. an astrolabe.

[2] Richard Abbatt, *The Elements of Plane and Spherical Trigonometry, and its Application to Astronomy, Dialling, and Trigonometrical Surveying*, London, 1832. Or, if this is too non-specific for the library you have in mind, then I'm afraid it will have to be Ernst Abbe, *Gesammelte Abhandlungen. Erster Band: Abhandlungen über die Theorie des Mikroskops*, Jena, 1904.

[3] OLIS is Oxford University's on-line library catalogue; I don't know what 'Dewey-decimal' is.

[4] See, for instance, my comments on the pattern of Irish science in *Annals of Science*, xlv, 1988, pp. 541–542.

[5] The most beautifully written of all English dictionaries, which of course is *Chambers Twentieth Century Dictionary*, Edinburgh, 1972, defines ramble as 'to go as fancy leads', and also 'to be desultory, incoherent, or delirious'.

From *Learning, Language and Invention: Essays presented to Francis Maddison*, ed. W.D. Hackmann and A.J. Turner. Published by Variorum, Ashgate Publishing Ltd, Gower House, Croft Road, Aldershot, Hampshire, GU11 3HR, Great Britain.

Incunabula instrumentorum

Scientific subjects as we should now call them were not the favourite themes of the early printers, and emerged only gradually from the presses. Furthermore, almost all of the early work put into print was either classical or medieval. The first scientific book ever printed was Pliny's *Historia naturalis* (Venice, 1469). Theophrastus (botany), Celsus (medicine), Euclid (geometry), and Aristotle (misc.) are notable among the Greek and Roman authors who followed Pliny into print in the incunable period.[6] Euclid's *Elements* is the oldest scientific textbook still in use, and established the laws of a geometry which has stood virtually unchallenged since. It was first printed in 1482 by the Augsburg Venetian Erhard Ratdolt, who illustrated it with little geometrical diagrams, the first printed use of such.[7] Most editions since have been similarly embellished.

While Euclid established the fundamental principles of geometry, the practical basis of astronomy was provided in the works of Claudius Ptolemy, principally his *Almagest*, which was first printed in 1496 in the form of the Latin 'epitome' prepared by Regiomontanus and Georg Peurbach.[8] Further printings and improvements of Ptolemy's works gradually followed.[9] They contained the essential tables and star catalogues without which no serious astronomical (or astrological) work could be done, including the making and using of astronomical instruments. Several treatises on instruments by Ptolemy, hitherto like all his writings known only to the few who had access to manuscripts, were also printed during the next century;[10] as were Regiomontanus's own writings on instruments, both new and Ptolemaic, which had remained unpublished long after his early death.[11]

The period saw many other important classical works put into print. The Ptolemy of the technologists was Vitruvius, whose *De architectura* (first printed at Rome in 1487) embraces many subjects in addition to the obvious one, including substantial parts of engineering, physics, and applied mathematics. It was a great influence on Renaissance engineers and architects, and some of the sixteenth-century editions were illustrated with their achievements, such as the Gothic cathedral of Milan.

The first printed books about scientific instruments were all short medieval treatises on the astrolabe: the *Opus preclarissimum astrolabii* by Andalus de Nigro (1270–1342), edited by the astronomer Petrus Bonus and printed at Ferrara in 1475; *De astrolabio*

[6] See Arnold C. Klebs, *Incunabula scientifica et medica: Short Title List*, Bruges, 1938, and also Bern Dibner, *Heralds of Science*, 2nd edn, Norwalk & Washington, 1980.

[7] *Preclarissimus liber elementorum Euclidis* ..., Venice, 1482. See Charles Thomas-Stanford, *Early Editions of Euclid's Elements*, London, 1926.

[8] *Epytoma Joannis de Monte Regio in almagestum Ptolomei*, Venice, 1496.

[9] Including the first complete *Almagest*, Venice, 1515; *Omnia, quae extant, opera, geographia excepta* ..., Basel, 1541; the Greek text of his astrological manual Τετραβιβλος, Basel, 1553, printed in conjunction with Philipp Melanchthon's Latin translation of it, *De praedictionibus astronomicis, cui titulum fecerunt quadripartitum*, Basel, 1553.

[10] *Ptolemaei planisphaerium* ..., Venice, 1558; *Liber de analemmate*, Rome, 1562.

[11] *Scripta clarissimi mathematici M. Ioannis Regiomontani, de torqueto, astrolabio armillari, regula magna Ptolemaica, baculoque astronomico, & observationibus cometarium* ..., Nuremberg, 1544, edited by Johann Schöner.

canones by Robert of Chester (*fl.* 1140), but usually ascribed to Robertus Anglicus, printed at Perugia about 1478; perhaps the most respectable (or respected, anyway), the *Magistralis compositio astrolabii* of Henry Bate (1246–*c.* 1310), written in 1274 and printed at Venice (by Ratdolt again) in 1485;[12] and *De astrolabo* by Gregoras Nicephorus (*c.* 1295–*c.* 1360), printed at Venice in 1498. A *De astrolabio* attributed to the fifth-century Neoplatonist philosopher and mathematician Proclus was printed in 1491, also at Venice.[13]

The position of the astrolabe as the pre-eminent scientific instrument of the medieval world is well attested by its standing alone in the early printed literature. No other instrument was treated to a printed book of its own until the appearance of works on sundials in the 1520s and '30s. Nothing else was said about scientific instruments in the incunable period, except in Regiomontanus's calendars (of which more anon), and in Gregor Reisch's encyclopaedia the *Margarita philosophica*, which appeared in various editions from 1496 onwards. Reisch describes and depicts the *baculus iacob* or cross staff, two types of quadrant, and the armillary sphere; and in his supplement of 1512 adds illustrations of the *torquetum* and the *polimetrum* (prototype of the theodolite), and a very thorough section on the astrolabe, based on an influential treatise traditionally attributed to the eighth-century Arabic astronomer Messahalla.[14] This pioneering encyclopaedia whetted the scholarly world's appetite for a more substantial literature on such instruments.

Regiomontanus had a few counterparts in the medical and biological fields, whose works began to appear in the 1490s, often mixing (as was usual) contemporary observation and experience with information compiled from earlier texts. A particularly interesting example was Hieronymus Brunschwig (*c.* 1450–*c.* 1512), who published works on anatomy and surgery in 1497 and on distillation of medicines in 1500, which were the first reliably contemporary treatments of practical surgery and practical medicinal chemistry to appear in print.[15] Another important medical and natural history work was the *Hortus sanitatis* (Mainz, 1491 and later editions), an encyclopaedia famous for its thousand and more woodcut illustrations in late medieval style.

But scientific writing by contemporaries was everywhere slow in entering the lists. Partly this was because of a backlog of two thousand years or so which confronted

[12] R.T. Gunther reproduces Henry's treatise in *The Astrolabes of the World*, Oxford, 1932 (hereafter cited as Gunther), pp. 368–376. Gunther obtained his text in true medieval style from his research assistant Miss E. G. Parker's handwritten transcription of the printed book in the Bodleian Library (Ashmole 465), which (Miss Parker's transcription, that is) was afterwards bound and placed in the library of the Museum of the History of Science as a substitute for the printed book.

[13] For a better list of astrolabic incunabula, see A.J. Turner, *The Time Museum: Catalogue of the Collection*, vol. 1, part 1, Rockford, 1985, pp. 52–53.

[14] Gregor Reisch, *Margarita philosophica nova*, Strasbourg, 1508 and 1512, and *Appendix matheseos in Margaritam philosophicam*, Strasbourg, 1512 (unpaginated); the astrolabe section in the latter is entitled, on its own title page, 'Tractatus de compositione astrolabii Messehalath'. For Messahalla's treatise, see R. T. Gunther, *Early Science in Oxford*, vol. 5, Oxford, 1929; and on its authorship, see Paul Kunitzsch, 'On the Authenticity of the Treatise on the Composition and Use of the Astrolabe Ascribed to Messahalla', *Archives internationales d'histoire des sciences*, xxxi, 1981, pp. 42–62.

[15] *Anathomia ossium corporis humani*, Strasbourg, 1497; *Buch der Cirurgia* ..., Strasbourg, 1497; *Liber de arte distillandi, de simplicibus* ..., Strasbourg, 1500.

the first printers and publishers. But I think it also took time to evolve the idea of doing what I am doing now, that is, frantically writing something for the specific purpose of participating in the printed word as the primary form of scholarly communication – instead of composing it by thinking and writing over a period of fourteen years, and then getting a young monk to copy it out neatly so that I can send it to a friend at Dieulacres.[16]

The Man in the Moon

The first contemporary scientist to confront the printing press was Johann von Königsburg or, in Latin, Johannes de Monte Regio, also known in his youth as Hans Müller, and whom posterity decided, for no very good reason, to call Regiomontanus (1436–1476). And he cultivated the new medium very directly and deliberately, not so much as an author but as the first scientific publisher. It is partly this which gives him his presiding reputation in the post-medieval world, for Regiomontanus was the first scientist who can be said to have become famous in his own lifetime. Historians often consider him (along perhaps with his own teacher and collaborator Peurbach) to be the founder of modern astronomy, in spite of the obviously more dramatic appeal of Copernicus.

Regiomontanus dragged astronomy out of the middle ages, setting himself the task of modernizing the mass of classical and medieval data and making it available and useful in the here and now, in spite of the conservatism of a world which (rightly) was in such awe of the knowledge of the ancients that it hesitated to correct it. He approached the *Almagest* in this spirit; and with a little editorial help the Bible of astronomy reigned supreme well beyond Copernicus's alleged destruction of the Ptolemaic universe, and into the epoch of Tycho Brahe and Kepler, who were the logical concluders of what Regiomontanus started. He has in fact been credited, not entirely incredibly, with making a remark which may indirectly have sown the seeds of heliocentric cosmology in Copernicus's brain.[17] It was, anyway, in Regiomontanus's epitome of Ptolemy that Copernicus was first alerted to the errors which inspired him to rethink Ptolemy's system.

A more subtle thing which I believe earns Regiomontanus the respect of posterity is the way in which he began to mediate scientific information to the (or a) public, approaching it as though it needed to be communicated, and perhaps even had a commercial value. He was an entrepreneur – quite a different fish from the cloistered and secretive Pole – and leapt at the potential of the printed book, which had been invented during his own lifetime. About 1472 he had a printing press installed in his house in Nuremberg, and began printing ephemerides, and then an astronomical calendar in book form covering a span of about 50 years. His *kalendarium*[18] was first printed

---

[16] I originally wrote Silverdale, but this would be an anachronism; so I've assumed that my friend in Silverdale, had he lived in the fifteenth century, would have been a monk at Dieulacres Abbey.

[17] See Edward Rosen's thoughtful consideration of this in *DSB*, vol. 11, 1975, pp. 351–352.

[18] Most editions are untitled; Klebs *Op. cit.* (n. 6) lists them, pp. 278–279. For a facsimile reprint with introduction by Ernst Zinner, see *Der deutsche Kalender des Johannes Regiomontan: Nürnberg, um 1474*, Leipzig, 1937.

Figure 1. Lunar volvelle with man in the moon which may be a portrait of Regiomontanus; from the 1482 Venice edition of Regiomontanus's kalendarium

about 1474, with various later editions (including at Venice from 1476 onwards); and he printed it not just in Latin but in German editions too, with strong hints of his Bavarian dialect and some newly-coined words (notably *sunnur*, ancestor of *sonnenuhr*) – for these things had never been discussed before.

This was a significant moment in the intellectual history of Europe, for not only did it push forward the expansion of language by introducing science into the vernacular, but it made calendrical, astronomical, horological, and such like data widely available for the first time. The calendar also established new conventions for setting out the information: its tabulation in columns and the pictorial indication of eclipses were, like Ratdolt's Euclidean diagrams, new (and successful) experiments in communication. Regiomontanus and Peurbach are also generally credited with inventing the modern system of astrological division into twelve houses, and the square horoscope diagram expressing this (first widely published by Reisch in his *Margarita philosophica*).

The German and Latin calendars contain text explaining the relevant astronomical calculations, and describing for the first time in print the making and use of a sundial. At the end they provide four actual working, or at least demonstration, instruments, printed on thicker paper: an unequal hours calculator, a lunar volvelle, a horary quadrant, and a *quadratum horarium generale*, known to us as a Regiomontanus sundial. As with much else in early printing, such things were not unknown in the manuscript tradition; but there of course they were unique products of an individual writer's ingenuity, and cannot have had more than the tiniest effect in scientific education or the dissemination of techniques. Their printed counterparts at once reached what we must assume were hundreds, and eventually thousands, of people. In short, Regiomontanus invented the printed paper scientific instrument, and his 1474 calendars are the first books on our scientific instrument bookshelf.

In 1518 two of the most interesting of Regiomontanus's followers, the astronomer Johann Stöffler and the scientific printer–publisher Jakob Köbel, issued their *Calendarium Romanum magnum*, a hugely expanded and lavishly printed imitation of the Regiomontanine calendars, with all the same ingredients writ large. These included the four printed instruments at the end. It shows that Regiomontanus's influence was very much alive over forty years later, and proves absolutely that Stöffler and Köbel were consciously continuing his work. They even issued it in a German as well as a Latin edition.

Volvelles and other instruments integral with printed books – moving illustrations, in fact – become common in the sixteenth century; while real instruments of printed paper also occur, though more rarely.[19] Peter Apian designed and published both between the 1520s and '40s. At the same time the instrument maker Georg Hartmann, who saw himself as Regiomontanus's successor in Nuremberg, made instruments not only of brass (and ivory) but also of paper, printed from copperplates which he had engraved,

[19] See Owen Gingerich, 'Astronomical Paper Instruments with Moving Parts', in R. G. W. Anderson et al. (eds), *Making Instruments Count* ..., Aldershot, 1993, pp. 63–74.

and mounted on wood and pasteboard. The paper astrolabe in the Museum of the History of Science has had its date altered on the original printing plate to 1542 from (probably) 1531, indicating that Hartmann used the plate for quite some years to produce a cheaper instrument on demand. Perfectly good scientific instruments (mainly astrolabes and quadrants) were later made in this way by other instrument makers.[20] Unfortunately their minimal survival leaves a distorted impression of their impact.

Regiomontanus was also associated with the making of instruments in the conventional sense, and there is no doubt whatsoever that a workshop operated under his direction in Nuremberg in the 1470s. There remains much to discover about German instrument making between the controversial astrolabe of 1462, supposedly made by Regiomontanus himself,[21] and the Hartmann workshop of the 1520s onwards. Existing lists of astrolabes, for instance, offer surprisingly little during this period.[22] Regiomontanus's name is associated more with sundials. The 'universal rectilinear' type of vertical sundial which is actually known by his name was made quite commonly by the mid-sixteenth century, often as a secondary feature on the back of a quadrant or nocturnal. The diptych sundial is also traditionally associated with him, and, from Hartmann's time, became the characteristic product of a considerable industry in Nuremberg.[23]

Regiomontanus's printed works are few;[24] but his revolution lay as much in the concept, and in the spirit, of the enterprise. After all, the printing press did not come with self-evident strategies for its use, nor indeed with warnings that it was going to democratize the élite world of knowledge. Yet at the same time that Caxton set about learning the new craft – to all intents and purposes in the first years of printing as a real business – Regiomontanus established his missionary headquarters in Nuremberg. His converts included printers as well as scientists, the earliest being Erhard Ratdolt. The first book that Ratdolt printed was an edition of Regiomontanus's calendar in 1476, the year of our hero's death. As we have seen, he went on to print other major scientific works, notably his Euclid, which was also a project he took over from Regiomontanus.

An edition of the calendar which Ratdolt printed in 1482 adds to the upper volvelle of the lunar instrument near the end of the book a very crisp man-in-the-moon face,[25]

---

[20] Examples include Philippe Danfrie's astrolabes, 1584, and Henry Sutton's quadrants, 1658; and see below.

[21] See Derek J. de S. Price, 'The First Scientific Instrument of the Renaissance', *Physis*, i, 1959, pp. 26–30.

[22] Sharon L. Gibbs et al., *A Computerized Checklist of Astrolabes*, New Haven, 1973; but see David A. King's comments in Anderson et al. (n. 19), pp. 60–61.

[23] See Penelope Gouk, *The Ivory Sundials of Nuremberg 1500–1700*, Cambridge, 1988.

[24] The other things he printed are Georg Peurbach, *Theoricae novae planetarum* and Marcus Manilius, *Astronomica*. For some of his own posthumously published writings, see nn. 8, 11, and 29.

[25] We probably tend to see a face in the moon, but the traditional myth, from Babylon to Baring-Gould (S. Baring-Gould, *Curious Myths of the Middle Ages*, London, 1901, pp. 190–208), is of a full figure carrying a bundle of sticks and with a little dog, usually stated in European folklore to have been banished thither for stealing or for Sabbath breaking. He is the old man gathering sticks whom King Wenceslas saw. Depicting sun and moon with faces is common in astronomical books and manuscripts, but this must be a convention linked to the apotheosis of the heavenly bodies, in which case the moon's face should be female.

in profile, far too unstylized to be anything but a real portrait [Figure 1]. It may be debated whether Regiomontanus's nose, which was certainly prominent, was as uneven as this; but I find it difficult to believe that so deliberate a representation could have been intended as anyone but Regiomontanus himself.[26] And perhaps there was some justice in translating him into the moon. He died in Rome in 1476 aged only 40. His calendrical expertise had taken him there, to advise on calendar reform, a topic of great moment in the later fifteenth century. He probably died of plague; though it was rumoured that he had been murdered – *Ha scelus infandum*! – by the sons of George of Trebizond, whose commentary on Ptolemy he had attacked.[27]

The last thing which Regiomontanus printed, in 1474 or '75, was a list of the works he had so far issued and, chiefly, those he intended to issue – in other words, the first scientific publisher's catalogue.[28] It included Ptolemy's *Almagest* and other works, Euclid's *Elements* (as published a few years later by Ratdolt), Archimedes on geometry and instruments, the *Conics* of Apollonius, the works of the great astrological authority Firmicus, and an ambitious list of his own projects (*opificis tentata*, the maker's essays), including commentaries on Ptolemy, Archimedes, and others, his epitome (*breviarium*) of the *Almagest*, and his important work on trigonometry *De triangulis omnimodis*.[29] This little *verlagsanzeige* is one of the great manifestos of intellectual optimism; and it stands, appropriately, on the western horizon of the Middle Ages, as the man in the moon goes down, and the age of humanism dawns.

The Many-Coloured Astrolabe

We adjourn now to the cradle of printing, Mainz, where Johann Gutenberg printed his famous Bible. Eighty years later, in March 1535, Peter Jordan printed here a special edition of Johann Stöffler's *Elucidatio fabricae ususque astrolabii*. It is not necessarily catalogued as such by our chronological librarian, for Stöffler's familiar title (familiar from perhaps a dozen other editions between 1513 and the 1590s)[30] does not appear in it. Instead, the title-page has been usurped by the printer–publisher, and the apparent title *Coelestium rerum disciplinae, atque totius sphaericae* ... is really the opening words of his preface, telling the 'candid reader' what a jolly good edition of Stöffler's work this is. If I were the librarian I think I would record it as *De compositione aut fabrica astrolabii, eiusdemque usu multifariisque utilitatibus*, which has been added as a fly-title at the head of the first page of the main text – albeit this is twelve

[26] As far as I'm aware, this has not hitherto been considered as a possible portrait of Regiomontanus. Zinner does not mention it: see Ernst Zinner, *Leben und Wirken des Joh. Müller von Königsberg genannt Regiomontanus*, 2nd edn, Osnabrück, 1968, pp. 267–274.

[27] The interjection is from the prefatory poem of Ratdolt's 1482 (man in the moon) calendar, which is the first documentary reference to the alleged murder.

[28] It is reproduced in Zinner *Op. cit.* (n. 26), fig. 45; and in Gouk *Op. cit.* (n. 23), p. 84.

[29] *De triangulis omnimodis*, Nuremberg, 1533, edited by Johann Schöner; for a facsimile reprint with translation by Barnabas Hughes see *Regiomontanus on Triangles*, Madison, 1967.

[30] See Gunther (n. 12), pp. 595–596.

folios into the book.[31] It is actually Stöffler's usual title translated from Latin into Latin!

At the end of the book an elaborate colophon occupies a full page, beautifully hand-coloured in the copy I have in mind. The central design is Jordan's trademark: a sandglass with a winged orb or sphere below and two disembodied hands grasping a large shell above.[32] Split around it is the word 'Fortuna'. This large and clear version of the symbol is useful not least as a representation of the sandglass, given that, unlike the astrolabe, it has no literature. The page is surrounded by the decorative border which was used on the title-pages of the first and second editions of Stöffler's book, and is therefore part of the *répertoire* of the original printer, Jakob Köbel. Before he died in 1533, Köbel formed some kind of partnership with Jordan, who inherited some of his materials, including the original woodcuts for the illustrations. The text of Jordan's version is essentially identical as well, though reset entirely in Roman type.[33]

Not only is this particular edition very rare, but the copy in my imaginary library is unique. For it contains a full-size printed paper astrolabe, the more impressive for being, like the colophon page, beautifully hand-coloured. The colouring uses pink, rust red, and turquoise, plus several shades of yellow, and gives a magnificent effect. The astrolabe consists of three woodcuts – front, back, and *rete* – printed on separate leaves and bound between the extensive preliminaries and the beginning of the main text.[34]

It should be clearly understood that this is not an illustration, nor merely a demonstration piece. It is a real astrolabe. It was meant to be cut out and mounted (or *vice versa*), to form a working instrument as real as its smaller contemporary by Hartmann, or the later paper astrolabes, more similar in size, by Danfrie or Krabbe.[35] It is included by R. T. Gunther as number 253 in *The Astrolabes of the World*, together with an illustration.[36] But he loses his nerve when it comes to giving his usual technical description of it. Since his survey also includes (for instance next door, at numbers 254 and 255) a number of mere book illustrations (of astrolabes which may or may not have once existed), subsequent students of the astrolabe have dismissed number 253 along with these illustration-only entries. It is deliberately omitted from the versions of the International Checklist of Astrolabes which I have seen. Since this wonderful object has been so undeservedly neglected, I include a description of it here.

The *rete* is 165 mm in diameter. There are 22 star pointers, which are flame shaped and mostly have little stars or asterisks at their points. All the stars are named using

---

[31] Folio 1^r, sig. A1^r as numbered, f. 12 *de facto*.

[32] Folio 78^r. The roughly shaped object held above the sandglass looks at first glance like a stone; but its shape and shading reveal it to be a shell of the scallop type, held with its concave side towards us. I wonder if it is connected with the use of ground shell as the material in early sandglasses?

[33] Apparently Jordan also published a black-letter version in June 1535.

[34] Folios 9^v, 10^v, and 11^r *de facto*, bound between prelims., f. 8 as numbered, and f. 1, sig. A1 as numbered.

[35] Krabbe's (though dated 1583) was issued in sheets bound with his book: Johann Krabbe, *Newes Astrolabium* ..., Wolffenbüttel, 1609 and 1625.

[36] Gunther (n. 12), pp. 427–428. It is also reproduced in colour in Turner (n. 13), p. 51.

their Arabic names. The ecliptic circle is divided into degrees and signs of the zodiac. The style of the *rete*'s tracery within the ecliptic circle does not conform to any of the conventional shapes by which astrolabes are ordinarily classified (e.g. Y-type, quatrefoil type, Fusoris type). It comes nearest, as one would expect, to the contemporary Hartmann astrolabes and their kindred, which have been termed Regiomontanus type.[37] The tracery in our specimen consists of five arcs attached within the upper part of the ecliptic circle, mirrored by five arcs attached within the lower part of the tropic of Capricorn circle. The central arc of the latter group is joined to the ecliptic by a pair of dolphins. The outer tropic circle has the conventional dragon's head and tail.

The front and back of the astrolabe are 195 mm. in diameter. The front is, of course, a single sheet bearing one projection. The edge is divided into hours and degrees, and within this is the standard astrolabe projection with all the usual lines. Its latitude is not stated; but it is in fact for 48°40'. The latitude cannot be calculated quite so precisely as this, but is readable to about 49 or a little less. This fits with the text and examples in the book (in all editions) which use the module 48 40, giving this as the latitude of Tübingen (and also as the middle latitude of the seventh region in the table of the climates of the world).[38] Stöffler was professor of mathematics at Tübingen. Mainz, where this edition of the book was printed, is at latitude 50°. So the intention is clearly to represent Stöffler's astrolabe exactly as described in the book.

The back of the instrument is typical of the astrolabes of the period. In concentric circles around the edge are scales of degrees and signs of the zodiac. Slightly offset within these are circular calendar scales of days and months. Within this, occupying the centre of the plate, are a shadow square in the lower left quarter, Jordan's sandglass trademark in the lower right quarter, and above in the upper half-circle a diagram of unequal hours on either side of a sun with rays and face. In the area of the central circle below the shadow square and sandglass are depicted clouds and a recumbent moon with a very distinctive profile face protruding from the concavity – not unlike our man-in-the-moon Regiomontanus.

Jordan's sandglass symbol proves that the astrolabe accompanied the book and was printed by him. Indeed, it can be considered a form of signature. Most of the astrolabe's technical features are copied from the partial and diagrammatic representations by which the book's description builds up. Hitherto Stöffler had not illustrated the complete astrolabe, nor shown the *rete* in realistic form. Incorporated in his diagrammatic *retes*, however, are a small number of star pointers, which are flame shaped, as here. What we end up not knowing, which is frustrating, is what *rete* design Stöffler might himself have offered us, and thus we cannot be sure that this is fully 100 per cent 'Stöffler's astrolabe', as Gunther dubbed it.

Stöffler died in 1531, and this posthumous edition of his *Elucidatio* may be taken as a kind of tribute to him. Soon after he died, his friend Köbel issued his own small astrolabe manual, which was also printed by Jordan.[39] Very much a summary of Stöffler's,

[37] King in Anderson et al. (n. 19), pp. 60–62.

[38] Stöffler (n.1)(1524), ff. 5^r, 10^v, 11^r, 45^v.

[39] Jakob Köbel, *Astrolabii declaratio* ..., Mainz, 1532.

it contained illustrations of the complete front and back of the instrument (for latitude 50°, and dated 1532) and of the *rete.* The former are very similar indeed to Jordan's, but the *rete* is not the same. Although Köbel died the following year, he may already have been helping Jordan to prepare the new edition of a book which he had himself originally designed and printed; and he may therefore have had a hand in designing the astrolabe. Of course, there probably existed brass instruments which Jordan could have copied. It goes without saying that Stöffler and Köbel had owned and used astrolabes, and were probably capable of making them with their own hands. But in fact, nothing survives which can be associated with them – apart from this.

International checklists notwithstanding, our many-coloured astrolabe has its own rightful place in the ultimate *verzeichnis* of real astrolabes. It links the literature and the instruments in a special and thought-provoking way. And along with Hartmann's work, which survives in fair quantities,[40] it emerges from a relatively fallow period in the archaeology of the instrument. Perhaps Jordan's paper astrolabe after Stöffler should be seen as part of a new era of astrolabe making (and using), of which Stöffler's book was the *primum mobile.*

The Swabian Enlightenment

The *Elucidatio* was evidently written, or finished, or as good as finished, in 1510. Stöffler's preface carries this date (in words: *sesquimillesimo decimo*); and several examples of horoscopes for dates in 1510 are used. One horoscope for March 10 1513 (2 minutes past 10 in the morning) is almost certainly a genuine prediction, perhaps for the intended date of publication.[41] The book was at the printer's in 1512, the colophon at the end of the first edition carrying this date, as do the dedicatory poems at the beginning. The title-page of the first edition is dated 1513, the year of final publication.

This was the first printed book devoted to a scientific instrument (or instruments) and written by a contemporary.[42] The time was ripe: 1512 saw the publication of the astrolabe treatise in Reisch's supplement to his *Margarita philosophica*; it also witnessed the last in a series of anonymous reprints of the Robert of Chester/Robertus Anglicus treatise.[43] Stöffler's work was not a dramatic break with this medieval tradition which had dominated the incunable period, and indeed they have things in common. But it was on quite a different scale, it set the instrument amidst a much more intense didactic programme, and it was written by a man of practical experience who had looked at the whole subject in a fresh and up-to-date way, in the Regiomontanine manner. It was not only the earliest modern book on the astrolabe but, as R. T. Gunther said, it was 'also the best',[44] and remains so to this day. And for over a century it was

[40] Six pages of it in Ernst Zinner, *Deutsche und Niederländische astronomische Instrumente des 11.-18. Jahrhunderts*, 2nd edn, München, 1967, pp. 362–368; 24 astrolabes in Sharon L. Gibbs et al. (n. 22).

[41] Stöffler (n. 1)(1524), f. 62r; other horoscopes ff. 59^v, 60^r, 63^r, 63^v.

[42] I follow Gunther (n. 12), and Turner (n. 13), in disregarding Johannes Angelus (or Engel), *Astrolabium planum in tabulis ascendens* ..., Augsburg, 1494, for although the author was a pupil of Regiomontanus and Ratdolt printed it, its content is entirely astrological.

[43] Reisch *Op. cit.* (n. 14), *Astrolabii quo primi mobilis motus deprehenduntur canones*, Venice, 1512.

[44] Gunther (n. 12), p. 567.

massively influential upon the whole range of scientific instrument literature, and the direct progenitor of a good deal of it.

The book is a folio consisting of 78 folios numbered in folios. In addition, ten folios of preliminaries precede this main text, most of them being an astonishingly long index. This description applies to the first and second editions of 1513 and 1524, printed and published at Oppenheim by Jakob Köbel; and more or less to a third edition, appended to Stöffler's commentary on Proclus's *De sphaera* (Tübingen, 1534), and to the fourth edition, printed by Peter Jordan at Mainz in 1535, with the special characteristics which we have been looking at. All contain the same text and exactly the same woodcut illustrations.

The title *Elucidatio fabricae ususque astrolabii* implies the two sections into which the book falls: the first (*fabrica aut compositio*)[45] covering in detail the structure and especially the mathematical structure and content of the astrolabe, with full-page illustrations of the scales and projections; the second (*multiplex eius usus*) explaining with examples the uses to which it could be put. Not much is said about astronomical observation, and I think one has to take this as confirmation that speculative observing and cataloguing, the great purposes of astronomy from Tycho's day onwards, were not part of the purpose of the astrolabe. The section on use is devoted largely to computing celestial positions, time, and latitude; to astrology and the casting of horoscopes; and to surveying, the attractive illustrations of tower measuring and other operations being the best known feature of Stöffler's book.

In his preface Stöffler tells us who his authorities on the construction (*constructio*, not *fabrica*) and use of the astrolabe were. The list is of some interest:[46] 'ab insignibus philosophis Hipparcho, Ptolemaeo, Amonio, Proclo, Philopono, Nicephoro, Hermanno contracto alemanno, Ioanne eligero, Hainrico bate'. It is an orderly sequence of classical and medieval authors, all of whom would be known to Stöffler in manuscript versions of their writings. Though three of our incunabula are present – Proclus, Nicephorus, and Henry Bate.

Stöffler's mentors begin not surprisingly with Hipparchos in the second century BC, who is still considered the most probable inventor of the astrolabe, and of stereographic projection.[47] Most of what we know of his astronomy comes down to us through Ptolemy, who needless to say is second on Stöffler's list. Ammonius, Proclus, and Johannes Philoponus were all late classical philosopher–astronomers (fifth and sixth centuries AD), Proclus holding a special interest for Stöffler, who wrote a commentary on his *De sphaera*.

These names are followed by four medieval authorities. Gregoras Nicephorus and Henry Bate we have already met. Johannes 'eligerus' I suspect is John of Lignères,

[45] *Fabrica* is the wrong word, in fact, for although it can mean fabric in the modern sense (which anyway is not what Stöffler intended), it really means a workshop or smithy, or the trade practised there. The correct word would have been *fabricatio*. In the text Stöffler uses *compositio* and *constructio*.

[46] Stöffler (n. 1)(1524), f. 1r; and I go through it to supplement rather than repeat Gunther's (n. 12) comments on it, pp. 403, 568–569.

[47] G.J. Toomer considers and supports this traditional view in *DSB*, vol. 15, 1978, p. 219.

whose fourteenth-century tables were of enduring importance to astronomers.[48] But special interest attaches to the oldest of the medieval names, Hermann the Shrivelled (1013–1054), usually called Hermann the Lame, *der Lahme*, or *le Boiteux* by polite non-Latin speakers.[49] He was not only, as Stöffler considered important, a fellow-countryman, but had lived not far away at the monastery of Reichenau, and designed his instruments for the classic 48° latitude. Hermann was one of the first in the Christian west to write in Latin on the instruments known from Arabic texts. From his writings on them stems the enduring love of European mathematicians for the astrolabe, the cylinder sundial, and the quadrant, each of which he made relevant to the interests of Latin-reading scholars. The borrowing of Arabic technical terms for the astrolabe was also Hermann's doing, and has affected every European language via his and, eventually, via Stöffler's Latin.

Later in the book Stöffler lists writers on the seven *climata* (latitudinal regions) of the world, taking his table of the climates from the best authority, one 'Joannes de Sacro Busto' (*sic*). He also acknowledges his astrological influences (authors, that is!), among them Marcus Manilius, whose *Astronomica* Regiomontanus had printed, and Julius Firmicus and Leopold of Austria, who were both on Regiomontanus's list of intended publications. It is also in the astrological section that he mentions Regiomontanus and Peurbach, in connection with the twelve-house system. Elsewhere he refers in passing to Messahalla.[50]

A group of people associated more immediately with Stöffler's book are the four who wrote commendatory verses to be printed in it: Johann von Wirsperck, Peter Günther, Georg Simler, and Philipp Melanchthon. Each has a poem on the page headed 'Epigrammata'; in addition, Günther contributes a verse to the title-page and Simler has one praising the printer, Köbel, on the colophon page.

The name which leaps out of this group[51] is that of Melanchthon (1497–1560), who was a 15-year-old student at the time. A few years later he went as professor of Greek to Wittenburg, where he immediately became the closest ally of Luther in the midst of Luther's rebellion. He became, essentially, the co-founder of the Reformation. Melanchthon was a prodigious scholar of Greek at a time when it (too) awaited revival, and at a young age (as this, perhaps his earliest publication, proves) adopted a Greek translation of his surname, Schwarzerd (μελαν χθων, black earth). He is known as an editor and translator of astronomical texts;[52] and his connection with Stöffler shows

48 Gunther (n. 12), p. 568, however, identifies him as John of Legnano (d.1382), a less well-known chap of similar sort.

49 *Contractus* means shrunk tightly or shrivelled. Dr A. Chapman diagnoses 'muscular dystrophy, congenital spinal curvature, premature osteomyelitis, or dwarfism: we can never know' (personal communication). To the fact that he was more than merely lame we owe his lifetime of sitting in his monastery writing about (and presumably making) some of the west's earliest scientific instruments. He deserves more honour from historians of instruments – though Gunther reproduces the text of his treatise on the astrolabe, pp. 404–422; and Claudia Kren's little article in *DSB*, vol. 6, 1972, pp. 301–303, is superb.

50 Stöffler (n. 1)(1524), ff. 10^{r-v} (climates), 60^{v} (astrology), 59^{v} (Regiomontanus), 52^{r} (Messahalla).

51 And did so for Gunther (n. 12) too, p. 568.

52 Such as Ptolemy's Τετραβιβλος (n. 9); Sacrobosco's *Libellus de sphaera*, Wittenburg, 1531; Peurbach's *Theoricae novae planetarum*, Wittenburg, 1535 (which had been Regiomontanus's first publication).

that this was no mere linguistic or textual exercise, but a genuine interest in the subject. He enrolled at Tübingen university on September 17 1512, more or less as Köbel's press was about to print the *Elucidatio*. While studying mathematics and astronomy under Stöffler, he also taught Greek, including to another, older linguist among Stöffler's pupils, Sebastian Münster, who came to Tübingen in 1514.[53]

Melanchthon belonged to a 'good Latin' club at Tübingen; and his poem is appropriately pithy and elegant:

Foelix Actiacae magister artis
Stofler nobilis: enitens probate
Orbes: quae iuga perferant volucres
Astrorum tacito graves meatu
Signas: hoc regimurque, vivimusque
Prognati imperio: lubente fama
Pennis munera gratiora librum
Phoebo Daedaleis novum reponis.

It is chastening to realize that a man famous in the annals of European religious history was present at the birth of the first modern book on a scientific instrument. It improves the standing of this very specialized phenomenon within the wider *gestalt* of the Renaissance, as well as indicating that the astrolabe and what it stood for were not irrelevant to that most dramatic consequence of the Renaissance, the Reformation.

Of the other versifiers, Simler is known to history as a humanist thinker on law, and was a friend of Köbel. Wirsperck tells us himself that he is an adviser (*consiliarius*) to the Elector of the Rhine Palatinate. Günther is unknown, in spite of the genealogical researches of our own R. T. Gunther. The Günthers were an old-established Swabian family, and may well have been related to Stöffler.[54] Peter's presence on the title-page is suggestive of a close link with the author.

Johann Stöffler (1452–1531) was born at Justingen, in the province of Swabia, not far from Tübingen. He studied at the university of Ingolstadt, and became a minister at Justingen in 1477. He gained a reputation as a clever mathematician with a particular expertise in mechanical and astronomical instruments. An elaborate celestial globe which he made (*fabrefecit*, the inscription says) in 1493 is preserved in the Germanisches Nationalmuseum, and is his only surviving instrument.[55] I do not know if he ever encountered Regiomontanus, who was only sixteen years his senior; but he moved in circles which were close to the master's influence, and became one of the first generation of his disciples. From 1499 onwards he compiled and published

[53] Viktor Hantzsch, *Sebastian Münster: Leben, Werk, Wissenschaftliche Bedeutung*, Leipzig, 1898, p. 10.

[54] Gunther (n. 12), indeed, succeeded in writing Stöffler into his own family tree, although not in identifying Peter Günther: an ancestor in the female line, Johann König, who was a professor and official of Tübingen university, married one Agnes Stöffler about 1530, and she is presumed to have been a daughter of our hero; granddaughter is more probable. The Günther family hailed from Möhringen near Tübingen, and R. T.'s father Albert Günther studied at the university of Tübingen, where the family was entitled to scholarships as distant kin of Duke Eberhard of Württemberg, who founded the university in 1477. See R. W. T. G[ünther] & A. G[ünther], *Günther Family Records*, London, 1910.

[55] Zinner *Op. cit.*(n. 40), p. 544 and plate 58.

ephemerides in direct continuation of those which Regiomontanus had issued.[56] Not until 1511, when he was 59, did he become professor of mathematics at Tübingen. By this time, as we have seen, he had already completed his great work on the astrolabe, which was published in 1513.

In the manner of the times, Stöffler is invariably described in his books as *Iustingensis* (from his home town), and sometimes *vir Germanus* as well. Peter Günther's verses go further and call him 'Stofler Germanus, origine Suevus' and 'Ioannes Alemannus & Suevus'. The medieval Duchy of Swabia had declined, losing its castle of Tübingen to the new Dukes of Württemberg and its southern town of Zurich to the Swiss Confederation. But like their neighbours the Bavarians, the Swabians were a proud people, descended as they deemed themselves to be from the Suevi of classical times, said to have been 'one of the greatest and most powerful peoples of Germany'.[57]

These regions of southern Germany in which we find ourselves are not backwaters: we are in the centre of Europe culturally, as well as geographically. As Gunther said, 'Much that was good flowed northward at that time, and the South German Astrolabe seems to have come as a child of the Italian Renaissance'.[58] The spirit of the Renaissance found a receptive second home amidst the civic pride and creativity of Germany's semi-autonomous towns, and the intellectual freedom of her universities. Some of the towns were highly developed trading centres with flourishing native crafts – Regiomontanus came to Nuremberg for this very reason. While out in the mountains and forests (of the Tyrol, or of northern Bohemia) were the most extensive mining and metal-processing industries in the world.

And we find ourselves also at a fascinating moment in European history. This is the height of the Renaissance. Stöffler's exact contemporary, Leonardo da Vinci, is alive; so too are the fathers of humanism, Linacre and Erasmus; so are Machiavelli, and Michelangelo. It is important for us to remember (whether it is Melanchthon who reminds us or not) that the mathematicians and astronomers, the printers and instrument makers, were fully part of this world; and the rise of scientific instruments and their literature is part (if I can put it thus) of the phenomenology of the age. If Stöffler's work is the true cornerstone (or should it be bracket?) of the scientific instrument bookshelf, then 1513 is the year in which the modern literature of scientific instruments was born, and its *heimat* was this region of southern Germany. Its godfather and guardian angel was the man in the moon. Its parents were the Swabian astronomer and his muse (Urania).[59] And its midwife was the town clerk.

The Town Clerk of Oppenheim

Jakob Köbel (*c.* 1465–1533) was the son of a goldsmith of Heidelberg, and studied at the famous university there. Legend has it that he also attended the university of

[56] *Ephemerides ab anno 1482 ad annum 1518* …, Ulm, 1499.

[57] William Smith, *A Smaller Classical Dictionary* …, London, 1870, p. 403.

[58] Gunther (n. 12), p. 567.

[59] She is depicted in the frontispiece of Köbel's *Astrolabii declaratio* …, Mainz, 1532.

Cracow in the early 1490s, studying mathematics alongside the young Copernicus. In 1494 he married the daughter of an alderman of Oppenheim, and became town clerk and surveyor. He was already involved in the book trade, and established his own printing press in Oppenheim; 96 works are known to have been printed and published there by him.[60] He generally described himself as *stadtschreiber zu Oppenheim*, and in his pictorial colophon at the end of Stöffler's *Elucidatio* as *prothonotarius et chalcographus* (chief clerk and printer),[61] projecting a confident self-image. He was the archetypal Renaissance urbanite: a respectable tradesman and municipal official, and proud of being both, yet also a scholar and a humanist, with many friends in the world of liberal learning. His particular passions were the law and mathematics.

In his own writings, which he only began printing about 1514, he took on the task of making Germany numerate and teaching her accessible calculating and measuring skills. They are all in the vernacular (except his *Astrolabii declaratio*), and have pleasant little illustrations, his aim being to reach – and win over – literate but not wealthy or scholarly German speakers. These charming *büchlein* include several elementary textbooks of arithmetic and reckoning, using both Roman and the then less familiar Arabic numerals; and *Eyn künstliche sonn-Uhr* (1532), reprinted several times under the less flattering title of *Bauren Compas* (peasant's compass), explaining how to use the human hand and a little stick as a sundial or compass.[62]

A compilation of writings in similar vein was published posthumously as *Geometrei. Von künstlichem Feldmessen und absehen Allerhandt Höhe* (Frankfurt, 1535 and later editions). It illustrates the primitive state of surveying instrumentation at the time, as its only tools are the measuring stick, the cross-staff, the mirror, and the quadrant. The approach is deliberately elementary and the language simple, explaining the necessary technical terms with patience and good humour – not least the overall term adopted: 'Dise Kunst des Feldtmessens wirt genent *Geometria* uff Griechisch, welchs heisst Messung der Erden oder Feldts'.[63] Special attention is paid to clarifying the names of measurements, like 'Füss oder Schüch' for Latin *pes*, reinforced by one of Köbel's most delightful illustrations, showing a row of motley Oppenheimers standing foot to foot to demonstrate exactly what a foot is (or rather what sixteen feet are).

His favourite instrument after the astrolabe (or even before it) was the Jacob's staff, as the cross-staff was universally known. He enjoyed the pun on his own name, and was employing it as a symbol as early as 1512: it is the 'mystery object' being held by the *putti* at the top of his title-page border (for instance, in Stöffler's book). In his

[60] Joseph Benzing, *Jakob Köbel zu Oppenheim, 1494–1533: Bibliographie seiner Drucke und Schriften*, Wiesbaden, 1962. Köbel has the singular honour, denied even to Stöffler, of inclusion in the *DSB*, vol. 7, 1973, pp. 418–420 (by Kurt Vogel).

[61] Late Latin *prot(h)onotarius*, chief clerk or secretary, is in fact in dictionaries of that catch-all language, English, as prot(h)onotary; so too is chalcographer, an engraver on copper. I assume that Köbel used the latter term loosely to mean printer; otherwise, there is no excuse for him not having made astrolabes.

[62] The use of the hand as a scientific instrument requires some attention from historians. It occurs again in Peter Apian, *Instrument Buch*, Ingolstadt, 1533.

[63] Jakob Köbel, *Geometrei* ..., Frankfurt, 1570, f. 2^{v}.

account of it, written in 1531 when he was in his mid or late sixties, he tells us that he now has to walk with a staff because he suffers from 'rampant gout': 'meines krancken Beins des wütenden Gegichts, Deszhalb ich auch am Stab gehen müsz'.[64] The implication, interestingly, is that his walking-stick and Jacob's staff were one and the same. This lends support to one hypothesis about this little-known instrument, that it originated in the medieval pilgrim's staff, which was known by the same name.[65]

Johann Dryander (1500–1560), a professor at Marburg, was a friend and follower of Köbel, and took over some of his woodcuts and unfinished writings. He added the section on the quadrant to the posthumous *Geometrei*, and later published it separately in Latin.[66] His completion of Köbel's vernacular account of *Das Nocturnal oder die Nachtuhr* (Frankfurt, 1535) and his own *Novi annuli astronomici* (Marburg, 1536) were the first separately published works on either instrument. He also edited for publication Stöffler's account of terrestrial projections, with folding plates which Köbel had already prepared.[67] As a printer Köbel was succeeded by Peter Jordan of nearby Mainz, with whom he had collaborated during his last few years, and then by Christian Egenolffs in nearby Frankfurt, who sustained Köbel's enterprise into at least the 1580s.

Regiomontanus's dream of scientific publishing – well served at first by Erhard Ratdolt – had no better champion than Jakob Köbel, in his combination of learning, a zeal to communicate science to the layman, and great skill in the art of printing. As well as his own writings, he printed many larger and more academic books. One was a beautifully printed version of the medieval treatise of Albertus Magnus on gemstones and alchemy, issued in 1518.[68] His *tours de force*, however, were Stöffler's *Elucidatio* (1513 and 1524), and Stöffler's Roman calendar in Latin and in German (1518 and 1522). These ambitious folios were printed with the highest standards of typography and page design, and unusually good illustrations. The pains and pride which Köbel took in them are obvious.

His part in the *Elucidatio* extends beyond that of mere printer and distributor. Its well-loved surveying and tower-measuring illustrations have the simple beauty and the humour that are undoubtedly Köbel's. Their boldness and clarity of line are unusual in book illustrations of the period, and a woodcut illustrator of great skill was responsible for them. But the concept was Köbel's. So too was the practical knowledge depicted, for he was the official surveyor to the town of Oppenheim. Several

64 Ibid., f. 20r.

65 There are two unreconciled etymologies for the term 'Jacob's staff': from the Biblical Jacob, referred to once as leaning on a staff; and from the patron saint of pilgrims St James (Latin *Jacobus*), lending his name to the walking-staff typically used by pilgrims. The latter is obviously the more likely, even if the cross staff did not actually evolve from the pilgrim's staff; though I think it did. On the history of the instrument, see John J. Roche, 'The Radius Astronomicus in England', in *Annals of Science*, xxxviii, 1981, pp. 1–32.

66 Johann Dryander, *Brevis et exquisita, quadrantis instrumenti geometrici, & horarii, explicatio*, Marburg, 1550.

67 Johann Stöffler, *Cosmographicae aliquot descriptiones*, Marburg, 1537.

68 Albertus Magnus, *Liber mineralium*, Oppenheim, 1518.

of the towers carry little messages from him. The first tower is actually inscribed with the word 'Oppenheim', and bears shields with symbols of the town and of the German Empire, including the head of the Emperor. The figure measuring the tower's shadow with an astrolabe is a fashionably dressed and well-to-do townsman – and this of course is the town clerk himself, somewhat youthful perhaps, but as he appears in his own imagination [Figure 2]. Another illustration shows a clock and bell tower, providing one of the earliest images of a clock face on a tower.[69] Another tower bears the initials (in monogram) ATH, probably those of Köbel's brilliant woodcutter.[70]

We do not know how the astronomer from Justingen and the town clerk of Oppenheim came to be involved in this venture together. There is no doubt, however, that their collaboration was thorough and warm. The book provided Köbel with an important component in his agenda of scientific printing. And Köbel provided Stöffler with what no author on a scientific instrument had ever before seen in his own lifetime – a printed version of his treatise. To all intents and purposes, and with due respect to their mutual mentor Regiomontanus, these two men founded the printed literature of the scientific instrument.

Lingua franca

The printed word was dominated by the Latin language for its first three centuries of existence. This was the *lingua franca* of academically educated Europe. Indeed, it was the most successful international language the human race has ever invented. And from the beginning of the translation of Arabic texts in the eleventh century AD, it was the language implied by the very word 'translation'; even Greek had to be rendered into Latin to be of use to European scholars. Hence Regiomontanus found himself learning Greek solely in order to edit a Latin version of Ptolemy's *Almagest.*

German is the only vernacular language to emerge with any credibility among our early technical and mathematical books. And it came closest to Latin in the size of its potential readership, being the dominant language across a huge swathe of central Europe. I assume that whatever dialect may have been spoken, elementary education would equip a subject of the German Empire to read a reasonably standard German. This offers Köbel's arithmetics, for instance, a market stretching from Brabant to Transylvania and from Denmark to Tuscany, as well as beyond this area in pockets where German diplomats, artisans, and scholars represented the cosmopolitanism of Teutonic society. It may be felt that I exaggerate the state of literacy at the time;[71] but the existence of so many works in German is itself compelling evidence of a high level of elementary

[69] The hours are the wrong way round, which is a common type of mistake in printmaking, but might also suggest the unfamiliarity of such a thing at the time. Which, the reversal corrected, is 10 o'clock at night!

[70] Stöffler (n. 1)(1524), ff. 69^v (Oppenheim), 72^v (clock), 74^r (ATH).

[71] I've never been a believer in the myth of mass illiteracy anyway: it was invented by middle-class historians to portray my ancestors in a bad light.

literacy (by which I mean literacy without Latin), and of course the publication of certain works in both languages, such as Regiomontanus's calendars, presupposes this.

The élite position of the astrolabe among instruments is reflected in its early literature being almost entirely in Latin – even when the great popularizer, Köbel, turned his hand to it. This reflects a perceived superiority of the study of (or inherent in) the astrolabe, which was meant for scholars, over other mathematical skills and instruments. The astrolabe was, however, vernacularized to a limited degree. One of the first works to bob up in Stöffler's wake was entitled *Wje man diss hochberumpt astronomischer und geometrischer kunst Instrument Astrolabium brauchen soll* (Bamberg, 1525). It is written – and printed – in simple style, with no illustrations (except for one little geometrical diagram), though in the course of explaining the word *astrolabium* it does include some scruffy Greek lettering, probably cut by the printer during his ale break.

This little-known but very fascinating book was written at Saint Joachimsthal, a mining town in Bohemia, by the official town physician (*leybartzt*), Johann Copp; his dedicatory preface is addressed to the *Burgermeyster*, Magistrate, and Council and dated 1524 (the year of Stöffler's second edition). Although our reference books know nothing of Copp himself, we are well acquainted with some of his patients, because he was the immediate predecessor at Saint Joachimsthal of one Georg Agricola, who became town physician in 1527, and wrote his famous *De re metallica* there.[72] In other words, the iconography of mining and metallurgy which has become our universal model for the period shows us the very area, and society, out of which this early astrolabe book comes. That is why its title-page goes on to claim that the use of the astrolabe is very enjoyable (*vastlustig*) and useful not only to physicians (obviously), but to master-builders, miners, master-gunners, and other technical professions ('Paumeystern, Pergklewren, Püchssenmeystern unnd andern künstlern').[73]

The use of the vernacular – and, as far as I can judge, a somewhat Bohemian version of it – thus arises from the book being written for these industrial and technical artisans, rather than for a scholarly audience. It is a more specifically defined part of the wider readership which Köbel had in mind. Though this readership, as Regiomontanus understood, was more likely to be interested in simpler time-telling instruments. The first German-language work on a scientific instrument which I can see on the shelf, only slightly earlier than Copp, is a pamphlet by a mathematics teacher, Heinrich Grammateus alias Schreiber, entitled *Ein kunstreich und behendt Instrument zuwissen am tag bey der Sonnen und in der nacht durch die Stern mancherley nutzberperkeit und aufgab in allen orten und endt der welt* (*sic*) (Nuremberg, 1522). The description of the instrument (which is not illustrated) is almost as inscrutable as the title, but it is evidently some kind of horological volvelle.

Peter Apian's *Ein künstlich Instrument oder Sonnen ur* (Landshut, 1524) has better claim to be the first separately printed work on sundials; while Sebastian Münster's

72 Basel, 1556.

73 I take the liberty of correcting Gunther's (n. 12) translation, p. 570, if only for the sake of the *büchsenmeistern*, who I feel sure would not be flattered to be called 'well-sinkers'!

début in the same vein and language was published by Köbel in 1528.[74] These two writers were the vanguard of the next generation. Münster (1489–1552) we met earlier, studying under Stöffler at Tübingen and being helped with his Greek by young Melanchthon. He is best known as a geographer and cosmographer, and was professor of Hebrew at Basel. But to us he is the founding father of the literature of sundials. The first really serious, full-length book on the subject was his *Compositio horologiorum* (Basel, 1531), which was successful enough to earn an expanded version in 1533 and a German edition soon after.[75]

Meanwhile Apian (1495–1552) was professor of mathematics at Ingolstadt, and installed his brother Georg as a printer there. His famous books with moving illustrations[76] have overshadowed his simpler and more accessible *Instrument Buch* of 1533, the illustrations in which were also available on separate sheets so that they could be cut out and made into working instruments. This book in some measure completed the Renaissance man's instrument library, dealing with a whole family of astro-horological angle-measuring instruments, all shown in use in the illustration on the title-page: quadrant, various *quadrata*, nocturnal, cross-staff, and human hand. In German, but in the academic folio format, the *Instrument Buch* was a seminal contribution to scientific instrument literature; and its title alone suggests a mature market interested in the subject of instruments *per se*. There is also some evidence that it was printed in larger numbers than most of the works I have mentioned, and that it was well used.[77]

Returning to the astrolabe for a moment, Stöffler's early folios generated a lineage of smaller format editions through the rest of the century, as did Köbel's *Astrolabii declaratio*, and the two were sometimes bound together. The enterprising Parisian publisher Guillaume Cavellat, who spearheaded a scientific publishing boom in Paris in and around the 1550s, issued the first editions outside Germany of the *Declaratio* in 1550 and of the *Elucidatio* in 1553; and a French translation of the latter followed in 1560.[78] Cavellat's volumes were small octavo in format and bound in soft vellum covers, the characteristics of a new style of book which emerged at this period, intended as cheap, portable textbooks rather than as tomes for the library shelf.

The *Elucidatio* and its little companion also inspired or influenced a vast bibliography of other books, both about astrolabes[79] and about other instruments and

---

[74] Sebastian Münster, *Erklerung des newen Instruments der Sunnen*, Oppenheim, 1528.

[75] *Horologiographia*, Basel, 1533; *Fürmalung und kunstlich beschreibung der Horologien oder Sonnenuhren*, Basel, 1537. On Münster, see Hantzsch (n. 53).

[76] *Cosmographicus liber*, Landshut, 1524; *Cosmographia*, Paris, 1531 and later editions, edited and augmented by Gemma Frisius; *Astronomicon Caesareum*, Ingolstadt, 1540. See Gingerich (n. 19).

[77] I know a library which has 4 (or 4½) copies, 2 (or 2½) of which are mutilated in some way. On Apian, see Siegmund Günther, *Peter und Philipp Apian, zwei deutsche Mathematiker u. Kartographen*, Prague, 1882.

[78] Johann Stöffler, *Traite de la composition et fabrique de l'astrolabe, & de son usage: avec les preceptes des mesures geometriques*, Paris, 1560, translated by Guillaume des Bordes and Jean Pierre de Mesmes.

[79] An impression of the cascade of astrolabe books in the century following Stöffler can be had from Gunther's (n. 12) bibliography (though it needs checking and revising), pp. 576–598.

techniques. Twenty years after Copp's work in German, Dominique Jacquinot's *L'usaige de l'astrolabe, avec un traicté de la sphere* (Paris, 1545) was the first in French. The famous treatise by Egnatio Danti, *Trattato dell uso et della fabbrica dell' astrolabio* (Florence, 1569), brought the astrolabe of Stöffler, Gemma Frisius, and Juan de Rojas, on whose Latin works it was based, to readers of Italian.[80] In English, apart from several printed versions of Chaucer's works, the first contemporary book on the astrolabe was John Blagrave's idiosyncratic description (with paper plates to make the actual instrument) of his own form of universal astrolabe, *The Mathematical Jewel* (London, 1585). The first printed account of a standard astrolabe in English was essentially a version of Köbel's, published as *The Travailers Joy and Felicitie* by Robert Tanner (London, 1587).

A new wave of scientific instrument books, and with it the Italian literature of instruments, began in 1537 with *La nova scientia* of Niccolo Tartaglia (*c.* 1499–1557). In spite of the title, there is something ever so familiar about his tower-measuring type of illustrations; and when he acknowledges a few authorities, one of them is none other than 'Giovanne Stoflerino'.[81] Tartaglia's *nova scientia* was ballistic geometry and gunnery instruments. A considerable literature subsequently pursues this theme, reaching a high point with the exquisitely illustrated works, in German, of Leonhard Zubler, the Zurich goldsmith, in the first decade or so of the seventeenth century.[82] The peaceful scenes of the Oppenheim tower measurers have been transformed into a world of canon aiming, range finding, and fortification designing: shooting holes in the towers is now the next step after calculating their height.

It was the kindred subjects of surveying and gunnery which more than anything else brought the vernacular into use in the scientific instrument literature. The primitive world of Köbel's 1535 *Geometrei* was essentially *status quo ante bellum*: its medieval picture dissolves in the second half of the century with the invasion of a newer style of instrumentation for land and military surveyors, an invasion accompanied at each step by a printed book, almost invariably in the native language of the author/inventor.[83] While it is correct to see this literature as successor to the highly influential surveying section of Stöffler, it has now found a new purpose – or rather two purposes: to teach the trade to the practising surveyor, and to announce and advertise new forms of instrument.

This is all, of course, another story for another night. Suffice it to say that a new *genre* of instrument books arises from the old, and that it largely employs the vernacular

[80] As well as something even more radical: Danti handed over his tower measuring to a woman (p. 124) – possibly the first picture of a mortal female using a scientific instrument?

[81] Niccolo Tartaglia, *La nova scientia*, Venice, 1537, prelims, f. 4^r.

[82] For instance Leonhard Zubler, *Nova geometrica pyrobolia* ..., Basel, 1614. For some other works of the kind, see n. 83.

[83] Examples include Abel Foulon, *Usage de l'holomètre* ..., Paris, 1555; Leonard Digges, *A Geometricall Practise, Named Pantometria*, London, 1571; Latino Orsini, *Trattato del radio latino* ..., Rome, 1583; John Blagrave, *Baculum familiare catholicon sive generale* ..., London, 1590; Bartolomeo Romano, *Protheo militare*, Naples, 1595; Philippe Danfrie, *Declaration de l'usage du graphometre*, Paris, 1597.

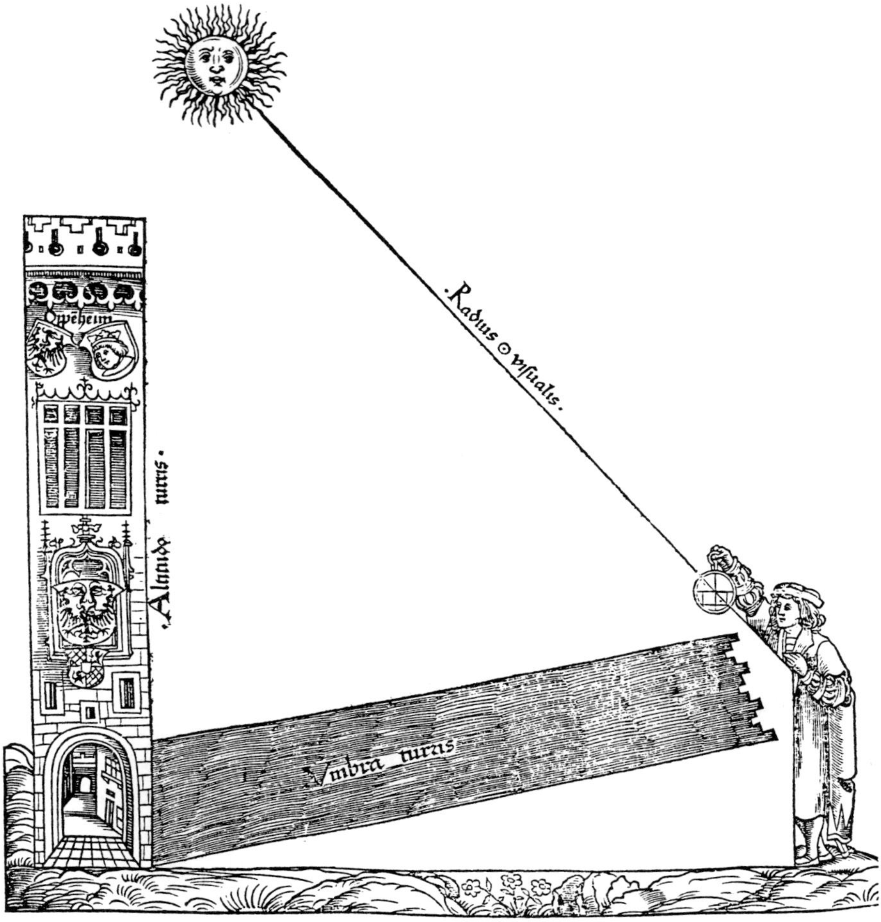

Figure 2. Illustration of the use of an astrolabe for measuring the height of a tower by means of its shadow; from Johann Stöffler, Elucidatio fabricae ususque astrolabii, *Oppenheim, 1524*

languages. Hence Leonard Digges's *Pantometria* in 1571 was still one of the few scientific books in the English language (its title notwithstanding). And as late as 1593 Thomas Fale's *Horologiographia* (its title also notwithstanding) was the first proper book on sundials in English. The Graeco–Latin titles evince a feeling of linguistic inferiority. But even with the less insular languages, it was only within this period that it became a natural thing for a Danfrie in France or a Gallucci in Italy[84] to do what Apian and Köbel in Germany found reason to do much earlier in the century, and Regiomontanus earlier still – to write in their native tongue rather than in Latin.

The Wiles of Girls

The thing about a ramble is that it will not necessarily lead you anywhere, nor have any particularly logical point at which to stop. If I stop now I know what you will say: Apian and Münster and Tartaglia have been given short *schrift*[85] – and what of Gemma Frisius, and what of Oronce Fine? Well, these are the flowers of the next generation, and when they blossom into print, chiefly in the 1530s, it is a sign that the beginnings are over and a mature and confident bibliographical *genre* has found its place on the bookshelf.

Anyway, I think I hear the wrecking-crew coming to rearrange the library, and put Stöffler back where he truly belongs, next to Stoke-on-Trent City Museum and Art Gallery,[86] and take away those books which we do not have – but only imagined we had, for completeness's sake. Jordan's *édition de luxe* of Stöffler, with its many-coloured astrolabe, for instance, is not in our imaginary library at all, but in the Museum of the History of Science.

The point of books is that, if we can find them, we learn things from them. Looking at their own history, of course, we learn different things than the authors meant to tell us. Yet perhaps the most difficult thing to gather from an old book is what the contemporary reader learned or wished to learn or was meant to learn from it, or what purpose or relevance that learning had. Luckily, one of Stöffler's epigraphists anticipated our question, and tells us down the centuries what the *Elucidatio fabricae ususque astrolabii*, the first true scientific instrument book, taught the reader of 1513.

'This book teaches you', he chants, 'what Stöffler and Teutonic excellence and genius have achieved; what pole the heavens are in, and the courses of the stars; to get to know the causes of fortune, and what your fates have in store; to seek one's times for

[84] For instance Giovanni Paolo Gallucci, *Della fabrica et uso di diversi stromenti di astronomia, et cosmografia* ..., Venice, 1597.

[85] I know that's not the form, I'm just joking.

[86] Stoke-on-Trent City Museum and Art Gallery, *Man and the Stars: Space Age Astronomy: An Exhibition Showing the Development and Role of Space Probes and the Ways they Benefit Mankind*, Hanley, 1980. Or, if that is too outlandish for the library you have in mind, then I'm afraid it will have to be John W. Stokes, *70 Years of Radio Tubes and Valves: A Guide for Electronic Engineers, Historians and Collectors*, New York, 1982.

prayers, and the wiles of girls; what the diversity of things is, their height and depth; and which constellations and courses you may depict with your hands.'[87]

Quid Iustingensis Stoflers, quid Teutona virtus
Possit, & ingenium: te docet iste liber.
Te liber iste docet, in quo sit vertice coelum
Astrorum cursus te docet iste liber.
Te liber iste docet fortunae noscere causas
Quid tua fata velint, te docet iste liber.
Te liber iste docet, sua tempora quaerere votis,
Furta puellarum te docet iste liber.
Te liber iste docet, quae sit distantia rebus
Altum & procerum, te docet iste liber.
Te liber iste docet manibus quae sydera lector
Pingas & cursus: tu mihi crede. Vale.

[87] I am grateful to Mrs R.E.W. Chapman for help with this and other Latin translations; and also to Mr F.R. Maddison and Mr A.J. Turner for information and comments which I have found useful in writing this essay.

Learning and Language in the Somerset Levels: Andrew Paschall of Chedsey

A.J. Turner

Andrew Paschall was a country parson. He is of interest because he was a learned parson possessed by the idea that 'There is infallibly a nearer way to knowledge than wt is in use'. 'Toward a discovery of yis', he continued in a letter to John Aubrey, 'are my general indeavours...'.[1] Although sometimes discouraged, Paschall never seriously wavered in his commitment to this vision despite passing the better part of his life in a remote and isolated village, Chedsey (now Chedzoy), about three miles north-east of Bridgwater, in the middle of the, then undrained, Somerset Levels. What Paschall sought was a method for the discovery of knowledge, natural, moral and divine, which because it guaranteed the knowledge found and locked it into the real, and therefore the true, explanatory system of the universe, automatically made all debate, discussion and contention superfluous. This Utopian aim, which Paschall hoped to realize through the agency of an effective universal language, was the guiding principal of his life and intellectual endeavours, but while his aim was universal reconciliation and harmony, it was contained in the social mould of a conservative Church and King Tory attitude. While Paschall's vision has its roots in the radical ideas of the Interregnum idealists such as congregated around Samuel Hartlib, Paschall's personal intellectual history illustrates how such ideas could move from a radical to a conservative context without losing their visionary, inspirational power.

Andrew Paschall was born in or close to London, perhaps in the family house in Soper Lane[2] *c.* 1631/2. We know nothing of his early years, but in 1647 he matri-

[1] Paschall to John Aubrey 18 July 1684, Bodleian Library, Oxford MS Aubrey 13 f. 67. Although the TALPA 2 teaching machine invented by Francis Maddison and his brother in 1962/3 was not exactly what Paschall had in mind when seeking 'a nearer way to knowledge', the common interest of the subject of this essay and the honorand of this volume in communication and language make it, I hope, an appropriate offering to Francis Maddison, whose learning has saved me from many a solecism and whose language and invention have offered countless hours of enjoyable conversation.

[2] In Cambridge matriculation records, Paschall is described as 'of Middlesex' (J.A. Venn, *Alumni*

From *Learning, Language and Invention: Essays presented to Francis Maddison*, ed. W.D. Hackmann and A.J. Turner. Published by Variorum, Ashgate Publishing Ltd, Gower House, Croft Road, Aldershot, Hampshire, GU11 3HR, Great Britain.

culated at Queen's College, Cambridge. Proceeding BA in 1650/1, MA in 1654 and BD in 1661, he held a Fellowship at Queen's from 1653 to 1663.[3] On 13 January 1657/8 he was appointed an university preacher.[4] At the Restoration he was obliged to demand re-admittance to his Fellowship which was granted in August 1660,[5] but in 1663 he resigned his Fellowship following his presentation to the living of Chedsey by Sir Francis Rolle.[6] Chedsey despite its isolation was a relatively well-endowed living worth £38 a year and among the six richest benefices in the county, its status underlined by the large and handsome church which still stands in testimony. Here, despite the inconvenience of being sometimes physically cut off by floods during the winter season, Paschall remained more or less contentedly until 1689. Then, having found 'a good curate' to supply his place, he moved to Bristol hoping to find there 'a fixed settlement' and 'to have much better opportunity of prosecuting usefull studyes, & doing some good by them than I had in Chedsey'.[7] On 25 March 1689 he was appointed Chancellor of Wells Cathedral,[8] requesting help towards further preferment from John Aubrey and William Holder in an undated letter of the same year.[9] It was perhaps thanks to their intervention that in July 1690, the Dean of St Paul's recommended him for the vacant prebend of St Decuman at Wells,[10] although the appointment was not made until the end of the year.[11] In September 1693 he was admitted canon residentiary at Wells and became the innocent cause of a new round of conflict between Bishop Richard Kidder and his chapter.[12] Appointed steward for two years in August 1694, he died soon after the term of this office before November 1696.[13]

From this brief account of Paschall's relatively successful, relatively undisturbed, and wholly undistinguished career, it would seem that he is typical of the hundreds of graduates who took over the livings of the Church of England in the course of the seventeenth century.[14] In fact he does not quite fit the stereotype. If 'there were, in

Cantabrigiensis...part I, Cambridge, 1922–27, vol. 3, p. 316), but Henry More, *Saducismus Triumphatus...*, 2nd edn, London, 1682, relation 24 mentions the Paschall house in Soper Lane, London. Soper Lane no longer exists, but there need not be any contradiction between the two remarks. I am indebted to Jennifer Drake-Brockman for drawing my attention to Henry More's statement.

[3] Venn *Op. cit.* (n. 2). According to Thomas Hearne in his edition of *Peter Langtoft's Chronicle (as illustrated and improved by Robert of Brunne) from the death of Cadwallader to the end of Edward the First's Reign ...*, 2 vols, Oxford, 1725, vol. 1, Preface p. lxxxxix, Paschall was admitted Fellow on 19 January 1652/3.

[4] John Worthington, *Diary and Correspondence*, ed. J. Crossley, 3 vols (Chetham Society Manchester, vols 13, 36, 114), 1847–86, vol. 1, p. 103.

[5] Hearne *Op. cit.* (n. 3), p. lxxxix.

[6] Paschall to Aubrey, 3 April 1676, MS Aubrey 13 (n. 1), f. 7.

[7] Paschall to Aubrey, 4 November 1689, MS Aubrey 13 (n. 1), f. 83.

[8] For details of Paschall's offices at Wells, see Aelred Watkins in 'Collectanea iii', *Somerset Record Society*, lvii, 1942.

[9] MS Aubrey 13 (n. 1), f. 88.

[10] Sir Robert Southwell to the Earl of Nottingham, 14 July 1690, *HMC* 71 Finch, ii, ff. 325, 357, 390.

[11] Paschall was installed on 23 December. *HMC* 12 Wells, ii, f. 176.

[12] *Ibid.*, f. 225. Paschall paid £100 caution money. For Paschall's difficulties, see 'The Life of Richard Kidder, Bishop of Bath and Wells', *Somerset Record Society*, xxxvii, 1924.

[13] *HMC* 12 Wells, ii, f. 237d.

[14] For a portrait of the English clergy in the seventeenth century, see Rosemary O'Day, 'The Anatomy

reality, two Clergies: a clerical elite [*sic*] drawn from among the high-flyers at the universities and protégés of the well-connected; and the rank and file, increasingly well-educated but possessing little hope of preferment outside the work of the parishes',[15] Paschall seems to belong to both. Though comfortable, his benefice was not exceptionally wealthy, but Paschall was not without connections. When at the end of his career he needed preferment following the failure of some mining speculations,[16] he obtained it relatively easily thanks to the relations which his active intellectual endeavours had brought him. Until then, despite his complaints, he seems to have stayed at Chedsey through choice rather than lack of opportunity. Only need provoked ambition. Paschall's inclination was always for retirement and solitude in which to pursue his intellectual work. Election to convocation, with the consequent interruption of a journey to London annoyed him in later life, and an illness at such times was not unwelcome. Until the last period of his life he waved aside efforts by his friends to procure his advancement. Gently rejecting Robert Nelson's efforts on his behalf, he writes 'I take it as an instance of yr friendship, but really I am of that make, that I choose rather to indulge my own genius though to my losse, than strive in a crowd'.[17] Fundamentally Paschall was a retiring; perhaps rather shy, intellectual displaced in a remote province. If in this he is typical of the kind of men who produced a rise in quality of the beneficed clergy during the mid–late seventeenth century, he is untypical in the depth of his commitment to philosophical investigation and Utopian reform orientated towards the goal of social harmony.

For the sources of Paschall's ideas we have no detailed information. At Cambridge Paschall moved in the orbit of the Cambridge Platonists. From 1664 until his early death in 1652, John Smith was a Fellow of Queen's College, and from 1650 Dean and Catechist. Paschall must certainly therefore have had some contact with him, perhaps even have been taught by him. The earliest reference to Paschall so far found is in the diary of John Worthington for January 1657/8, and on 31 July Worthington further records sending a copy of his comments on a manuscript of Mr Naylor's to Paschall.[18] Presumably either through Worthington or Smith, Paschall came into contact with Henry More and Benjamin Whichcote, supplying the former in later years with a ghost story used in the *Saducismus Triumphatus.*[19] If he did not personally know Ralph Cudworth he approved his work, reading the *True Intellectual System of the Universe* on its appearance in 1678 'with great satisfaction' and longed 'to heare of the publication of what remaines of this first part, as also of the other two which he doth promise but not with so much assurance as I could wish.[20]

of aProfession: the Clergy of the Church of England' in Wilfred Prest, ed., *The Professions in Early Modern England*, London, New York & Sydney, 1987, pp. 25–63.

[15] *Ibid.*, p. 51.

[16] See letters to Aubrey in MS Aubrey 13 (n. 1), ff. 71, 73, 74, & 83.

[17] Paschall to Robert Nelson 31 August 1686, British Museum Add MS 30277, f. 34.

[18] Worthington *Op. cit.* (n. 4), vol. 2, i, p. 115.

[19] Relation 24 (see n. 2).

[20] Paschall to Aubrey 8 July 1678, MS Aubrey 13 (n. 1), f. 32.

From the Cambridge Platonists Paschall perhaps derived his insistent rationality, his orientation towards natural philosophy, and his belief that the object of human enterprise should be to recover the true, ancient, pre-lapsarian knowledge – knowledge which could heal the schisms of mankind. For the origins of his commitment to the universal language as the method which could enable this ancient knowledge to be recovered there is as yet no precise evidence, nor for the origins of his enthusiasm for co-operative work. Both were of course relatively widespread in mid-seventeenth-century England, but Paschall's explicit acceptance of the means towards restoring religious harmony and allaying intellectual strife, when placed beside passages in his letters treating of alchemy and Utopian republic schemes, suggest some curiously radical associations for a man who was to become a firm Church and King Tory.

But it would be unwise at present to speculate too far on these matters. Of Paschall's life at Cambridge, and at Chedsey before 1669 we know nothing directly, and can only extrapolate backwards from the letters that have survived from after this date. Of these the most important are a long series of nearly eighty letters sent to John Aubrey between 1672 and 1694, and smaller groups sent to Joseph Glanvill, Robert Hooke, Francis Lodowik and John Ray. As these names indicate, despite his isolation in the countryside, Paschall was known to, and in regular contact with, some leading members of the Royal Society as he was with all the leading exponents of the universal language. That Paschall thus enjoyed some intellectual exchanges resulted mainly from his friendshp with John Aubrey. Aubrey[21] was crucial to him. He was Paschall's 'universal intelligencer', sending him news and books, copying manuscripts for him, bringing him into contact with some of the intellectual leaders in London. In particular it was he who orchestrated the group interested in developing John Wilkins' universal language which included Paschall, Lodowik, Hooke, Thomas Piggott (Fellow of Wadham College), John Ray and Aubrey himself.[22]

The driving force of Paschall's life was a commitment to the advancement of learning. This was a manifestation of his constant desire to be useful. Whether in society, to his friends, or to other individuals, Paschall felt a duty and a need to contribute what he could. In the letters to Aubrey he can be seen indefatigably at work; interceding for a parishioner unjustly prosecuted at law;[23] endeavouring to promote a scheme for draining the moors around Chedsey;[24] or trying to advance John Ray's career.[25] No personal involvement was needed to provoke Paschall's activity – his concern was for

---

[21] For whom see Michael Hunter, *John Aubrey and the Realm of Learning*, London, 1972.

[22] For a detailed technical study of the work of this group see Vivian Salmon, 'John Wilkins' *Essay* (1668): Critics and Continuators', *Historiographia linguistica*, i, 2, 1974, pp. 147–163.

[23] Paschall to Aubrey 31 October 1671 & 20 November 1677. MS Aubrey 13 (n. 1), ff. 25 & 26.

[24] *Ibid.*, 18 September 1683, MS Aubrey 13 (n. 1) f. 59 *et seq.*

[25] Paschall had suggested Ray as a tutor in France for the son of the patron of his living, Sir Francis Rolle (MS Aubrey 13 (n. 1), f. 7), since Rolle was especially concerned that his son should have a botanical education. Ray refused the offer as he had a similar proposal in 1670. Charles. E. Raven, *John Ray Naturalist, his Life and Works*, 2nd edn, Cambridge, 1950, pp. 165, 179. Paschall to Aubrey 3 April 1676, MS Aubrey, 13 f. 7.

the general welfare. An advertisement for a new method of bee-keeping is enough to set him off acquiring more information and suggesting a way in which a cheaper contrivance 'of more profit to the public' might be made.[26]

But activities such as these represented mere marginal alleviations in the condition of man. At the root of Paschall's thought was the pessimistic theological commonplace that thanks to the Fall, man is in a state of darkness and ignorance. To overcome this a great reformation was needed in education and learning, which would incline men 'to an unprejudicate converse with things, and from thence forme their notions & govern their reasonings'.[27] Though Paschall is never explicit, it seems clear that underlying this is the quite irrational assumption that as the world of things is reduced to intellectual order, so will the world of man, by some undefined process, settle into social order and harmony. Certainly he states the object of his investigations into the universal language in terms which suggest this. The language was to be 'a universal remedy against errour & Bigotry, a panacea for ys epidemicall disease of human nature,...'[28]

Belief in the degraded condition of mankind and in the limitations of human knowledge after the Fall was common to seventeenth-century intellectuals. Allied to it were a diversity of curative programmes. The most effective of those put forward by Interregnum idealists tended to envisage a planned society wherein men trained according to their abilities fulfilled the rôle natural to them. Such an aim clearly implied fundamental alterations in social structure and the practice of politics. It also tended to value knowledge 'as it hath a tendency to use'.[29] Paschall's ideal was quite different. His was a vision of an ordered society, where men fulfilled the duties belonging to the position in society in which they were naturally placed at birth by a God whose providence meant that the position decreed at birth was the proper position.[30] Where the social structure proposed by Interregnum radicals allowed for change and movement of personnel within itself, Paschall's structure was essentially static, though this could be obscured, or its effects mitigated, by the fact that it was also paternalistic.

In the structure of society and political life, Paschall's ideal demanded no fundamental readjustment, though much reformatory work, and a high sense of duty at the top

---

[26] *London Gazette*, 987, 991, 1675. Paschall to Aubrey 16 June 1675 & 18 January 1675/6, MS Aubrey 13 (n. 1), ff. 4, 6. A copy of part of the second letter is preserved in Royal Society MS LB.P, f. 56. The bee-keeping method in question was that of John Geddes described in *A New Discovery of an Excellent Method of Bee-houses and Colonies*, London, 1675. For the non-originality of the method to Geddes, see D.J. Bryden 'An Excellent Method of Bee Houses and Colonies: John Geddes and the Royal Society, a Seventeenth-Century Exercise in Marketing' (forthcoming), *Notes and Records of the Royal Society of London.*

[27] Paschall to William Musgrave 3 March 1683/4, Bodleian Library MS Ashmole 1813. Printed in R.T. Gunther, *Dr Plot and the Correspondence of the Philosophical Society of Oxford*, vol. 12 of his *Early Science in Oxford*, Oxford, 1939, pp. 272–273.

[28] Paschall to Aubrey 2 March 1679/80, MS Aubrey 13 (n. 1), f. 39.

[29] Boyle to Hartlib October 1646. *Works*, vol. 1, p. xxxiv.

[30] '...I consider yt although it be not so delightfull to be ingaged in ye warre against evill, as to injoy a free conversation with ym who live in ye exercise of not cathartic onely, but perfective vertues; yet tis best for us to be in yt station in wch our supreme ruler places us, and to do willingly & diligently our master's worke'. Paschall to Aubrey 20 November 1677, MS Aubrey 13 (n. 1), f. 26.

levels of society.[31] In relation to learning, it recognized the value of practical knowledge, and insisted that techniques should be investigated in order that they might be improved. This however was secondary to a view of learning as a means to constructing the real structure of the world. Everything whether natural, moral or 'of common Humane life' had a proper place within an ordered framework. True knowledge was that framework, and the investigation into all activities that Paschall supported was an investigation intended to make evident this true framework and to reduce all subjects to it. He writes to Aubrey:

> I am glad that Mr Moxon goes on in his exercises. I wish others may be gotten to do ye like yings in other matters. That some in Theology, Medecine, Mathematiques, &c will attempt what you say *that* gentleman hath done in our law.[32]

All investigation into natural philosophy is seen by Paschall as part of a process of classification, and his major efforts – creating a machinery for scientific communication in his attempt to found a philosophical correspondence club in Somerset, and trying to make viable the universal character of John Wilkins are informed by this purpose. So too were his long labours over the philosophical language, which if perfected, would at once embody the real structure of the world, and give instant understanding of it.

Compared with most other aspects of Paschall's life, his attempt to found 'a Clubb for Philosophical correspondence' in Somerset, is relatively well known. At least two copies of printed *Propositions* setting out the aims of the correspondence have survived[33] and other documents relating to the affair have been published.[34] The aim of the correspondence was to overcome the sense of isolation which deterred all but the most committed men from pursuing natural and philosophical activities in isolated, local areas. Allied to this was a desire to 'carefully inquire into all the *rarities* of Nature that may be in or near the places of [the correspondents'] respective Residences such as the more extraordinary *Phænomena* of *Minerals, Plants, Animals, Waters, &c* and whatever else may serve to promote and inlarge the *History of Nature*'.[35] Communication of moral and philosophical ideas was encouraged as specifically were ideas concerning the universal language, but '*Speculative* matters, or *Controversies* of Religion' dogmatically expressed were ruled out, everyone being expected to 'propose his Conceptions

[31] See for example his views on education set out in a letter to Aubrey discussing the latter's 'Idea of Education of Young Gentlemen'. Paschall's letter is printed in Aubrey's *History of Surrey*, Preface. The original, probably as a result, appears to have been lost.

[32] Paschall to Aubrey 16 August 1678, MS Aubrey 13 (n. 1), f. 34. On the bottom of f. 31 Aubrey has scribbled 'Mr Merret's method of the laws of England', presumably a note to himself to mention it in his reply to Paschall's letter of 11 June 1678, which is presumably the method to which Paschall here refers.

[33] London, printed for John Collins. There is one copy in the British Library, London, another in the Bodleian Library, Oxford.

[34] A.R. & M.B. Hall, *The Correspondence of Henry Oldenburg*, vol. 7, 1669–70, Madison, 1969, pp. 137–139 letters 1248, 1248a. Nicholas Steneck, '"The Ballad of Robert Crosse and Joseph Glanvill" and the background to *Plus Ultra*', *British Journal for the History of Science*, xiv, 1981, pp. 59–74 (65–66).

[35] *Propositions*... I.

modestly in the way of *Queries* and *Hypotheses*, for the preventing and excluding of *Wrangling disputes*'. All communication between members of the club was to pass through the hands of the secretary who was also charged to maintain contact with the Royal Society, to organize a foot-post weekly between Bath and Bristol, and to obtain for members such books, journals and instruments as they might need. For these services members would pay 2s 6d per quarter.

Because the secretary of the Correspondence was Joseph Glanvill, discussion of it has tended to associate the correspondence with him.[36] It is clear however from the very first words of a letter writen to Glanvill by Paschall on 18 June 1669[37] that the idea for the correspondence came from the group around Paschall who invited Glanvill to join them because of his known contacts with, and easy access to, the Royal Society:

> Our proposition aims first at the gratifying Proficients in the best knowledge with a probable way of communicating wt they have for the good of others, then at serving ingenuous inquiryes with means of coming to satisfaction in wt they seek after. Wee fancy yt this may be done in this countrey, if but a few persons shall resolve to meet in place and times convenient, and so write as frequently as ther shall be occasion; this body from a lesse considerable beginning may grow greater and some valuable proportions of that truth, whiche alone can repaire mankind, may freely circulate in it, for the good of others, who are estranged from wt is true because it is not duly made present to their mindes.

The similarity of this aim with the declared aims of the early Royal Society and with Hartlibian efforts at information exchange, and its congruence with the needs of isolated local intellectuals such as Paschall or William Cole, surveyor of the customs at Bristol and dedicated naturalist who quickly joined the correspondence circle, will be evident. Paschall wished local materials and discoveries to serve the higher ends of the Royal Society, but he also wished, as an university intellectual in exile, to benefit from a two-way exchange:

> ...if this succeed, we shall have the Hardyness to hope, as well as to wish, that some of those incomparable persons in London, who labour so much in ye service of truth, and are knowne to you, wil joyne, and take charge of giving directions & resolutions to us in our doubts, and importing what occurres to them worth our notice. *I fear that I am now fallen into pleasant dreame.*

His vision went even further:

> ...it seems no hard matter for such a company in London to set such correspondencyes agoing, not only in the two Universityes, but in most countreyes of England. I will not

[36] Thus Steneck (n. 34) makes no mention of Paschall and ascribes authorship of the *Propositions* to Glanvill. No doubt the latter had a say in their formulation, but it seems clear from their phrasing and the special attention paid to the universal language that Paschall also played an influential part. Since Glanvill in his letter to Oldenburg of 19 July 1669 states that 'I am to meet next weeke with some of ye cheif in ye Designe, in order to ye methodizing of it', it seems clear that initiative for the Correspondence did not come from Glanvill and that the *Propositions*, at least in their final form, were the work of a committee.

[37] Royal Society, London, MS P, no. 53, printed in *Old Corr.*, vol. 6, 1248a, pp. 140–141.

undertake to say what happy effects are likely to ensue such a union and commerce among ye sons of truth, could it bee brought to passe.[38]

The project for the Somerset philosophical correspondence was well received by the Royal Society[39] and Oldenburg even told Glanvill that he wished to join it.[40] In Somerset itself however the project quickly ran into difficulties. Glanvill an over-enthusiastic exponent of new philosophy, new styles in preaching, and new styles of behaviour, was already a controversial and unpopular figure in many parts of the county. Involving him in the correspondence as secretary doomed it. In a letter to Evelyn 20 November 1669, John Beale, rector of Yeovil, dismissed the project out of hand as 'a ridiculous foppery' in 'ye Air & Dirt of the Country'. For Beale the 'soyle was baren, ye Country unripe, the Designe Ridiculous & intolerably Romantick'. The root of Beale's objections however was personal dislike of Glanvill and perhaps, as Glanvill suggested, the fact that he had not been consulted at the very beginning.[41] Once Beale thought that the Royal Society might have given official countenance to the project, he moderated his tone considerably and says that he 'will conforme to my vtmost' despite the fact that he had asked Evelyn to try to obstruct any such official recognition being given.[42]

The correspondence did in fact begin operations: the names of at least six members are known, together with the descriptions of one or two others, and in late 1669 and 1670, information, queries, and replies passed between Somerset and London in letters from Glanvill to Oldenburg. But conditions in Somerset were not propitious: the personal unpopularity of Glanvill, the social uneasiness of the county thanks to the large, active and disruptive non-conformist community, the difficulties of communication between the few but scattered members, all meant that by the end of 1670 the correspondence was effectively dead. As Paschall put it later 'that wel-meant designe failed then'.[43] If Oldenburg had seen it as a means towards 'ye so much desired *Historia Naturalis*; for which I have excited many in foreign parts',[44] for Paschall this natural history was only the basis for establishing a greater classificatory philosophical structure of which the truth would be guaranteed once it could be expressed in the universal language [Figure 1].

The fifth head of the *Propositions for a Philosophical Correspôndence...* desired:

that if any of the Corespondents who studies the *Universal Character* of the learned *Dr Wilkins*,... shall have any *Objections* or doubts about it; or else have any thing to suggest

[38] *Ibid.*, p. 141.

[39] *Old Corr.* (n. 37), vol. 6, p. 155; 24 July 1669, letter 1256.

[40] *Ibid.*

[41] Glanvill to Oldenburg 17 December 1669, see *Old Corr.* (n. 37), vol. 6, p. 373, letter 1345.

[42] Beale to Evelyn 20 November 1669 and 6 December 1669. Evelyn Papers Christchurch College, Oxford. Corres. 90 & 92. Since Glanvill told Oldenburg on 17 December that he had sent *Proposals* to Beale, who had made no reply, 'above a Fortnight since' it must have been receipt of the *Proposals* that provoked Beale's angry reaction.

[43] Paschall to Aubrey, 7 April 1676, MS Aubrey 13 (n. 1), f. 8^{v}.

[44] Oldenburg to Beale *c.* 17 September 1669, see *Old Corr.*(n. 37), vol. 6, p. 245, letter 1286.

for the facilitating, or promoting the learning or spreading of it; he shall impart the same to the Secretary of the Correspondence...who will endeavour to procure an *Answer* to such *Doubts*, or spread any such Communication, as may tend to the advantage of that noble *Design*.[45]

This is a striking testimony to the importance that Paschall ascribed to the real character and the universal language so amply described by John Wilkins in his *An Essay towards a Real Character, and a Philosophical Language* (1667). The book made an enormous impression on Paschall. Here it seemed at last was an unambiguous language in which two men could discourse about God and nature clearly and without fear of confusion, since they would be using words which were isomorphic with things and so indubitable, and sounds which were at once congruent with the meanings being expressed, but independent of all existing individual languages. Paschall was impatient that 'the establishment of this great ying' be 'so managed that in yis yere [1676] might be nothing of Schisme, nothing of Babell'.[46] However much Paschall had a genuine interest and skill in linguistic studies, his excitement over the universal language seems to have stemmed from the rôle he ascribed to it as a universal panacea. If words are made up just as things are, then disagreement would have to cease. This however was a very simplistic view. As Thomas Piggot pointed out to Aubrey:

I will not deny but that it would be most highly philosophicall to compound words just as things are compounded this I confesse would be a language proper for Angells whose knowledge is infinite, but I dare be bold to say it will never be plausible with men and we justly expect a second confusion when we strive to build such a babel.[47]

Although, as Paschall studied the universal language more carefully he gradually came to understand the problems it posed and the fundamental differences between John Wilkins' ideas and those of Seth Ward, he never entirely abandoned his belief in the remedial possibilites of artificial language. He did however sometimes despair. On 2 March 1679/80, he wrote to Aubrey that:

...I am bewildered in my thoughts about yis subject, and fallen into a rationall despair of ever bringing them to any good issue, in a ying yat I doubt is too great for any man in yis state of imperfection...Perhaps tis better for us (I think it is for me) instead of seeking after universall remedy against errour & bigotry, a panacea for yis epidemicall disease of human nature; to make the best use we can of the known languages....[48]

The technical content of Paschall's investigations into universal language have been studied elsewhere,[49] and here is not the place to rehearse details of the many efforts that he made to popularize the subject.[50] What may be underlined is the vast

45 *Propositions...* (n. 35) V.

46 Paschall to Aubrey 12 December 1676, MS Aubrey 13 (n. 1), f. 14.

47 Piggot to Aubrey 25 February 1676/7, MS Aubrey 13 (n. 1), f. 16.

48 MS Aubrey (n. 1), 13 f. 39.

49 By Vivian Salmon (n. 22 above).

50 For one of them, see A.J. Turner, 'Andrew Paschall's Tables of Plans for the Universal Language 1678', *Bodleian Library Quarterly*, ix, 6, 1978, pp. 346–350.

Sr

I received & read ye Letter wch was written in ye univ. char. with much pleasure: & (if I may declare my opinion freely) I seeme to discover ye difference between this wch was written by a man Educated in Learning, & those other papers. In ym is not ye right notion of many words: but not so in this Letter: & yt there my doubts are few & triviall. & yet I write them freely bec. I am willing yt my friend would so deale wth me. I thinke yt yrselfe in London; & Mr Pigot in Oxford & Mr Lodowick in Holland & ye gentleman in Denmarke in his own country may promote ye Knowledge & use of this language, And intercourse in Letters should be continued among them. but this is desireable, that my Ld Bishop of Sarum will finish & publish his designe speedily: yt if it may

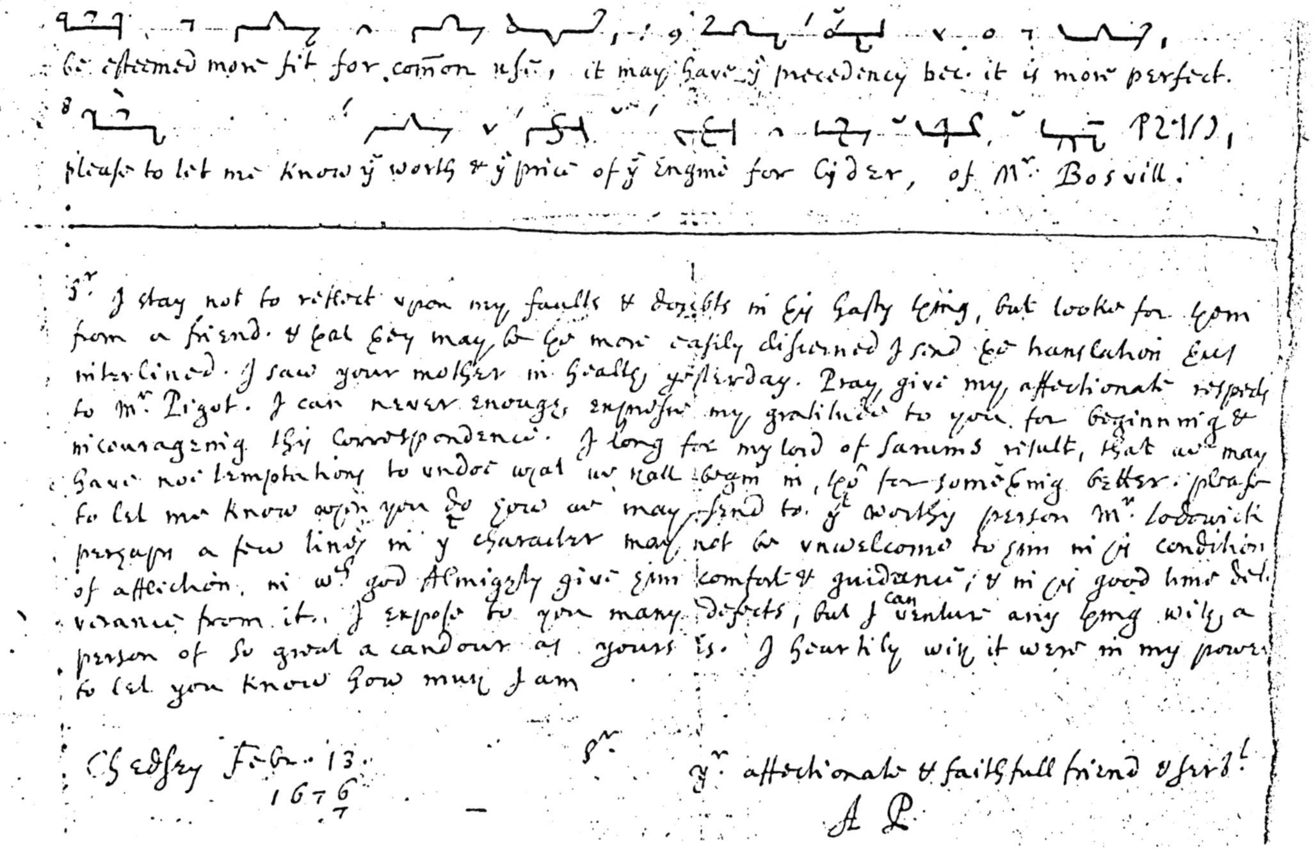

be esteemed more fit for comon use, it may have ye precedency bec. it is more perfect.

please to let me know ye worth & ye price of ye Engine for Cyder, of Mr. Bosvill.

Sr. I stay not to reflect upon my faults & defects in this hasty thing, but looke for them from a friend. & that they may be the more easily discerned I send the translation here interlined. I saw your mother in health yesterday. Pray give my affectionate respects to Mr. Pigot. I can never enough expresse my gratitude to you for beginning & incouraging this Correspondence. I long for my lord of Sarums result, that we may have noe temptation to undoe what we shall begin in, tho for something better. please to let me know when you doe for we may send to that worthy person Mr. Lodowick perhaps a few lines in ye character may not be unwelcome to him in his condition of affliction, in wch god Almighty give him comfort & guidance, & in his good time deliverance from it. I expose to you many defects, but I can venture any thing with a person of so great a candour as yours is. I heartily wish it were in my power to let you know how much I am

Chedsey Feb. 13 1676/7

Sr

Yr affectionate & faithfull friend & servt

A P

Figure 1. *A letter written by Andrew Paschall to John Aubrey using the 'real character' and 'universal language' of John Wilkins, 13 February 1676/7. Bodleian Library, Oxford, MS Aubrey, 13 f. 14*

effort that he invested in it, an effort fuelled by his belief that the language could be the means to universal redemption and, even if not, certainly something that could only be for the public good. For Paschall public good and universal harmony are intertwined. 'God Almighty Guide, Bless, Succeed all good indeavours for the benefit of mankind' he prays at the end of the short letter which accompanies his own new form of universal language sent to Aubrey in November 1682.[51]

Two years later in a letter to William Musgrave, secretary of the Oxford Philosophical Society he summarized his own position succinctly, 'I feel to be true; that I have a desire to learn; and a sincere willingness & zeale to serve, according to my power, them who devote themselves to advance the good of mankind, by indeavouring to promote usefull Knowledge'.[52] It is this mixture of delight in learning and social commitment that makes him so eager to institutionalize, to co-ordinate, to ease the machinery of intellectual communications and to put his hope in '...ye conjoined labours of learned men'. It was this attitude which led him once again in the 1680s to attempt to organize a further collaborative work, this time with a largely antiquarian bias, in association with the Oxford Philosophical Society. The advantages of collaboration would be '...a shortening of the way to the attainment of that reall knowledge allready gotten; and consequently a gaining of advantage towards a further progresse. It is supposed that the thing may be done by the united labours of some men'.[53]

In Paschall then intellectual delight and curiosity conjoined with a social ideal and a belief in the utility of natural philosophy. The vision served was firmly conservative and pressed natural philosophy into service as a pillar of the restored social structure of Caroline England. Paschall was a firm Church and King Tory. Thanks to his influence not a man left Chedsey to join Monmouth's rebellion, and even in 1687 he could deliver himself of the judgement that under James II there was 'one of ye best governments in the world, could Englishmen but know when they are well'.[54] In the same letter he quoted with evident approval a statement of the extreme Church Tory view of non-resistance. Underlying this attitude was not a thought but an emotional desire, that although the king's power over the Church was unlimited, he would never abuse it. The total unreality of this view was exposed by the actions of James II. But this wilful failure to face up to social realities underlies Paschall's intellectual vision. His thought is all of a piece. The world is well-ordered if we only see it aright (which through collaborative investigation and the use of universal language one might eventually hope to do), and so too is society. It seems mere wilfulness to Paschall that men should run against the natural order. He has difficulty in believing that such men are really human. 'As your dissenters so ours, what shall I call them? for they are of another strain perhaps.'[55] From the disorders of man Paschall turned with relief, and for relief, to the order of the natural world.

[51] MS Aubrey 13 (n. 1), f. 58.

[52] Gunther *Op. cit.* (n. 27), p. 272.

[53] Paschall to William Musgrave 3 March 1684/5 in Gunther *Op. cit.* (n. 27), p. 273.

[54] Paschall to Aubrey 23 April 1687, MS Aubrey 13 (n. 1), f. 81.

[55] *Ibid.*

A propos de deux petits quadrants indiens

Jean-Pierre Verdet

En 1982, le Musée d'Histoire de la Science de l'Université d'Oxford et l'Observatoire de Paris ont acquis deux instruments indiens dont l'identité morphologique oblige à penser qu'ils proviennent d'un même gabarit, ou que l'un a servi de gabarit à l'autre. Ce sont deux quarts de cercles astronomiques en laiton de 12 centimètres de rayon, munis d'une alidade de visée, d'une étroite fente et de deux index, l'un rectiligne et l'autre cruciforme. Sur l'une des faces, que nous qualifierons de diurne, on trouve l'index rectiligne, cinq quarts de cercles concentriques, et deux inscriptions. Sur l'autre face, que nous qualifierons de nocturne, on trouve l'index cruciforme, quatre cercles concentriques, un tableau et une inscription.

Description de la face diurne [Figure 1]

L'inscription longue donne, avec une redondance dont les Indiens sont coutumiers, la date de fabrication de l'instrument et la latitude à laquelle il est utilisable. La date est donnée dans deux systèmes chronologiques différents. D'abord en années vikrama (*saṃvat* 1872), puis en années çaka (*çaṃke* 1738), avec la précision qu'il s'agit d'années courantes (*pravartamâna*), et non d'années révolues, habituellement notées *gata*. Les années vikrama commencent en 58 avant notre ère; leur usage est habituel en Inde Centrale. Les années çaka commencent en 78 de notre ère; ce sont les années les plus usitées dans l'ensemble de l'Inde. Cet instrument a donc été fabriqué en l'an 1815 de notre ère, le 11 de Jyeshtha (mois indien luni-solaire à cheval sur nos mois de mai et de juin), 3 ghatî avant que le soleil ne se couche. Cette dernière précision n'étant destinée qu'à souligner que l'artisan a terminé l'instrument à un moment faste, il note d'ailleurs qu'il a été protégé par les dieux (*sadaiva*). Les unités de temps traditionnelles de l'Inde sont le ghatî, qui vaut 24 de nos minutes, et le vighatî, ou soixantième de ghatî, qui vaut 24 de nos secondes. La latitude d'utilisation peut être déterminée à partir de l'ombre méridienne d'un gnomon de douze doigts, le jour de l'équinoxe, (*akshabhâ 5,0*), puis

From *Learning, Language and Invention: Essays presented to Francis Maddison*, ed. W.D. Hackmann and A.J. Turner. Published by Variorum, Ashgate Publishing Ltd, Gower House, Croft Road, Aldershot, Hampshire, GU11 3HR, Great Britain.

à partir de l'hypoténuse du triangle rectangle formé par cette même ombre et le gnomon (*akshakarṇa 13*), la latitude est de plus donnée directement (*22° 30'*). Cette valeur est évidemment confirmée par le calcul du triangle rectangle donné précédemment, en effet 5/13 vaut 0,38 qui est le sinus de l'angle 22° 33'.

L'inscription courte indique que « l'instrument a été fait pour le littoral du Surat» région de l'ouest de l'Inde Centrale. Il existe, dans cette région du Surat, une ville elle-méme dénommée Surat dont la latitude est environ 21°.

Le quart de cercle extérieur donne une graduation divisée en degrés numérotés de trois en trois. Les deux quarts de cercle suivants donnent la longueur de l'ombre des gnomons de 12 et 7 doigts respectivement, en fonction de la hauteur du soleil. Les petites barres verticales associées aux nombres indiquant des quarts d'unité, et celles horizontales des dixièmes d'unité [cf. Tableau I]. Les deux quarts de cercles intérieurs donnent, toujours en fonction de la hauteur du soleil et pour le jour du solstice d'été, le temps écoulé en ghatî et en vighatî depuis son lever, ou le temps à venir avant son coucher.

Cette face montre que ce quart de cercle s'apparente à un autre instrument également conservé au Musée d'Histoire de la Science de l'Université d'Oxford [Figure 2]. Ce quadrant (*turîya yantra*), signé Vuhlo Varmma mais non daté, a été construit pour la latitude 32°. Il est constitué de huit arcs de cercles concentriques. L'arc extérieur est une division du quart de cercle en degrés, numérotés de six en six. Puis viennent sept arcs de cercles de longueurs décroissantes, divisés en parties inégales. Le quatrième arc de cercle est divisée en quinze parties, numérotées de un à quinze. Il s'agit d'une division en ghatî du demi-jour équinoctial. A ces arcs de cercles sont associés les constellations du zodiaque traversées par le soleil au cours de l'année, excepté pour le premier arc de cercle où le signe des Gémeaux – *Mithuna* – sont remplacés par *âshâdha*, mois luni-solaire à cheval sur nos mois de juin et de juillet, mois qui contient bien le jour du solstice d'été. Ceci, joint au fait que les deux petites excroissances constituent une alidade et que cet instrument est muni d'un fil à plomb, indique clairement qu'il s'agit d'un cadran solaire portatif. L'emploi en est simple. Une visée du soleil à travers l'alidade donne le nombre de ghatî écoulés depuis le lever du soleil, ou le nombre de ghatî qui s'écouleront jusqu'à son coucher. Il est à noter que la visée ne se fait pas à l'œil mais par la recherche d'une image parfaitement ronde du soleil, par exemple, sur la main. On sait alors que les deux parties de l'axe cylindrique constituant l'alidade sont alignées avec le soleil, et que la main est orthogonale à cet axe.

Transcription du quadrant de Vuhlo Varmma:

90, 84, 78, 72, 66, 60, 54, 48, 42, 36, 30, 24, 18, 12, 6
Ashâdha: 17;40, 16, 15, 14, 13, 12, 11, 10, 9, 8, 7, 6, 5, 4, 3, 2, 1
Vṛshakarkatau: 17;15, 16, 15, 14, 13, 12, 11, 10, 9, 8, 7, 6, 5, 4, 3, 2, 1
Meshasimhau: 16;15, 15, 14, 13, 12, 11, 10, 9, 8, 7, 6, 5, 4, 3, 2, 1
Mînakanye: 15, 14, 13, 12, 11, 10, 9, 8, 7, 6, 5, 4, 3, 2, 1
Kumbhatule: 13;45, 12, 11, 10, 9, 8, 7, 6, 5, 4, 3, 2, 1
Makaravṛçikau: 12;45, 10, 9, 8, 7, 6, 5, 4, 3, 2, 1
Dhanuḥ: 12;20, 10, 9, 8, 7, 6, 5, 4, 3, 2, 1

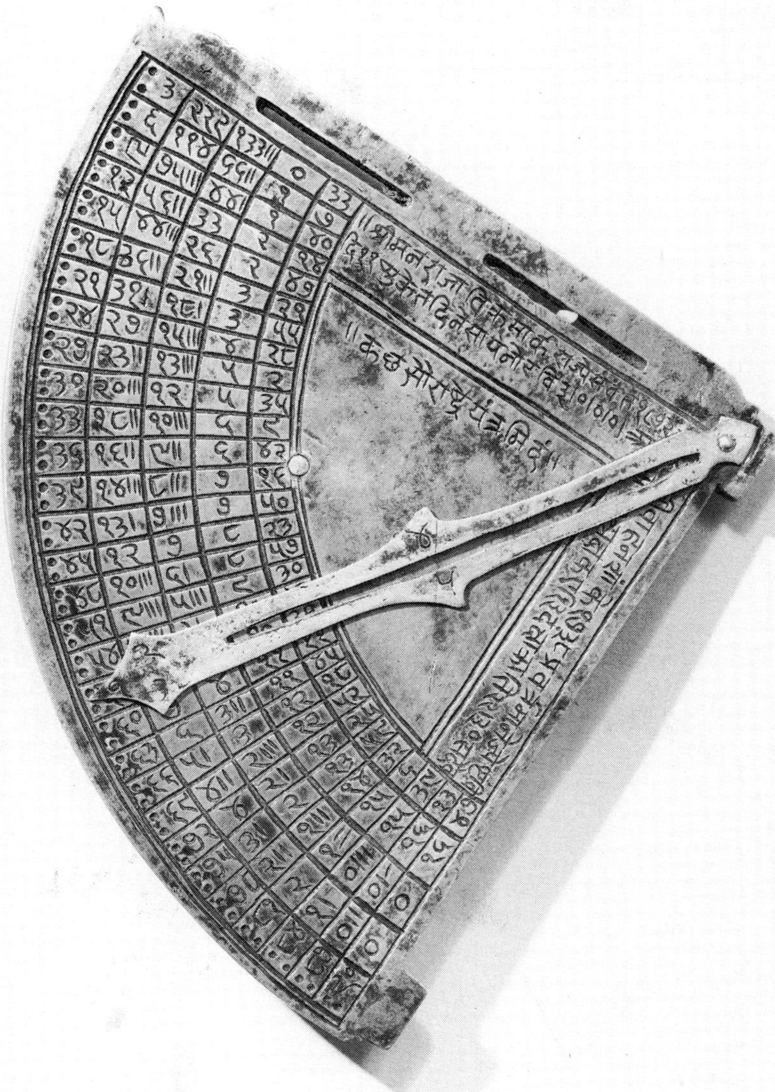

Figure 1. Face diurne d'un quart de cercle daté 1815 du Museum of the History of Science, Oxford (inv. no. 81–19)

Figure 2. *Face d'un quart de cercle* (turīya yantra) *du Museum of the History of Science, Oxford (inv. no. 57–84/191)*

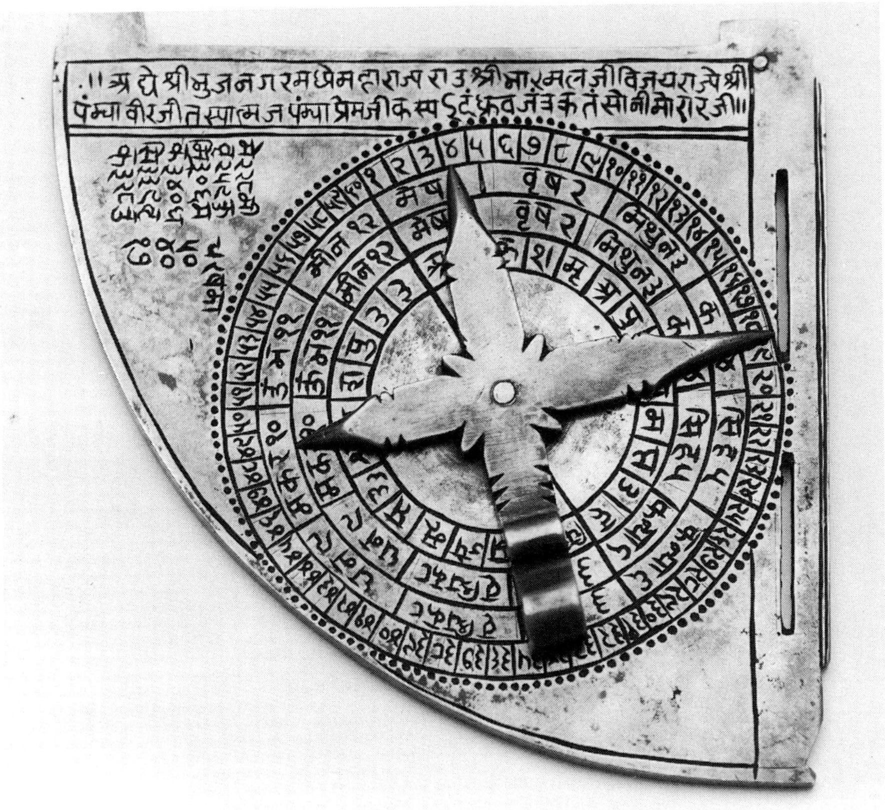

Figure 3. Face nocturne du quart de cercle de 1815 du Museum of the History of Science, Oxford (inv. no. 81–19)

« Après s'être prosterné au pied de l'époux de la déesse montagne, Vuhlo Varmma a fait ce quadrant pour que le sage connaisse le temps » La déesse montagne est *Pârvati* et son époux est *Çiva*, parfois aussi appelé *Kâla*: le temps.

On peut toutefois s'étonner que le quart de cercle que nous présentons ici ne donne le nombre de ghatî écoulés depuis le lever du soleil que pour le jour le plus long de l'année. En réalité, donnant ce renseignement pour ce jour-là et pour une latitude très basse, l'instrument est pratiquement utilisable toute l'année avec une précision suffisante au regard de celle de l'instrument lui-même, sauf autour du passage du soleil au méridien. On remarquera, sur l'instrument de Vuhlo Varmma, que les intervalles de temps écoulé en fonction de la hauteur du soleil diffèrent assez peu d'un mois à l'autre. Ce qui est déjà vrai pour ce quart de cercle, tracé pour une latitude de 32°, l'est encore plus pour celui que nous décrivons, tracé pour une latitude de 22° 30'. Nous donnons, dans le Tableau II, les intervalles de temps en ghatî et vighatî gravés sur l'instrument, puis calculés pour des déclinaisons du soleil valant respectivement +20°, 0°, -20° et +23° 30'. On voit que l'accord entre les valeurs gravées et celles calculées pour une déclinaison du soleil de +23° 30' – celle correspondant au jour le plus long de l'année – est assez bon. Les Indiens, commes les Grecs, choisissant plus souvent d'arrondir la valeur de l'obliquité de l'écliptique à 24° plutôt qu'à 23° 30', on peut être tenté de comparer les valeurs gravées avec les valeurs calculées pour une déclinaison maximale du soleil de 24°, malheureusement l'accord est moins bon que pour 23° 30'.

Description de la face nocturne [Figure 3]

Le cercle extérieur est divisé en soixante parties égales, numérotées de 1 à 60. Il s'agit de la division du nychtémère en ghatî, unité indienne de division du jour valant 24 minutes. Chacune de ces divisions est subdivisée en trois parties de huit minutes.

Les deux cercles suivant comprennent l'un et l'autre 12 divisions, il s'agit des 12 signes du zodiaque, un système étant décalé par rapport à l'autre. Le nom de chaque signe est suivi de son numéro d'ordre. On lit successivement sur ces deux cercles:

> *Mesha* 1 (le Bélier); *Vṛsha* 2 (pour *Vṛshabha*, le Taureau); *Mithuna* 3 (les Gémeaux); *Karka* 4 (pour *Karkata*, le Cancer); *Siṃha* 5 (le Lion); *Kanyâ* 6 (la Vierge); *Tulâ* 7 (la Balance); *Vṛçika* 8 (le Scorpion); *Dhanu* 9 (pour *Dhanus*, le Sagittaire); *Makara* 10 (le Capricorne); *Kumbha* 11 (le Verseau); *Mîna* 12 (les Poissons).

Le cercle intérieur comprend 26 divisions plus ou moins égales. Chacune de ces divisions contient une syllabe, à l'exception de la quatorzième qui en contient deux. On lit successivement: *A*, pour *Açvinî*; *Bha*, pour *Bharaṇî*; *Kṛ*, pour *Kṛttikâ*; *Ro*, pour *Rohiṇî*; *Mṛ*, pour *Mṛgaçiras*; *A*, pour *Ardrâ*; *Pu*, pour *Punarvasu*; *Pû*, (au lieu de *Pu*), pour *Pushya*; *A*, pour *Açleshâ*; *Ma*, pour *Maghâ*; *Pû*, pour *Pûrvaphâlgunî*; *U*, pour *Uttaraphâlgunî*; *Ha*, pour *Hasta*; *Ci*, pour *Citrâ*; *Sva*, pour *Svâti*; *Vi*, pour *Vishakhâ*; *A*, pour *Anurâdhâ*; *Jy*, pour *Jyeshṭhâ*; *Mu*, pour *Mûla*; *Pû*, pour *Pûrvashâḍhâ*; *U*, pour *Uttarâsâdhâ*; *Çr*, pour *Çravaṇa*; *Dha*, pour *Dhanishṭhâ*; *Ça*, pour *Çatabhishaj*; *Pu* (au lieu de *Pû*) pour *Pûrvabhâdrapadâ*; *U*, pour *Uttarabhadrapadâ*; *U* (au lieu de

Re, répétition de l'initiale précédente par faute d'inattention), pour *Revatî*. Il s'agit donc des 27 nakshatra, tels qu'on les trouve dans le *Paitâmahasiddhânta.* Dans les textes védiques, on a 28 nakshatra (*Abhijit* est inséré entre *Uttarâsâḍhâ* et *Çravaṇa*) et ces nakshatra commencent par *Kṛttikâ*. Les nakshatra sont des astérismes (c'est-à-dire des étoiles ou des groupes d'étoiles) dans lesquels la lune est censée séjourner chaque nuit, d'où le nombre de 27 ou 28 puisque la durée de la révolution sidérale de la lune est comprise entre 27 et 28 jours. Dans le cas des 27 nakshatra, chacun couvre 13° 20'. Nous donnons les étoiles directrices des différents nakshatra dans le Tableau III, selon Kirfel (*Die Kosmographie der Inder*), puis selon Pingree, à partir du *Paitâmahasiddhânta*, traité d'astronomie remontant au Ve siècle. ('History of mathematical astronomy in India', *Dictionary of Scientific Biography,* vol. 16, p. 565). Le fait que la quatorzième division contienne deux initiales montre que cet instrument n'a pas été gravé avec beaucoup de soin. Le graveur a d'abord tracé un diamètre comme axe de symétrie pour ses deux systèmes de signes zodiacaux, se trouvant ainsi piégé pour la division du cercle en 27 nakshatra.

La première partie du tableau est composée d'une série de nombres insérée entre deux colonnes d'initiales. Ce sont les lettres initiales des signes du zodiaque (*Râçi*), c'est-à-dire, en descendant pour la colonne de gauche et en montant pour celle de droite: *Me,* pour *Mesha*; *Vṛ,* pour *Vṛshabha*; *Mi,* pour *Mithuna*; *Ka,* pour *Karkata*; *Si,* pour *Siṃha*; *Ka,* pour *Kanyâ*; *Tu,* pour Tulâ; *Vṛ,* pour *Vṛçika*; *Dha,* pour *Dhanus*; *Ma,* pour *Makara*; *Ku,* pour *Kumbha*; *Mî,* pour *Mîna*. Et, au milieu, on trouve la série de nombres: 228, 259, 306, 340, 339, 328. La deuxième partie est composée de trois nombres: 50, 40, 17, placés sous le terme *carashaça* qui signifie différence ascensionnelle. Il s'agit donc des temps de lever des signes du zodiaque et des différences ascensionnelles, pour une latitude donnée, qui permettent de trouve ces temps en corrigeant les ascensions droites. Ces temps sont donnés en vighatî, c'est-à-dire en soixantième de ghatî, soit 24 secondes pour un vighatî. Les durées de lever des signes confirment que l'instrument a été fabriqué pour une latitude de 22° 30'. En effet les projections des trois premiers signes du zodiaque sur l'équateur valent respectivement 27° 50', parcourus en 278 vighatî; 29° 54', parcourus en 299 vighatî; 32° 16', parcourus en 323 vighatî. Les corrections pour une latitude de 22° 30' valent respectivement 4° 49', soit 48 vighatî, 3° 56', soit 40 vighatî, et 1° 36', soit 16 vighatî, valeurs de correction en bonne accord avec celles du tableau: 50, 40 et 17. D'où les ascensions obliques données par l'instrument: 228 (= 278 – 50); 259 (= 299 – 40); 306 (= 323 – 17); 340 (= 323 + 17); 339 (= 299 + 40) et 328 (= 278 + 50).

Il s'agit donc de l'instrument que les Indiens appellent le *Dhruvabhrama Yantra*, c'est-à-dire « l'instrument qui tourne autour du pôle ». C'est un instrument ingénieux pour trouver l'heure de nuit et la position des signes du zodiaque. Le principe de l'instrument est simple. Près de l'étoile Polaire, à environ 17 degrés de celle-ci, on trouve la deuxième étoile la plus brillante de la Petite Ourse, β *Ursae minoris, Markati* pour les Indiens. La ligne qui joint la *Polaire* à *Markati* tourne, comme le fait tout le ciel en 24 heures sidérales, ou en 23 heures 56 minutes de temps solaire moyen. Par conséquent l'estimation de l'angle que cette ligne fait avec la verticale d'un lieu donne une indication sur la position de la sphère des étoiles et donc sur celle des signes. D'autre part, la position du soleil dans les signes étant connue, de mémoire ou à partir d'éphémérides, le temps est facilement trouvé par addition ou soustraction des ascensions respectives.

Rappelons. Si l'on suppose les pôles célestes déterminés, à la fois sur la sphère locale et sur la sphère des fixes, la position relative des deux sphères – c'est-à-dire l'aspect du ciel étoilé pour un observateur local – ne dépend plus que d'un seul paramètre. Les astronomes ont décidé de choisir l'angle horaire du point γ (équinoxe vernal), on a donné à cet angle horaire le nom de temps sidéral T_s. L'angle horaire est la valeur de l'arc qui mesure l'angle dièdre formé par le cercle horaire d'un astre et le demi-méridien sud. L'angle horaire H d'un astre A est lié à son ascension droite α et au temps sidéral T_s par la relation: $H = T_s - \alpha$. Un cas important est celui où l'on a $H = 0$, c'est-à-dire lorsque l'astre est à son passage supérieur au méridien du lieu. On a alors $T_s = \alpha$, d'où une nouvelle définition du temps sidéral: c'est l'ascension droite des astres en passage supérieur au méridien à l'instant considéré.

Ainsi l'aspect du ciel montré par la Figure 4 correspond à $T_s = 0$ heure. Or si nous simulons la position correspondante de *ß Ursae minoris* par rapport à la Polaire sur notre quart de cercle, l'index Est indique 15 ghatî, c'est-à-dire 6 heures. Ceci parce que les Indiens prennent l'origine du temps sidéral non pas au passage supérieur du point γ au méridien mais à son lever. On a donc entre le temps sidéral indien $T_{s(i)}$ et le temps sidéral défini par les astronomes $T_s = T_{s(i)} - 6$.

Trois des branches de l'index cruciforme servent effectivement d'index, on peut les noter Est, Ouest et Nord, la quatrième beaucoup plus lourde, que l'on peut qualifier de Sud, se comporte comme un fil à plomb. C'est-à-dire que lorsque l'instrument est tenu dans un plan vertical, les index Est et Ouest se trouvent à l'horizontale alors que l'index Nord et l'index Sud indique la verticale. Les trois index vrais ont des longueurs différentes déterminées par le rayon des cercles qu'ils doivent décrire. L'index Est est le plus long; il décrit la division du nychtémère de huit minutes en huit minutes. L'index Nord décrit le cercle du lever des signes pour la sphére droite. L'index Ouest parcourt le cercle du lever des signes pour la sphère oblique correspondant à la latitude 22° 30'.

Utilisation du *Dhruvabhrama yantra*

Prenons la configuration de la Figure 3, pour, par exemple, le 15 septembre 1815. L'instrument est tenu dans un plan vertical, de telle sorte que la Polaire et *ß Ursae minoris* soient vues toutes les deux dans la fente AB, la Polaire étant vers le centre du quart de cercle. L'index Est donne alors le temps sidéral en ghatî, à la façon indienne, soit 18 1/3 ghatî, c'est-à-dire 18,33gh x 24m = 439, 92m = 7h 20m. Il est donc en réalité 1h 20mn, en temps sidéral tel qu'on le définit aujourd'hui. L'index Nord montre le signe qui franchit le méridien à ce moment-là, *Mesha*, le Bélier, de plus il permet d'évaluer ce temps sidéral de 1h 20m avec une assez bonne précision, bien qu'il soit trop court pour parcourir la division du nychtémère de huit en huit minutes. L'index Ouest montre le signe qui se couche, *Makara*, le Capricorne.

L'astronome, ou l'astrologue, peut alors déterminer l'heure de la nuit, ou combien de ghatî doivent s'écouler avant le passage du soleil au méridien du lieu. A partir d'un almanach, il connaît le signe et le degré du signe dans lequel se trouve le soleil, ce jour-là. Le 15 septembre 1815, le soleil est à 22° 30' dans le Lion, son ascension droite est donc 11h 30m. C'est-à-dire que le soleil passera au méridien 10h 10m après *ß Ursae minoris*, il est donc environ 2 heures du matin [Figure 4].

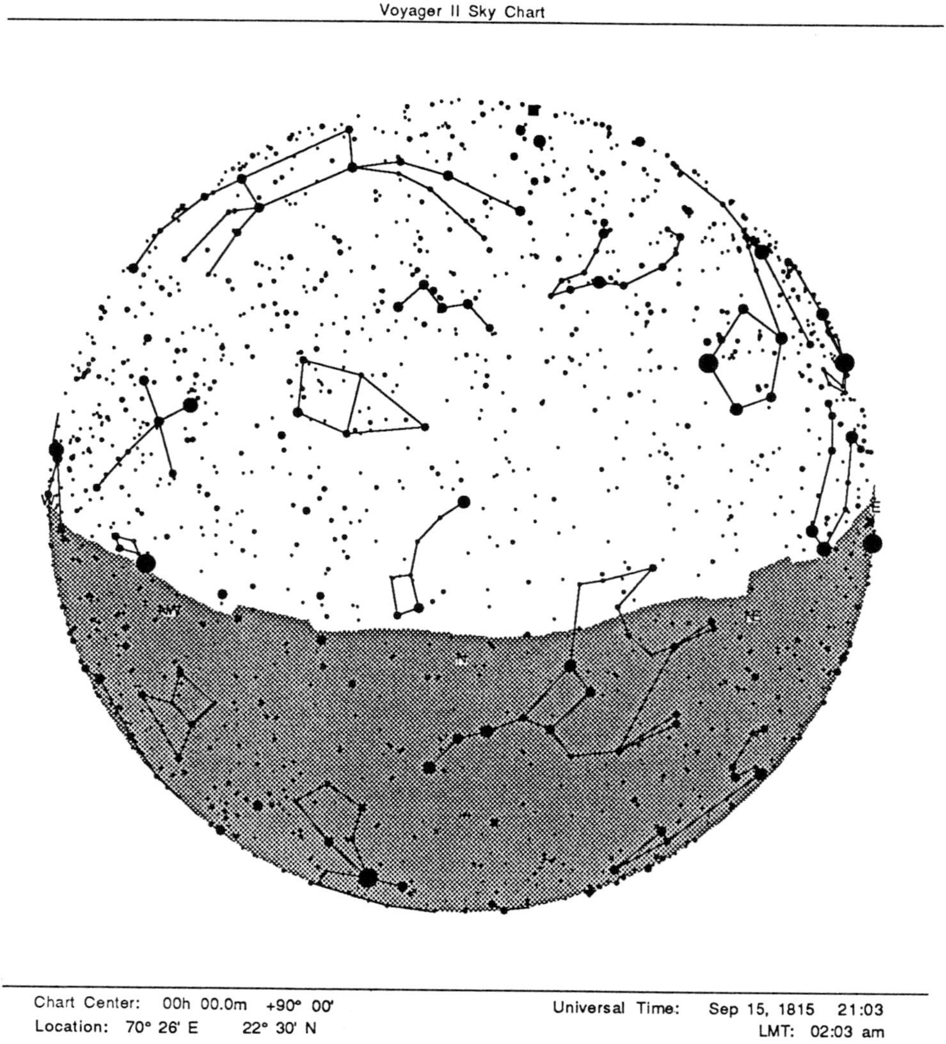

Figure 4. Carte céleste d'après Voyager 2

Tableau I

Hauteur	**Ombres 12 d. gravées**	**Ombres 12 d. calculées**	**Ombres 7 d. gravées**	**Ombres 7 d. calculées**
3	229	228,97	133 ǀǀ	133,57
6	114	114,17	66 ǀǀ	66,60
9	75 ǀǀǀ	75,76	44 ǀ	44,20
12	56 ǀǀ	56,46	33	32,93
15	44 ǀǀǀ	44,78	26	26,12
18	36 ǀǀ	36,93	21 ǀǀ	21,54
21	31 ǀ	31,26	18 ǀ	18,23
24	27	26,95	15 ǀǀǀ	15,72
27	23 ǀǀ	23,55	13 ǀǀǀ	13,74
30	20 ǀǀǀ	20,75	12	12,12
33	18 ǀǀ	18,48	10 ǀǀǀ	10,78
36	16 ǀǀ	16,52	9 ǀǀ	9,63
39	14 ǀǀǀ	14,82	8 ǀǀ	8,64
42	13 ǀ	13,33	7 ǀǀǀ	7,77
45	12	12,00	7	7,00
48	10 ǀǀǀ	10,80	6 ǀ	6,30
51	9 ǀǀǀ	9,72	5 ǀǀǀ	5,67
54	8 ǀǀǀ	8,72	5	5,08
57	7 ǀǀǀ	7,79	4 ǀǀ	4,55
60	7	6,93	4	4,04
63	6	6,16	3 ǀǀ	3,57
66	5 ǀ	5,34	3	3,12
69	4 ǀǀ	4,61	2 ǀǀǀ	2,69
72	4	3,90	2 ǀ	2,27
75	3 ǀ	3,22	2	1,88
78	2 ǀǀ	2,55	1 ǀǀ	1,49
81	2	1,90	1 =	1,11
84	1 ǀ	1,26	0 ǀǀǀ	0,74
87	0 ǀǀ	0,63	0 ǀ –	0,37
90	0	0,00	0	0,00

Tableau II

Haut.	Δ*t* **gravés**	Δ*t* (δ = +20°) **calculés**	Δ*t* (δ = 0°) **calculés**	Δ*t* (δ = -20°) **calculés**	Δ*t* (δ = +23,5°) **calculés**
3	0;33	0;35	0;32	0;35	0;36
6	1;07	1;09	1;08	1;11	1;11
9	1;40	1;44	1;37	1;47	1;46
12	2;14	2;18	2;09	2;23	2;21
15	2;47	2;51	2;70	3;00	2;56
18	3;21	3;25	3;15	3;37	3;30
21	3;55	3;59	3;48	4;17	4;04
24	4;28	4;32	4;21	4;56	4;38
27	5;02	5;06	4;54	5;37	5;12
30	5;35	5;39	5;27	6;20	5;46
33	6;09	6;12	6;02	7;05	6;19
36	6;42	6;45	6;36	7;52	6;52
39	7;16	7;18	7;10	8;45	7;26
42	7;50	7;51	7;45	9;46	7;59
45	8;23	8;24	8;20	11;03	8;32
48	8;57	8;56	8;56	13;56	9;05
51	9;30	9;29	9; 33		9;38
54	10;04	10;02	10;12		10;11
57	10;37	10;34	10;52		10;44
60	11;11	11;07	11;37		11;17
63	11;45	11;40	12;28		11;50
66	12;18	12;13	13;37		12;22
69	12;52	12;46			12;55
72	13;25	13;19			13;28
75	13;59	13;52			14;00
78	14;32	14;26			14;33
81	15;06	15;02			15;06
84	15;39	15;43			15;39
87	16;13				16;12
90	16;47				16;43

Tableau III

Nakshatra	**Selon W. Kirfel**	**Etoiles directrices selon D. Pingree**	**Longitude du début**
Açvinî	β, γ Ari.	β Ari	0°
Bharanî	35, 39, 41, Ari.	33 Ari.(?)	13°20'
Kṛttikâ	η Tau. etc. (Pléiades)	η Tau.	26°40'
Rohiṇî	a, ν, γ, δ, ϵ Tau.	α Tau.	40°00'
Mṛgaçiras	λ, ϕ_1, ϕ_2 Ori.	15 Ori.(?)	53°20'
Ardrâ	α Ori.	λ Ori.(?)	66°40'
Punarvasu	β, α Gem.	β Gem.	80°00'
Pushya	δ, γ Canc.	ϵ Canc.	93°20'
Açleshâ	ϵ, δ, σ, η Hyd.	β Canc.(?)	106°40'
Maghâ	α, η, γ, ζ, μ, ϵ Leo.	α Leo.	120°00'
Pûrvaphâlgunî	δ, ν Leo.	δ Leo.	133°20'
Uttaraphâlgunî	β, 93 Leo.	β Leo.	146°40'
Hasta	δ, γ, ϵ, α, β Corv.	η Corv.	160°00'
Citrâ	α Virg.	α Virg.	173°20'
Svâti	α Boot.	α Boot.	186°40'
Vishâkhâ	ι, γ, β, α Libr.	ι Libr.	200°00'
Anurâdhâ	δ, β, π Scorp.	δ Scorp.	213°20'
Jyeshṭhâ	α, σ, τ Scorp.	α Scorp.	226°40'
Mûla	λ, υ, ι, ν, η, μ, ϵ Scorp.	δ Oph.	240°00'
Pûrvâshâḍhâ	δ, ϵ Sag.	γ Sag.	253°20'
Uttarâshâḍhâ	α, ζ Sag.	o Sag.	266°40'
[Abhijit]	α, ϵ, ζ Lyr.	α Lyr.	
Çravaṇa	α, β, γ Aquil.	α Aquil.	280°00'
Dhanishṭhâ	β, α, γ, δ Delph.	α Delph.	293°20'
Çatabhishaj	λ Aquar.	λ Aquar.	306°40'
Pûrvabhâdrapadâ	α, β Peg.	α Peg.	320°00'
Uttarabhâdrapadâ	γ Peg., α Andr.	α Andr. (?)	333°20'
Revatî	ζ Pisc.	ζ Pisc.	346°40'

The Contributors

J.W. Allan	Keeper of Eastern Art, the Ashmolean Museum, Oxford & University Lecturer in Islamic Art.
Giles Barber	Librarian, The Taylorian Institution, Oxford.
Silvio A. Bedini	*Emeritus*, The Smithsonian Institution, Washington DC.
J.A. Bennett	Curator, Whipple Museum of the History of Science, Cambridge.
J. Bergsagel	Professor of the History of Music, University of Copenhagen.
A. Chapman	Wadham College, Oxford.
Charles Dowsett	Sometime Calouste Gulbenkian Professor of Armenian, University of Oxford.
J.A.Drake-Brockman	Milton, Oxfordshire.
W.D. Hackmann	Assistant Curator, Museum of the History of Science, Oxford.
R.M. Harré	Reader in Philosophy of Science, University of Oxford.
D.R. Hill	Deceased.
J.D. North	Professor of Philosophy, University of Groningen.
E. Poulle	Sometime Director, Ecole de Chartes, Paris.
E. Savage-Smith	Oxford.
Brian Scott	Sometime Reader in Latin, The Queen's University, Belfast.
A. Segonds	CNRS, Paris.
A.V. Simcock	Librarian, Museum of the History of Science, Oxford.
A.J. Turner	Le Mesnil-le-Roi, Les Yvelines. Gruppo Beltrame Research Fellow in the History of Time, Florence.
J.-P Verdet	Observatoire de Paris.
Colin Wakefield	The Bodleian Library, Oxford.

Index